FUNDAMENTALS OF WASTE AND ENVIRONMENTAL ENGINEERING

FUNDAMENTALS OF WASTE AND ENVIRONMENTAL ENGINEERING

S N Mukhopadhyay

teri

THE ENERGY AND RESOURCES INSTITUTE

Creating Innovative Solutions for a Sustainable Future

ISBN 978-93-8653-010-3

Suggested citation

Mukhopadhyay, S. N. 2020. *Fundamentals of Waste and Environmental Engineering.* New Delhi: TERI

Published by
The Energy and Resources Institute (TERI)
TERI Press
Darbari Seth Block
IHC Complex, Lodhi Road
New Delhi – 110 003
India

Tel. 2468 2100 or 7110 2100
Fax 2468 2144 or 2468 2145
India +91 • Delhi (0)11
E-mail teripress@teri.res.in
Website www.teriin.org
Bookstore https://bookstore.teri.res.in

Printed in India

Preface

Every nation is constantly looking for solutions to problems of pollution. With the progress of human civilization, all genes- and genome-regulated life forms (LFs), particularly living human genomic body/bioreactors (LHGBRs), are experiencing a wide range of health effects owing to exposure to pollutants present in the environment. Pollutants are generated heavily by activities of LHGBRs in pursuit of human progress. Construction-related activities are significantly contributing to pollution, posing a risk to human health and the environment beyond borders and on a global scale. Pollutions of all forms and dimensions are putting LHGBRs at great risks. Global climate change and the issue of persistent organic pollutants are posing threats to all LFs. Likewise, in recent years, e-waste has emerged as a global threat as it contains hazardous heavy metals. Noise pollution is also becoming a major public health concern. There is an urgent need of abatement efforts for all forms of pollution by proper disposal of wastes and garbage generated by LHGBRs activities whether on land, in water, or in air. Awareness of all stakeholders of waste treatments to minimize pollution and environmental engineering applications to keep the environment clean are now imperative. This book deals with these knowledge industry (KI) materials. The dimensions of pollution crisis are enormous. Understanding these dimensions appropriately requires adequate educational background.

I would like to mention the names of people who have contributed in the preparation of this book. I will always be indebted to Dr Nilay Choudhury, who taught me waste processing technologies during my undergraduate years at the Jadavpur University. He was a renowned expert in environmental engineering and waste treatment and later became the founding chairman of Central Pollution Control Board (CPCB) of India. I learnt a lot from him through class lectures and tutorial exercises. Later, during my teaching career at Biochemical Engineering Research Centre (BERC), Department of Biochemical Engineering and Biotechnology DBEB, Indian Institute of Technology (IIT) Delhi, I had the opportunity to teach biological waste treatment and biochemical engineering courses to undergraduate and postgraduate students. Also, I had the opportunities to organize a CPCB-sponsored short-term training course on biotechnology treatments of biological waste and wastewaters at DBEB under the aegis of Foundation of Innovation and Transfer Technology (FITT), IIT Delhi. I wish to sincerely acknowledge the support of CPCB and FITT for the same. I also recognize the sincere support of all my well-wishers and my colleagues and staff involved in the event. Encouragements received from Prof. T K Ghose, the founding head of DBEB, and Prof. Prem Vrat, the then Deputy Director (Faculty), IIT Delhi, are also sincerely acknowledged.

I was inducted as a lifetime member of Kali Mandir Society, Chittaranjan Park, New Delhi by its former president Dr Ananda Mukherjee, who also provided valuable inputs on domestic wastewater treatment. I sincerely acknowledge his help. In all my endeavours to complete this book, I have received unending support from my son Parag, daughter-in-law Pooja, my daughter Paromita, son-in-law Arvind, and love of my grandchildren Ayan and Anika. My wife Sakuntala has always encouraged me to pursue my goals and is there whenever I need her. I sincerely acknowledge her support throughout my journey of book writing. Also, I thank TERI Press for publishing this book. Suggestions from readers are most welcome and will be gratefully acknowledged.

Contents

1 Introduction to Waste and Environmental Crisis

1.1 INTRODUCTION

Humans and all life forms on the earth are victims of various kinds of pollution. Macromolecular biopolymeric living human genomic body/bioreactors (LHGBRs) emerged as a species about 200,000 years ago. They came into existence through creation of its nature/environment by almighty supranature (ASN). The emergence of LHGBRs is mainly attributed to the naturally acquired design engineering (NADE) and recombinatorially acquired design engineering (RADE) principles. Interaction between NADE and RADE resulted in creation of anatomical/structural and physiological designs of LHGBRs. Activities of LHGBRs on earth can be categorized into creators, discoverers, developers, innovators, builders, spiritual leaders, and many others. Emergence of LHGBRs on the earth can be further understood with the help of flowchart given in Figure 1.1.

During such activities, LHGBRs, besides being involved in the creation of populations by cell fusion process biotechnology (PB), produced many visible and invisible products and co-products/wastes by using nature-gifted knowledge. Although these contributed to the progress of life forms, they were also accompanied by many negative consequences because various activities performed by LHGBRs resulted in the generation of wastes, wastewaters, and pollutants and made environment/atmosphere toxic to life forms.

1.2 LHGBRS AND OTHER LIFE FORMS IN RELATION TO WASTE TYPES

Wastes and wastewaters are of different types and are largely resource based (Figure 1.2). Agricultural biowastes, constructional waste/garbage from shelters/houses of LHGBRs, household waste foods/food wastes (WF/FW), and organic and inorganic solid refuses/effluents from municipal/corporation and industries are major solid and semi-solid waste resources. The sizes and amounts of these waste materials may vary considerably and range from large/macroparticles to micro/nanoparticles.

Volatile suspended solid (VSS) concentrations in wastewaters are the most important solid components in water supplies. Depending on the available forms and sizes in the environment/atmosphere, these wastes may cause environmental problems for life forms,

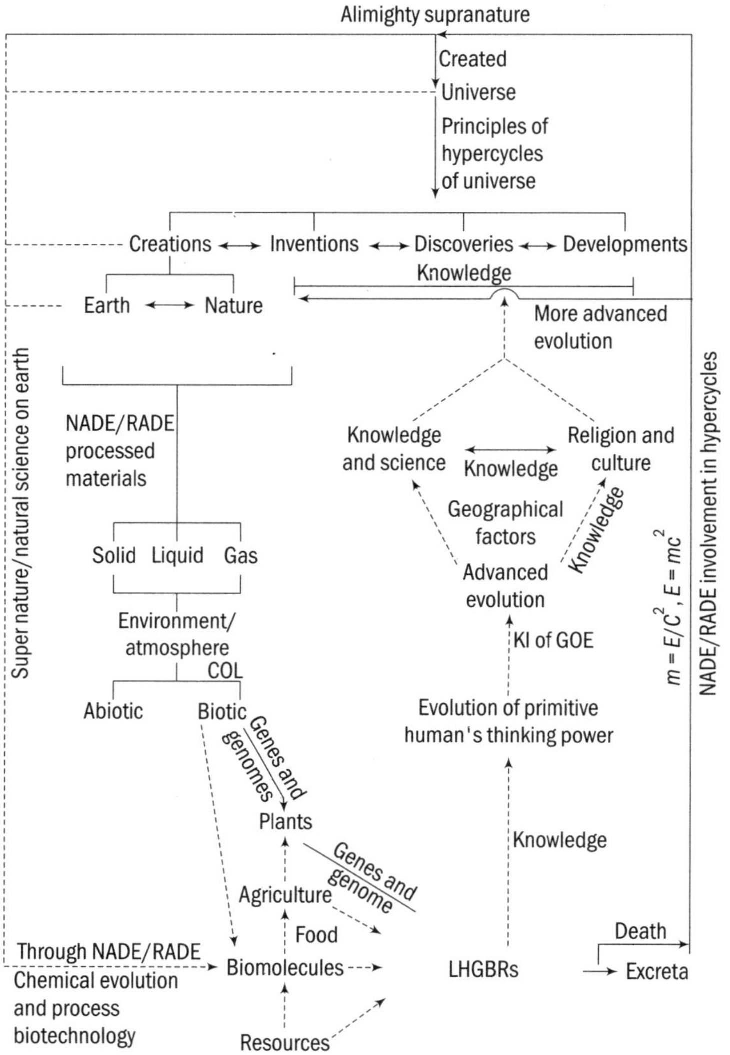

Figure 1.1 Emergence of LHGBRs on the earth

COL - chemicals of life; GOE - Gods on Earth; LHGBRs - living human genomic body/bioreactors; NADE - naturally acquired design engineering; RADE - recombinatorially acquired design engineering

including LHGBRs, and may also be hazardous. They cause ailments/diseases of various types in genes and genomes regulated LHGBRs and other life forms.

1.3 LHGBR DESIGNS

1.3.1 Delivering Products and Pollutants

Macromolar biopolymeric (MBP) standing LHGBR designs are shown in Figure 1.3. In terms of normal physiology, they are primarily of two types, that is, males and females.

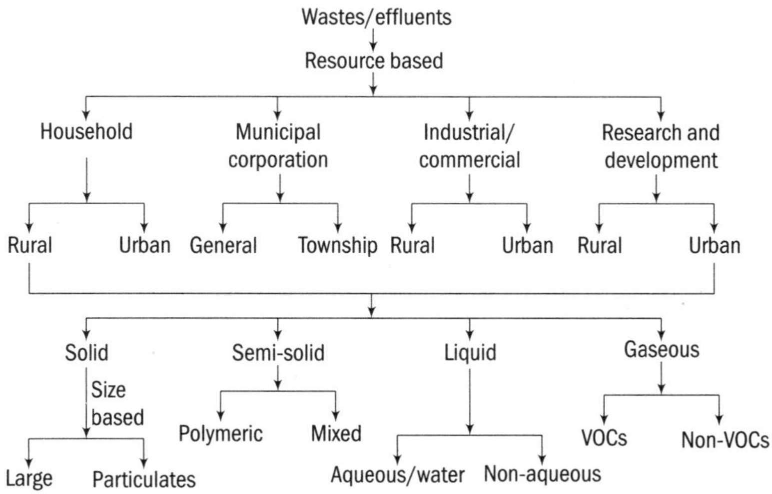

Figure 1.2 Waste/effluent types generated by LHGBRs and other life forms
VOCs - volatile organic compounds

They are fabricated by chemicals of life (COL) of genes of food macromolecules and their supermacromolecular polymeric genomes, called human genome (HG). HG/genes and genomes regulate growth, activities, and functions of individual LHGBRs through KI participated brain/mind to release products and co-products/wastes (Table 1.1). The first macromolecular physical map of HG is shown in Figure 1.4. Internal core parts of LHGBRs, where major bioreactions occur, are shown in Figure 1.5.

1.3.2 LHGBRs as Supreme Creatures/Super Organisms

Products of LHGBRs are many. These make LHGBRs to act as creators, innovators, and so on as stated earlier. Their activities and performances led them to be known as Gods on

Table 1.1 Products and co-products/wastes of LHGBRs

Products	Co-products/wastes
Visible	
Proteinous flesh/biomass	Faeces/wastes/pollutants
Brain, blood and bones for KI	Urine/liquid wastes
Anatomical/structural and physiological design forms	Waste heat
Work power	Sweat
Deeds—good or bad between birth and death limits	
Invisible	
Thinking/knowledge power	Gaseous excretions/off smell

KI - knowledge industry

Figure 1.3 Anatomical/structural standing physiological designs of modern LHGBRs

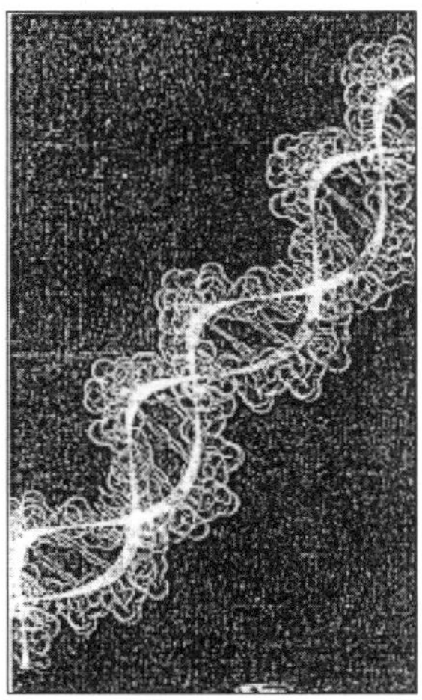

Figure 1.4 An LHGBR's genes' structure

Earth (GOE). The following are some of the reasons for terming LHGBRs as GOE:

(i) Galaxy's outstanding designs
(ii) Geographical occurrence discoverers
(iii) Gender opportunity detectors
(iv) Grow on diets
(v) Governors of diseases
(vi) Genomic of diseases
(vii) Genomic order detectors
(viii) Genomic order decoders
(ix) Go on discovering
(x) Go on dieing
(xi) Genome order designer

However, among HOE, those who tend to do bad deeds/sins may be termed as Sinners on Earth (SOE).

As supreme creatures on the earth, LHGBRs, and developed global-level KI. Activities of LHGBRs through genes and genome regulations and KI participations of communities produce solid, semi-solid, liquid/water, and gaseous wastes and release them into the

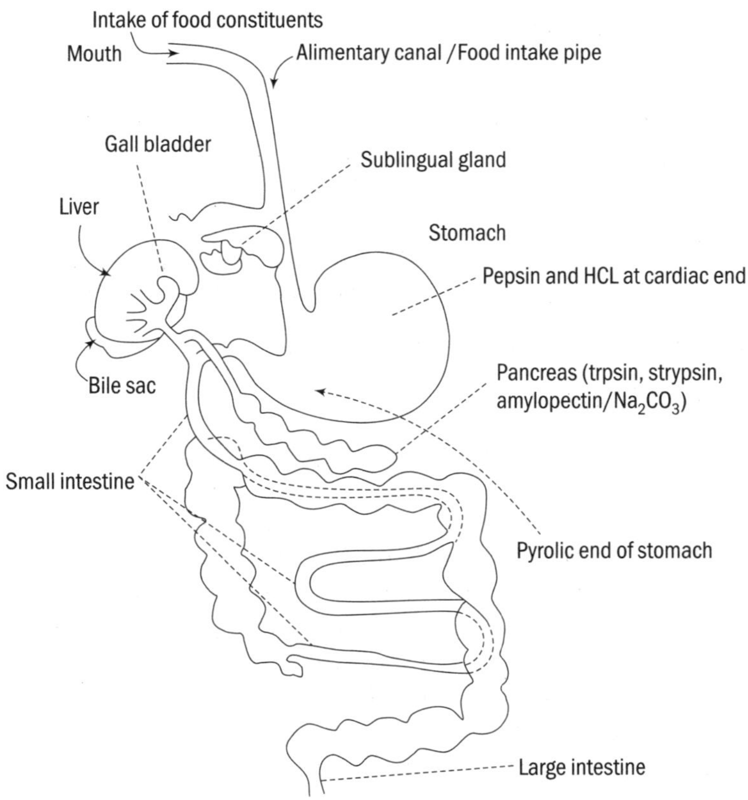

Figure 1.5 Internal design of LHGBR as the consumer of food using NADE/RADE

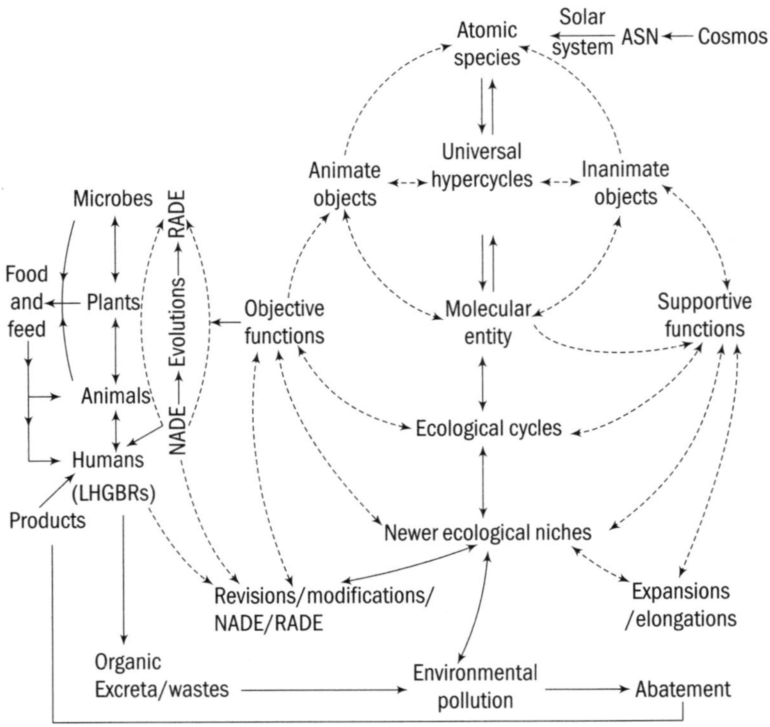

Figure 1.6 Participation of LHGBRs in universal hypercycles in causing pollution of the nature
ASN – almighty supranature; LHGBRs – living human genomic body/bioreactors; NADE – naturally acquired design engineering; RADE – recombinatorially acquired design engineering

ecological niches of universal hypercycles (UH) (Figure 1.6). Exposure of LHGBRs and other life forms to unwanted garbage/waste materials may be hazardous to their health.

1.3.3 Involvement in Universal Hypercycles

1.3.3.1 Universal hypercycles

Figure 1.6 shows cooperative interactions of quasispecies via cyclic coupling. These interactions are called universal hypercycles (UH) and in them ecological cycles are involved. In ecology, wastes have been defined as materials ranging from macro to subatomic particle sizes that are not directly beneficial to LHGBR activities and functions; rather, they may be hazardous or harmful to them and cause diseases. However, through activities of LHGBRs on the earth, these materials are generated in surroundings/environment. The involvement of ecological cycles in the creation of life forms, including microbes, plants, LHGBRs, and food, using NADE–RADE principles is a unique feature of UH. In this, water, air, food/feed, and shelters play important roles for normal physiological functions in the form of metabolism of LHGBRs in the environment (Figure 1.7).

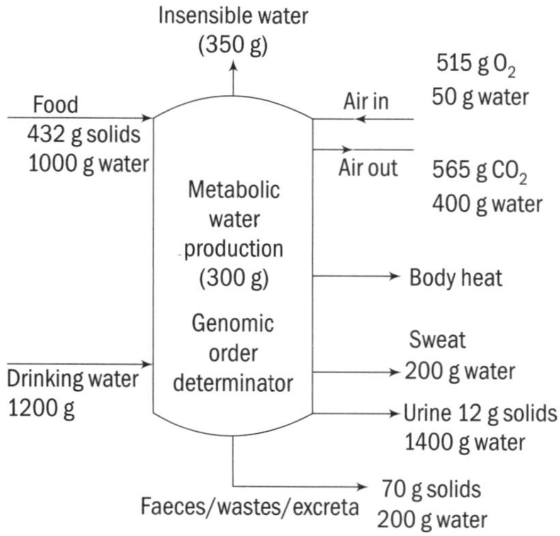

Figure 1.7 Performance of LHGBRs in the environment

Note The correlation is between physiological variables in LHGBRs and steady-state mass balance on the body for an average person at rest per day (neglecting solid content in sweat less than 2 g).

The progress of LHGBRs is associated with environmental pollution caused by pollutants/ wastes. For protecting life forms from hazards/ill effects of pollution in environment, efforts should be made to reduce activities that contribute to pollution. Thus, the definition of environment needs to be understood properly.

1.3.3.2 Environment

Environment of habitation of life forms, including LHGBRs, may be defined as an unbounded/ open cosmic enclosure, containing all states of matters and energy, for example, gaseous/ air, liquid/water, solid, and plasma in nature. "Environ" is a German word consisting of two parts: "En" meaning inside and "viron" meaning circuit. Thus, environment is a cosmic circuit enclosure. Gaseous state in this enclosure is generally referred to as air and liquid (primarily referred as water). Solids are the condensed forms of gaseous and liquid materials, that is, chemicals that form the major groups such as rocks/stones, trees/ plants, LHGBRs, animals, insects, and other life forms. Plasma state is the finest form of all matters that may exist in subatomic/atomic forms of particles in the environment under suitable, specific conditions. As pollution causes changes in nature by generating toxicants, different life forms in the environment, including LHGBRs, become threatened. So, pollution generated hazards calls for due attention and prevention.

At present, environment is getting polluted heavily by particulate matter (PM) of size 1 μm (PM_1) or less in diameter, called ultrafine particulate matter, compared to size 2.5 μm ($PM_{2.5}$) in air. PM_1 is the most dangerous of all suspended particulate matters because it penetrates deep into genomes of life forms, developing ailments in organelles such as lung, bloodstream, and other parts of LHGBRs and consequently causing disorder in their functions related to diseases. A Chinese study reveals that PM_1 is the primary cause of cardiovascular diseases linked to pollution and human health hazards. Experts have warned that PM_1 can cause heart problems as well.

1.3.3.3 Pollution hazards and problems

Pollution caused by wastes and effluents in air and water, that is, in nature, is hazardous to agricultural food production. So, to save the earth it is essential to cut down generation of wastes/effluents and wastewater and reduce emissions. In a recent report, the World Health Organization (WHO) has called for an urgent action to combat climate/environment change by reducing emissions of black carbon, ozone, methane, CO_2, sulphur compounds, and so on that result from human activities. Accordingly, all countries must endeavour to drastically reduce all types of pollution, that account for health hazards, causing over several million premature deaths annually. Therefore, provision for control of pollutions of all types has been included in Millennium Development Goals (MDGs). Thus, reduction in discharges of all types of pollutants, as well as creation of wealth from wastes and value-added products (VAPs), has become a major task and the fundamental responsibility of citizens of all countries in order to protect LHGBRs and all other life forms. Treatment of wastes by environmental systems engineering, along with PB process engineering, has always been effective in reducing pollution level and health hazard possibilities caused by the presence of undesirable toxic materials in the surroundings.

1.4 ENVIRONMENTAL CRISIS AND ALARMS

It is a common dilemma of all LHGBRs that has underlined so clearly as environmental crisis and alarms are now unfolding. It is affecting everyone, regardless of rich and poor, old and young, where they are from, and born or unborn. Ecology is getting fouled every day because there are no ecological sanctuaries. In the present-day scenario, it is much more than just a crisis of physical dimensions. Foremost on everyone's mind is the mitigation of these burning problems. It brings into question the values and virtues now widely held by individuals and the principles by which global society without border is organized. Most important, for national sovereignty and continuous economic growth in the finite biosphere of earth, there is a need to find solution to pollution, taking care of PB facets of all LHGBRs. That is why, all nations need solution to pollution. Unfortunately, the solution approaches designed by LHGBRs to deal with the adverse effects of environmental pollution alarms are becoming increasingly costly.

1.4.1 Global Climate Change and Warming Effects

Persistent organic pollutants (POPs) are chemical substances that persist in the environment, accumulate in the food web, and pose risk of adverse effects on LHGBRs and wildlife. In the late 1960s, polychlorinated biphenyls (PCBs) were detected in high concentrations in wildlife in Sweden. PCBs continue to threaten the survival of marine predators. Concerted efforts are thus still needed to mitigate PCB pollution. As a result, there have been urgent efforts to review current methods of PCB mitigation in the marine environment in Europe and elsewhere. In addition, Teheran in Iran was issued multiple "red status" warnings for poor air quality. It is time to return blue skies to Iran. Iranian gasoline consumption is attributed for poisonous air quality. It appears that global warming is preventing tomorrow's clean climate. In a study, it was found that climate change fuelled the beginning of Syria's Civil War by resulting in severe and longer regional drought and causing unrest in the country (Sills 2016). The unrest turned to civil war when the regime began suppressing protesters and that civil war allowed ISIS, also known as the Islamic State of Iraq and Syria, to rise.

For decades, scientists have warned on global warming and proved that it is no longer theoretical. Global warming caused by LHGBR activities has led land ice to melt and ocean water to expand. Scientists warned that the rising sea level would eventually imperil the coastlines in the United States. Now, those warnings are manifesting the realities. The inundation of the coast has begun. The sea has crept up to the point that a high tide and a brisk wind are all it takes to inundate streets and homes. For example, Highway 80, the only road to Tybee Island, Georgia, is forced closed several times in a year owing to high tides.

1.4.2 Interventions of Pollutants

In the prehistoric period, LHGBRs used to obtain food for livelihood by hunting environmental life forms such as animals.

As and when LHGBRs learnt to use agriculture for their food production, the interventions in the natural systems became progressively more complex and widespread. Thus, agriculture can be regarded as an effort of LHGBRs to shape the environment to suit their needs. Effectively LHGBRs selected and favoured those biological species of animal and plants that were useful to their needs for food, clothing, shelter, and agriculture. However, in the process, the global composition and distribution of both plant and animal species got substantially altered. By use of such materials, LHGBRs started generating both biological and non-biological wastes and wastewaters. Since the beginning of LHGBR civilizations, many thousands of years ago, human race has developed large numbers of chemical compounds to help protect against insects, pests, and microbial diseases, thus enhancing prey–predator interactions and fouling the environment. Over time LHGBRs learnt to produce and use chemicals to restore and enhance soil fertility for agriculture. These chemical interventions , primarily designed for the purpose of enhancing and ensuring food supply and animal feed for LHGBRs and other purposes led to far-reaching consequences. It generated agricultural residues/wastes and chemical wastes/effluents and resulted in contamination of drinking water supplies. Figure 1.8 shows the categorization of these water pollutants.

Drinking water supply of LHGBRs needs to fulfil its chemo-physical and biogenic properties. This is because the chemistry and biochemistry of life forms on this planet are

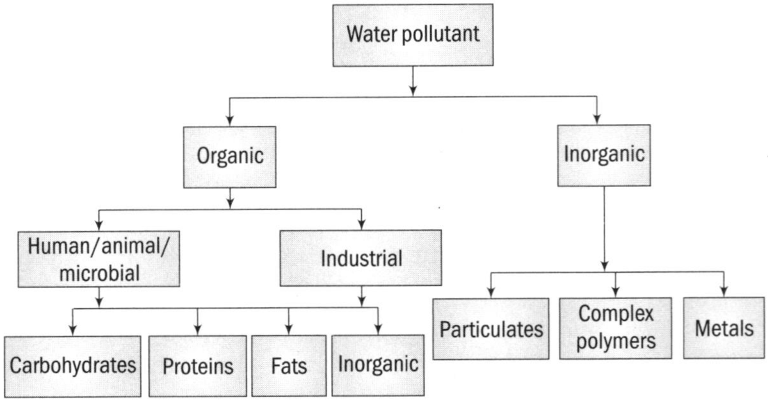

Figure 1.8 Categories of water pollutants

essentially related to the chemistry of water, carbon, and nitrogen. In fact, water makes up to 70%–90% of the weight of most life forms on earth.

1.4.2.1 Water pollution

Life began in the sea and and therefore properties of water have direct influence on the chemistry and biochemistry of all living organisms, including LHGBRs, on the earth. Any segment of nature in which there is an exchange of materials between organisms and environment is termed as ecological system or ecosystem. An ecosystem that normally exists in a natural body of water consists of four groups of components: (i) abiotic substances/ pollutants, (ii) producers, (iii) consumers, and (iv) decomposers. Interventions of these four groups in waterbodies always need special attention.

1.4.2.1.1 Abiotics

These are non-living polluting components of the system. These groups include CO_2, O_2, and detritus (non-living organic and inorganic particulate matters and all other inorganic ions).

1.4.2.1.2 Producers

This group comprises photosynthetic organisms of the following three types:

(i) Rooted or large floating bodies of plants such as water hyacinth
(ii) Microscopic floating plants, usually algae, called phytoplankton
(iii) Pathogenic eukaryotic and prokaryotic microorganisms (these are generally more common in large ponds, lakes, and slow streams such as rivers compared to rooted vegetation)

1.4.2.1.3 Consumers

This group includes heterotrophic organisms, mostly animals (aquatic), that feed on other organisms or organic detritus. Herbivores and carnivores are included in this group. LHGBRs are the largest component of this group. In LHGBRs, one of the key roles of biomembranes is to limit the passage of small molecules between aqueous compartments in a cell. Understanding the interactions between membranes and small molecules at the molecular level is of significant importance in a number of areas of research. For example, passive partitioning and the subsequent location and movement of drugs in membranes are believed to affect their interactions with membrane-bound proteins, and thus knowledge of such interactions is crucial in the field of drug design. Distribution and solubility of molecules in the cell membrane are important determinants of their bioavailability, which is important for normal growth and toxic effects to the cell. In the case of physical chemistry, it is of interest to understand the partitioning of small molecules in an inhomogeneous environment, such as the lipid bilayer, that cannot be explained by regular bulk solution theories. In a membrane, a solute is not distributed homogeneously, but rather a gradient is formed that varies with density, polarity, and chain order in different regions within the membrane. It has been suggested that different molecules have different effects on the membrane structure based on their location within the membrane.

Computer simulation methods were used by Mukhopadhyay (2014) to obtain information about the equilibrium and dynamical properties of solutes in model membranes. Although standard equilibrium molecular dynamics (MD) simulations can provide information on the distribution of solute molecules in the lipid bilayer, thermodynamics quantities such as free energy and entropy cannot be accurately determined from such simulations. This is because equilibrium simulations do not adequately sample those regions of the configurational space, in particular the high-energy regions that make important contribution to these quantities.

Researchers have described MD simulations using the umbrella sampling technique to determine the free energy of partitioning of pentachlorophenol (PCP), an environmental pollutant, in palmitoyl-oleoyl-phosphatidylcholine (POPC) lipid bilayer. In an earlier equilibrium MD simulation, the distribution of PCP in model phospholipid bilayers was studied. The simulation results were in qualitative agreement with the "bilayer effect", in which partitioning of non-polar molecules into lipid bilayer at room temperature appears to be enthalpy driven rather than entropy driven. Here, a quantitative analysis showed that the favourable enthalpy of partitioning of PCP in lipid bilayer was due to highly favourable van der Waals interactions between the PCP and acyl chains of lipid molecules in a region of high tail density and low free volume in the bilayer.

Figure 1.9 shows a snapshot of the starting configuration of the PCP–POPC lipid bilayer system with 128 lipids, 2460 water molecules, and 2 PCP molecules placed in the lipid bilayer, one per leaflet. A series of 10 ns MD simulations were performed at 298 K in which PCP molecules were constrained to lie in several overlapping ranges, referred to as windows, at different depths in bilayers (0.125 nm apart along the direction perpendicular to the bilayer surface) by means of a biasing (umbrella) potential with a force constant of 1800 kJ/(mol nm^2). In total, these simulations represent 210 ns of simulation time. The distribution of molecules in the presence of this biasing potential was obtained for each window. By combining together the results from all windows spanning the lipid bilayer, the bias was removed and the potential of mean force $G(z)$, which is related to the distribution of PCP in

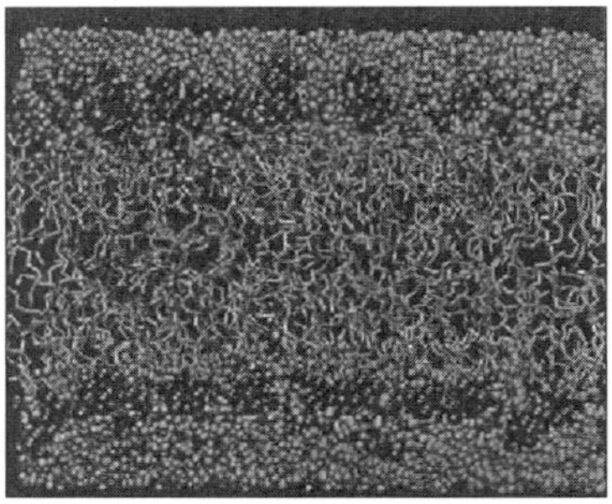

Figure 1.9 Starting configuration of the PCP–POPC bilayer system

the bilayer through the equation $\rho(z) = \exp[-G(z)/RT]$, was constructed using the weighted histogram analysis method (WHAM).

Figure 1.10 shows the potential of mean force (PMF) of PCP in the lipid bilayer. The PMF averaged over two leaflets of the lipid bilayer (Figure 1.10 (b)) shows a clear minimum

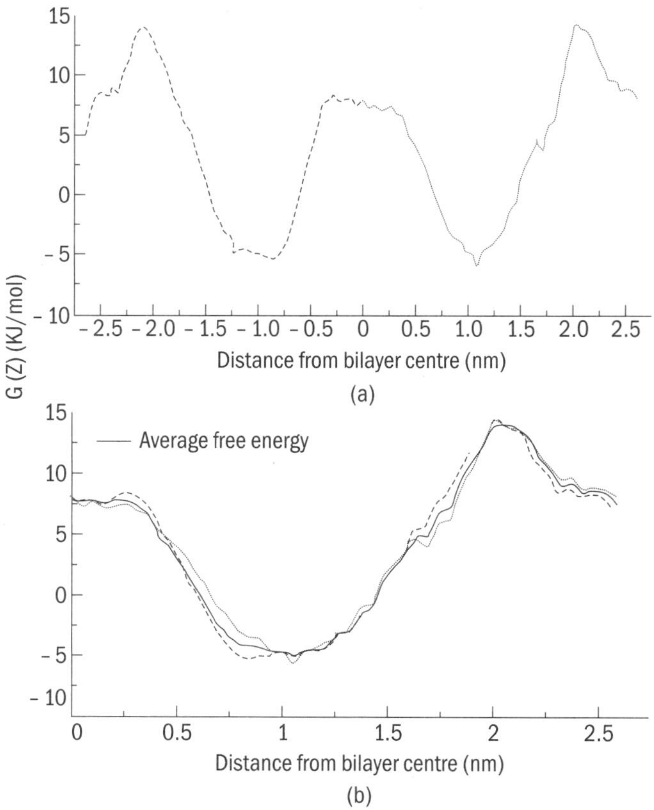

Figure 1.10 (a) Potential of mean force of PCP in the lipid bilayer and (b) its average over the two leaflets of the bilayer

(about –5 kJ/mol) between 0.75 and 1.25 nm from the centre of the bilayer. PCP clearly shows unfavourable free energy of partitioning at the centre of the lipid bilayer, which has the maximum amount of free volume. The lipid/water interface, which is between 1.75 and 2.5 nm from the bilayer centre, acts as a physical barrier (~14 kJ/mol) for the partitioning of PCP into the lipid bilayer.

Figure 1.11 shows the electrostatic and van der Waals components of PCP interaction with the lipid bilayer. PCP has more favourable van der Waals interactions, compared to electrostatic interactions, with POPC in the lipid bilayer (Figure 1.11 (a)). The most favourable van der Waals interactions occur between 0.75 and 1.25 nm from the bilayer centre, similar to the minimum in the PMF profile (Figures 1.11 (a) and 1.11b)). In a previous equilibrium MD simulation of the distribution of PCP in POPC lipid bilayer, the upper acyl chain region of the lipid bilayer was shown to attain optimum packing in the high-density region. This is due to an increase in the van der Waals interaction between PCP and lipids in the dense region of the lipid bilayer as corroborated by the dependence of the van der Waals component of the interaction energy on the membrane position (Figure 1.11 (a)). Figure 1.11 also shows that PCP has favourable electrostatic interactions with POPC and water molecules

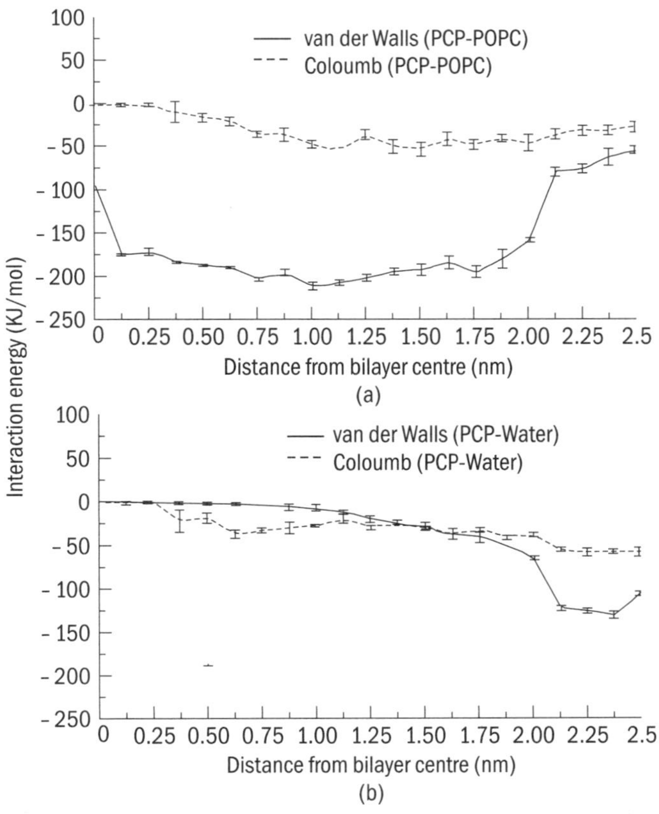

Figure 1.11 Electrostatic and van der Waals components of PCP interaction with the environment: (a) POPC; (b) water

in the lipid bilayer, which is due to the hydrogen bonding between the hydroxyl group of PCP and oxygen atoms in the lipid/water interface.

As mentioned, partition of PCP in the lipid bilayer at the room temperature seems to be driven by a favourable enthalpy (Figure 1.11); however, interaction energy indicates the enthalpic component of the change in free energy and ignores the entropic contribution. PCP increases the order parameter of the acyl chain in the lipid bilayer. The insertion of PCP in the already constrained upper acyl chain region of the bilayer at the room temperature results in an entropy loss, which hinders solute insertion into the lipid bilayer. Thus, a favourable enthalpy, as shown in Figure 1.11, is responsible for the observed partitioning of PCP into the aligned-chain acyl part of the bilayer, which is attributed to tight packing of the solute in that region.

To summarize, the thermodynamics of small molecule–membrane interactions and the atomic level interpretation of such interactions can be obtained by using a combination of umbrella sampling and WHAM methods. The combination of umbrella sampling and WHAM allows computational efforts to be focused on the most difficult part of the problem, that is, sampling the high-energy region of the configurational space, which is not adequately sampled in traditional equilibrium MD simulations, and thus enabling a better estimate of thermodynamic quantity such as free energy. In addition, MacCallum has recently shown that simulations using the umbrella sampling technique performed at different temperatures can be used to decompose PMF of a solute in a bilayer into enthalpic and entropic contributions.

1.4.2.1.4 Decomposers

This group consists of saprophytic organisms, chiefly bacteria and fungi, which decompose soluble organics and organic detritus. Although the body engineering of LHGBRs has contributed to the growth of civilization, it has resulted in the disposal of used waters and wastewaters/effluents in natural waterbodies. This discharge poses hazards to aquatic environment. For preventing hazards, regulation and control of pollution are crucial. Thus, the need of treatment of wastewaters/effluents to bring down their pollution strength to a safe limit before disposal is imperative. In India, Central Pollution Control Board (CPCB) is the deciding authority for setting limits to wastes/effluents. Water pollution control for safe environment involves the protection of the aquatic environment and the maintenance of good water quality in reservoirs, ponds, lakes, streams, rivers, estuaries, seas/oceans and so on for safety of aquatic life and transport. LHGBRs are not the only species that are adversely affected by environmental deterioration owing to pollution, other life forms (for example, animals and plants) also get affected. Owing to rise in pollution and LHGBR population explosion, an average of one species per year gets extinct. As population on earth goes up, the number of extinct species goes down, thus increasing the pollution strength of the environment.

1.4.3 Air pollution

Cities air all over the world are getting overpolluted due to LHGBRs' activities, increased use of transport vehicles, construction activities, and coming up of high-rise buildings. It is forcing city population to breathe air that is almost as noxious as the air during festival seasons. As an example, air quality index based on data collected/calculated at US Embassy as per India standards by SAFAR@MoES-IITM-IMD (based on 1 station data per city at 4 p.m.) is in 'very poor' category, that is, close to 'severe' zone witnessed during 'Diwali' in Delhi. It is the level of air pollution that often prompts the government to take emergency measures such as shutting down schools in many cities. So the gravity of the problem of air pollution is imaginable. The list of threatened and endangered species in the world owing to air pollution, although it is not yet complete because of the lack of details from areas such as China and the Amazon basin, now includes 275 spices of mammals and 300 species of birds. As LHGBRs move upwards along the steeply inclined curve of population growth and air pollution, new species such as transgender LHGBRs are added in the ecological domains due to genetic alterations.

1.4.4 Land pollution

Huge quantities of solid wastes are generated by activities of LHGBRs. For example, daily waste/garbage generation in Delhi NCR is reported to be nearly 8000–8500 metric tonnes. In terms of composition, 51% of urban municipal solid waste (MSW) is of organic nature of which nearly 17.5% is recyclable, and 31% is inert waste. This huge garbage may pose health risks if it is left untreated or unhygenically disposed. A part of this garbage is dumped at landfill sites for developing residential colonies or industry sites/commercial complexes. In residential areas of most cosmopolitan cities of developing countries, large quantities of organic/biodegradable garbages/wastes are accumulated. Food waste (FW) and waste food (WF) garbages are typical examples. Such waste need not be termed as waste as it can be recycled or converted into reusable materials through treatments for generation of VAPs. More recently, Delhi Metro Rail Corporation (DMRC) has commissioned a new facility for recycling of waste.

Innovative recycling of WF/FW is a newer thinking in recent times. In 2015, in Manhattan, United States, a chef turned FW into cuisine. He reformed/reformulated burger buns from stale rye bread topped with rye bread crumbs and also used discarded cheddar cheese with a patty of wasted carrot and beet pulp. In recent years, many upscale restaurants in the United Kingdom and United States have begun using WF ingredients such as fish ribs and pulp of vegetables and fruits to create fresh dishes. Certain chefs in India have started following the practice in small scale and this trend is only to grow with the emphasis on being eco-conscious without causing harm to health. This approach of WF/FW recycling may also promote innovative diet design calculations (IDDC) in food formulation/design.

1.5 POLLUTION ABATEMENT NEEDS

1.5.1 In Water

1.5.1.1 Natural means

The role of algae in wastewater treatment is well known. Nature has been changing since its creation. Water contents/resources on the earth are also changing. As water supplies are getting polluted owing to LHGBRs activities, efficient water management is essential. An economical, environment-friendly, and efficient water purification and management is regarded as one of the key strategies for environmental protection in almost every country to combat crisis resulting from the alteration of global climate.

1.5.1.2 LHGBR-designed means

- **Economic returns:** Many European countries have adopted E4 water project. This is for "economically and ecologically efficient water management in the European chemical industry" and started in 2012. It aims at optimizing water consumption in the chemical industry in Europe. Another objective of this project is to develop new integrated techniques for greater efficiency and sustainability in industrial water treatment for purification and management. For several years, nearly 19 international water treatment partners have been engaged in developing new technologies to lower water consumption, effluent volumes, and power consumption, inferring that the use of integrated engineering is increasing. In addition, it has necessitated simultaneous development of supporting analytical and sensor technologies, which are also on the rise. Water crisis and treatment of wastewater are persistent challenges faced by water engineers. The kinds of problems encountered in the areas of pretreatment and post-treatment of water need to be properly understood for tracing pollutants. It requires rapid pollution tracing and water quality testing. For such a need, the enhanced chemiluminescent oxidase (ECLOX) probe system was developed at the University of Birmingham in more recent years. Highly efficient wastewater treatment units provide better effluent quality and lower space requirements.
- **Greener image:** The efficient use of water in treatment and production processes is not only a potential source of economic return but also helps create a greener image. Gaining greener image is one of the objectives of all nations. New technologies are being developed in various countries. In India, the government has taken multiple initiatives to reinforce its green image. Clean Ganga mission is one such initiative. In this regard, efforts of Bhabha Atomic Research Centre Technologies to develop membrane-based technology to rid Ganga of toxins are highly appreciable. This technology can help remove heavy metals from water in rivers, reservoirs, or ponds. Successful implementation of Ganga Action Plan by the government will be an extraordinary achievement and it will enhance the country's green image. Gangajal Delivery Scheme launched by the government to deliver "Gangajal" at the doorsteps of people through Indian Postal Services is another initiative of the government. Water reuse is gaining significance in many parts of the world owing to water shortage. It has become almost mandatory in many parts of the world to explore ways to reuse or recycle water for creating green image.

1.5.2 In Air/Climate

1.5.2.1 Natural means

Air pollution caused by increase in fine respirable particle levels does get subsided because of natural wind blows that may be moderate to strong. In winter season, relatively calm weather and falling environmental temperatures usually bring down the boundary layer (that is, the lowest part of the atmosphere), trapping pollutants near the earth's surface. It may, however, result in the formation of grey to heavy smog, particularly over cities. If there is no moisture in the climate/atmosphere, fog may not be formed. As huge amounts of trash and particulates pollute air/environment/climate, majority types of pollution occurs near almost all landfill areas, providing means of toxicity. It poses a serious crisis by alteration of global climate. According to experts, the presence of particulates of 1 μm size or less in diameter (PM_1), also called ultrafine particulate matter (PM), in air can cause heart problems. Such ultrafine PM appears in air because of heavy construction activities and vehicular exhausts in almost all parts of the world. The shifting of natural wind blow results in movement of polluted air masses in multiple directions. So LHGBR-designed means are required to monitor and control/regulate such particulates.

1.5.2.2 LHGBR-designed means

Usually people stay healthy in winter season. The moisture-laden winds in winter causes the accumulation of nano-sized PMs near the earth's surface. It may make the climate/air very smoggy. Moreover, a large number of vehicles on the road are also the major source of air pollution in winter. Worsening air quality in winter is also due to emissions from industry, thermal plants and other sources if they are not equipped with installed eco-friendly LHGBR designed devices. In almost all cosmopolitan cities in the world, NO_2 and SO_2 levels have been found to exceed 100 μg/m^3 in air in most areas although the permitted standard is less than 80 μg/m^3. This highlights the role of constructions, demolitions, vehicles, and traffic in causing environmental pollution. Also, rapid build-up of air pollutants in winter months may result because of the inversion effect. Attempts have been made by various countries to reduce air pollution resulting from the activities of humans/LHGBRs. For instance, car-free day observed in Delhi helped to reduce air pollution (Table 1.2). This demonstrates that such measures make a difference.

Table 1.2 Comparison of ambient air quality of Dwarka: 2 days before and on car-free day observed in 2015 in Delhi

Particulate	Panchsheel Apartment, Sector 10, Dwarka		Pollution reduction (%)	Ayushman Hospital and Dwarkadheesh Apartment, Sector 12, Dwarka		Pollution reduction (%)		Near Police Chowki, Sector 3, Dwarka	Pollution reduction (%)
	Nov 20	Nov 22		Nov 20	Nov 22		Nov 20	Nov 22	
SO_2	5	4	20	7	5	28.6	9	8	11.1
NO_2	57	44	22.8	123	85	30.9	318	140	56.0
PM_{10}	511	375	26.6	454	421	7.3	544	497	8.6
$PM_{2.5}$	315	276	12.4	280	254	9.3	311	282	9.3

Notes (i) Standard as per CPCB norms: SO_2: 80, NO_2: 80, PM_{10}: 100, $PM_{2.5}$: 60 (all figures in μg/m^2)
(ii) Air quality monitored from 8 a.m. to 4 p.m. on both the days

In Delhi, every year air quality becomes worse and worse in winter as can be seen in Figure 1.12. Therefore, it becomes necessary to take preventive measures from the public health point of view. Also, prevention of pollution caused by FW/WF through using or recycling sustainably produced biogas by anoxic digestion will greatly reduce the net flow of CO_2 emissions in the atmosphere/air, thereby contributing immensely to the reduction in global warming.

1.5.3 On Land

1.5.3.1 LHGBR-designed means

1.5.3.1.1 Economic returns

Solid wastes/garbages are one of the major contributors to pollution. Their recycling by LHGBR process designs can give economic returns (as discussed previously).

1.5.3.1.2 Greener image

All over the world, reuse of solid wastes, garbages, and sludges is assuming significance. Preventing pollution by reuse of huge solid garbages accumulated in many areas will aid in creating green image if it can be carried out sustainably using a suitable process design (Figures 1.13 and 1.14). This reuse approach will also prevent health hazards. The reuse of garbages by construction of solid state biogas plants is a feasible approach. 'Make in India' concept has been successfully implemented in the case of the development of solid state Deenbandhu biogas plant technology in India. Cost of installation of such a technology is low to moderate depending on the model and size of a plant. The effort of Ministry of Non-conventional Energy Source in this regard is praiseworthy. It is a great aid in creating green image in many localities.

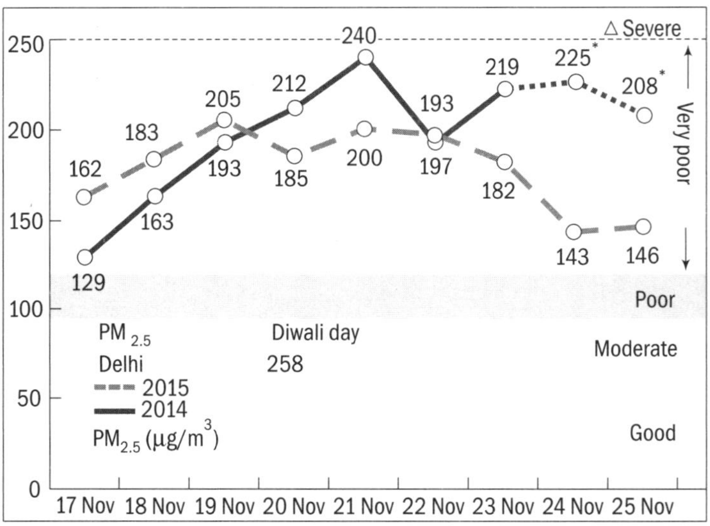

Figure 1.12 Variations of AQI in Delhi
* Forecast

Figure 1.13 Accumulated garbage in a locality

Figure 1.14 Children exposed to open garbage sites are highly vulnerable to various diseases

1.5.3.1.3 Preventing epidemics

Accumulated garbage is a serious health hazard. Seasonal changes further aggravate the situation. This is of concern particularly in rainy season when the number of breeding places for mosquitoes and other harmful insects grows, thus increasing the possibility of epidemic diseases. Malaria, dengue, leptospirosis, typhoid, cholera, jaundice, gastroenteritis, and other epidemic diseases are likely to break out in such localities. For preventing such epidemics, abatement needs are obvious. For immediate prevention, spraying of disinfectants may be useful. Other preventive measures include the following:

- Self-protection from mosquitoes
- Spraying of insecticides and microbicides in the area

Figure 1.15 High-rise building construction activities cause dust pollution

- Keeping watch on the rodent population in the area and preventing infection by vaccination
- Washing of hands frequently with disinfectants

1.5.3.1.4 Better town planning/high-rise buildings

Everybody needs shelter. The concept of "housing for all" aims to ensure that all people have access to adequate housing. This concept has rejuvenated the housing sector/business and increased demolition/construction activities in most parts of the world, including India. These activities generate ultrafine particles that remain suspended or float in air and are swept by blow of winds from one region to another. However, for better town planning, high-rise building constructions have become essential to provide housing for more and more families and fulfil the objective of 'housing for all' (Figure 1.15).

CONCLUSION

Wastes and wastewater should be recycled or reused through treatment/purification. This will offer a great scope for the transformation of wastes into VAPs. The valuable materials are generated by process integration for ecology balancing and minimizing environmental pollution. Realization of this concept will create opportunities to earn, reduce incidence of ailments and diseases in LHGBRs and other life forms, increase human resource development, and enhance and establish knowledge. It will also increase the scope for discoveries, innovations, and developments, The ritual performances of LHGBRs are associated with generation of huge quantities of pollutants on land, in water resources, and in atmosphere/air and this needs proper management and measures to ensure there are no adverse environmental impacts.

Bibliography

Alcamo, I. E. 2001. *DNA Technology: The Awesome Skill*, 2nd edn. San Diego, CA: Harcourt Academic Press

Bailey, J. E. and D. F. Ollis. 1986. *Biochemical Engineering Fundamental*, 2nd edn. New York: McGraw-Hill

Brown, L. R. 1972. *World without Borders*. New Delhi: Affiliated East West Press Pvt. Ltd

De Young, L. R. and K. A. Dill. 1990. Partitioning of nonpolar solutes into bilayers and amorphous *n*-alkanes. *J. Phys. Chem.* 94: 801

DeVido, D. R., J. G. Dorsey, H. S. Chan, and K. A. Dill. 1998. Oil/water partitioning has a different thermodynamic signature when the oil solvent chains are aligned than when they are amorphous. *J. Phys. Chem. B*. 102: 7272

Eckenfelder, W. W. and D. J. O'Connor. 1961. *Biological Waste Treatment*. Pergamon Press

Green, D. W. and R. H. Perry. 2008. *Perry's Chemical Engineers' Handbook*, 8th edn. New York: McGraw-Hill

Humphrey, W., A. Dalke, and K. Schulten. 1996. VMD: visual molecular dynamics. *J. Mol. Graph.* 14: 33

Hurst, C. J., R. L. Crawford, G. R. Knudsen, M. J. McInerney, and L. D. Stetzenbach (eds). 2002. *Manual of Environmental Microbiology*, 2nd edn. Washington, DC: ASM Press

Jepson, P. D. and R. J. Law. 2016. Persistent pollutants, persistent threats. *Science* 352: 1388–1389

Kumar, S., D. Bouzida, R. H. Swendsen, P. A. Kollman, and J. M. Rosenberg. 1992. The weighted histogram analysis method for free energy calculations on biomolecules. I. The method. *J. Comput. Chem.* 13: 1011

Koubi, L., M. Tarek, S. Bandyopadhyay, M. L. Klein, and D. Scharf. 2001. Membrane structural perturbations caused by anesthetics and nonimmobilizers: a molecular dynamics investigation. *Biophys. J.* 81: 3339

Leach, A. R. 2001. *Molecular Modeling: Principle and Application*. Harlow, UK: Prentice Hall

Loehr, R. C. 1974. *Agricultural Waste Management*. New York: Academic Press

MacCallum, J. L., P. Mukhopadhyay, H. Luo, and D. P. Tieleman. 2003. Large scale molecular dynamics simulations of lipid-drug interactions. In *Proceeding of the 17th Annual International Symposium on High Performances Computing Systems and Applications and the OSCAR Symposium*, edited by D. Senechal. Ottawa, Canada: NRC Research Press

Marqusee, J. A. and K. A. Dill. 1986. Solute partitioning into chain molecule interphases: monolayers, bilayer membranes, and micelles. *J. Chem. Phys.* 85: 434

Mezei, M. 1989. Evaluation of the adaptive umbrella sampling method. *Molecular Simulation* 3: 310

Mukhopadhyay, S. N. 2018. *Process Biotechnology Fundamentals*, 4th edn. New Delhi: Viva Books

Mukhopadhyay, S. N. 2012. *Process Biotechnology: Theory and Practice*. New Delhi: TERI Press

Mukhopadhyay, S. N. 2016. *Food Engineering: Process and Technology*. New Delhi: Viva Books

Mukhopadhyay, S. N. 2016. Genes and genome regulated LHGBRs products and facets, p. 48. BCGGR-2016, DBS, BITS Pilani, Rajasthan

Mukhopadhyay, S. N. 2016. Macromolecular LHGBRs products and coproducts: users and curses, Invited Lecture Abstract, ICM-2016, MGU, Kottayam, Kerala

Mukhopadhyay, P., H. J. Vogel, and D. P. Tieleman. 2004. Distribution of pentachlorophenol in phospholipid bilayers: a molecular dynamics study. *Biophys. J.* 86: 337

Mukhopadhyay, S. N. 2017. Products, facets and curses of LGHBRs. *INAE monthly e-Newletter* 8(7), July 1

Rich, L. G. 1974. *Environmental System Engineering.* New York: McGraw-Hill

Seelig, J. and P. Ganz. 1991. Nonclassical hydrophobic effect in membrane binding equilibria. *Biochemistry* 30: 9354

Sills, J. (ed.). 2016. *Science* 352 (6292): 1404

Spellman, F. R. 2015. *Handbook of Environmental Engineering.* Boca Raton, FL: CRC Press, Taylor & Francis Group

Tchobanoglous, G. (ed.). 1979. *Wastewater Engineering: Treatment, Disposal, and Reuse.* New York: McGraw-Hill

Tieleman, D. P., S. J. Marrink, and H. J. Berendsen. 1997. A computer perspective of membranes: molecular dynamics studies of lipid bilayer systems. *Biochim. Biophys. Acta* 1331: 235

2 Water, Wastewater, and Non-aqueous Liquids

2.1 INTRODUCTION

In all cities of the world, water is supplied for domestic, public, and industrial uses. Surface and groundwater resources are the major sources of water supply. Surface water resources include lakes, rivers, streams, ponds, and artificial reservoirs. Groundwater sources include tubewells, borewells, and so on. Absolutely pure water, consisting only of H_2O and free from any dissolved matter, is most difficult to obtain. Such pure water would be soft, colourless, and odourless and would have a pH value of 7.0. It is the last requirement that is difficult to meet because traces of natural CO_2 lower the pH. Rainwater/natural water resource collected in reservoirs/cisterns would be soft; it is still the practice in rural areas to build such cisterns in order to have water resources that saponify easily for washing purposes. In Bermuda, rainwater collected on roofs of dwellings is used as potable water and water for washing. Surface waters are comparatively soft (that is, low in mineral content) and most likely contain easily soluble salts and sediments. Groundwater (for example, from deep wells) is usually free from suspended matters, but is much harder (that is, rich in calcium and magnesium salts) than surface water in the same vicinity. In regions of heavy rainfall (for example, Cherrapunji), surface water contains less mineral matter because of dilution. Ocean water and sea water differ from the water discussed above because of being saline. It contains nearly 3.5% salts, of which 80% is common salt (NaCl). Therefore, ocean water is unfit for common uses until purified to a specific usable limit (Table 2.1).

Table 2.1 Classification of water

Type	Concentration of total salts (ppm)
Soft	<100
Medium hard	100–200
Hard	200–500
Saline	>500

ppm - parts per million

2.2 WATER RESOURCES, USES, POLLUTION, AND PURIFICATION

Water is classified into different types based on the total salt concentration (Table 2.1).

2.2.1 Foreword

Water is used in biochemical and chemical reactions and in downstream processing for product purification in industries. Also, it is the heat transfer medium for both heating and cooling processes. The quality of life depends largely on the quality and purity of water available. Availability of good quality water is not only crucial for domestic daily life and reaction but also vital to industry and agriculture. Water is a unique natural resource. Unlike other natural resources, the total amount of water in the Earth's water cycle is constant. It can be neither increased nor diminished. The total volume of natural water on earth is about 1400 million km^3. Its weight is nearly 1.3×10^{18} tonnes. In maintenance and management of aquatic system and waterways, pollution should be minimal as far as possible. The presence of dissolved or suspended pollutants such as silt, chemicals, faecal matter, metallic elements, organic materials/nutrients, or oil (due to oil spill) in concentration higher than normal in natural waterways (lakes, rivers, and oceans) may cause pollution of aquatic transport system by eutrophication, dissolved oxygen depletion followed by death of aquatic life, fire hazards, and so on. In water management, it is very important to address these water pollution problems taking into account the current aqua-biotic phenomena and their interactions with chemical, physical, and biochemical phenomena. So, in waterway transportation system, prevention of pollution and fire hazards by oil spills may become essential. In order to protect environment from pollution, therefore, prevention of water pollution is a must keeping in view the industrial process development.

2.2.2 Safety and Fitness Concerns

Safety during water purification process is important. In water industries, the process biotechnology of water purification involves chlorination. The practice of chlorination is highly undesirable. A high concentration of chlorine in water becomes a health hazard. Epidemiological studies have shown linkage between water chlorination and various forms of human cancer. Scientific results in recent years led scientists to claim that the practice of water chlorination accounts for nearly 30% cases of human bladder tumours. This has been attributed to chromosomal aberration of living mammalian cells. Care should be taken to ensure that there is no formation of low concentration of genotoxic compounds from humic substrates during chlorination. Biological safety of water purification system is, therefore, an important concern.

Also, in many cases, over-chlorinated water has been used in food biotechnology industries (for example, canning) for cooling purposes. Cooling water containing more than 100 organisms per mL is usually unfit. In such cases, water should not be over-chlorinated in an attempt to make the water impurity free. This is because, under certain conditions, an excess of chlorine in water may lead to corrosion which can have a serious effect on the processed food.

Besides hazardous metallic contaminants, water and wastewater usually contain organic nitrogenous materials. In wastewater purification, microbial nitrogen rich and carbon poor (or low C/N ratio) environments of denitrification process may serve as a key to clean water. When C/N ratio is high, causing *short circuiting of the nitrogen cycle*, the objective of wastewater purification process is defeated.

In water avenues, for the bioremediation of hazardous substances and protection against fire hazards due to oil spills and leakage of oil tankers, technologically redesigning of biocells with the aid of biomolecular design engineering science has been suggested. The redesigned biocells would be capable of degradation of multi-component oil, thereby preventing water pollution.

Oxygen deprivation hazard in water by pollution reduces free oxygen radicals. To protect aquatic life, an adequate supply of oxygen in water resources is necessary. This is because living biocells use water molecules, which are now regarded as radicals, and each molecule possesses two unpaired electrons. As urine and faeces of mammals/animals contain large quantities of oxygen scavengers, they are known to be hazardous in terms of reducing oxygen availability.

2.2.3 Resources and Withdrawal

Water is a multipurpose prime resource. Natural water resources are commonly grouped under four categories: (i) atmosphere or rainwater, (ii) surface water (ocean, sea, river), (iii) stored water, and (iv) groundwater (Table 2.2). Besides these, another important source of water is the wastewater generated from domestic, municipal, and industrial activities. Water is not equally distributed in nature and among people in the world. Nearly 97% of water on earth is sea water. Of the rest, 22% is groundwater and 77% is ice locked in the form of glaciers and polar ice caps. It implies that less than 1% of the supply of freshwater takes part in the hydrological cycle. Nearly half of that 1% is found in rivers, lakes, and swamps. In India, freshwater withdrawal is very important because it is primarily an agricultural country. Annual internal renewable water is estimated to be about 1850 km^3 and the per capita data for 1990 is 2.17 m^3. Of this resource, annual withdrawal percentage of water resource is 18%, whereas per capita withdrawal is about 612 m^3. Table 2.3 shows that the largest consumption of freshwater in India occurs in the agriculture sector.

2.2.4 Purity, Pollution, and Hazard

Natural water resources (for example, groundwater) are, in general, relatively pure and free from bacteria and other contaminants because of the filtering action of the earth. This filtering action removes not only most of the bacteria but also many suspended particles and organic materials. Deep wells usually contain fewer microorganisms compared to water from shallow wells. It is due to the deeper layers of filtering materials. Water resources receive microbial load, organic nitrogen, and other polluting substances from soil, air, sewerage, organic wastes, dead plants and animals, process industry effluents, municipal sewage, and so on. To define the water quality standards, various specifications have been adopted in different countries (Table 2.4). Groundwater is replenished by the hydrological/ water cycle. This replenishment may be due to rainfall, less evapotranspiration, and surface run-off.

Drinking water supplies in almost all urban metropolitan cities in our country are known to contain pollutants. This is primarily because sewerage is absent or partial. In such situations, microbial pollutants may predominate sometimes and pose a major health hazard. Sewer leakage also results in contamination. In developed countries, as well as mountainous health resorts and industrialized areas, road deicing and other urban processes also cause pollution in groundwater supplies.

Many of the organic contaminants are very site specific. In some urban metropolitan cities in our country, two groups of compounds may also be detected in water supplies from time to time. These are chlorinated hydrocarbon solvents and aromatic compounds resulting from the presence of petroleum fraction in water owing to chlorine disinfection. Industrial organic chemicals in groundwater may arise from the same source responsible for contaminating land, for example, industrial premises, fuel storage, transport, and pipeline spillages. Also, water resources become polluted and contaminated by sewerage discharge and human excrement. Effluent loads discharged into natural streams or stored water receptacles at various disposal points cause water resource pollution. Major effluent load discharging systems may emanate from process industries.

Pollution by sewerage and human excrement may be the greatest threat associated with drinking water. They spread infectious diseases, such as typhoid fever, dysentery, or jaundice. In such cases, water contains living microorganisms or viruses that cause diseases, thereby posing a health hazard. Organisms such as *Escherichia coli* and *Streptococcus* sp. are indicators of faecal pollution, and they result in the exertion of biological oxygen demand or BOD (carried out for 5 days at 20°C) and chemical oxygen demand (COD) in potable water as well (Table 2.4).

2.2.5 Purification of Environment and Hazard Protection

2.2.5.1 *Bioprocessing*

Chlorination is a highly effective method for maintaining purity of municipal/corporation water supplies. It purifies water from microbial contamination, thereby preventing spread of infectious diseases. Not only municipal/corporation water supplies but also other water supplies must pass through a series of purification steps before pipeline supply to make the water safe either for human consumption or for process industry and agriculture requirements. In some cases, however, the water resource may be sufficiently pure and may require treatment only with a disinfectant such as chlorine and chlorocresol, depending on the quality of treatment desired and the purpose. In many instances, the condition of water resource is such as to require normally three stages of primary purifications, namely, sedimentation, filtration, and disinfection. However, in recycling of used water in industries or for their disposal in natural waterways, stringent treatment methods are needed to make wastewater/liquid effluent safe for aquatic life and to prevent the filling of water basin with metallic or other solids.

2.2.5.2 *Bacterial biotechnology*

2.2.5.2.1 Oxic

There are limits to the self-purifying ability of a natural body of water. This limit pertains to both capacity and speed of self-purification. In industrialized societies, rapid man-made water purification has become imperative. In a more recent process, a mix of various strains of bacteria has been used to assimilate organic matter, natural and man-made, consuming oxygen and producing chiefly CO_2 and H_2O. The technological success largely depends on the type of water and pollutants, their amount and concentration, the consistency of these determinants, and last, but not the least, the speed of bacterial assimilation and the presence of inhibitory ingredients. Many biotechnological processes have been developed in laboratories and pilot plants. Their operational reliability is of utmost importance to ensure both continuous supply and necessary quality of treated clean water after the purification process. Oxic biological nitrification of organic matter present in polluted water is also one of the major concerns in this bioprocessing to avoid eutrophication.

2.2.5.2.2 Anoxic

Many microbes not only survive but even grow rapidly when starved of oxygen. For survival, these microbes avail gene machines that code for enzymes essential for anoxic growth. The most successful bacterial groups are those that can use nitrate as a substitute for oxygen/air. In anoxic stage of water purification/treatment, less energy is derived from nitrate reduction than that from conventional cell respiration, and so genes for nitrate reduction are activated only when nitrate is present but oxygen is unavailable. However, the product of nitrate reduction is very toxic. Actually, nitrate-respiring bacteria in water purification do make provision for removal of nitrate so formed.

2.2.5.2.3 Integrated oxic–anoxic process biotechnology

In the water purification, bacterial nitrogen metabolism is an important process to clean water. Bacterial metabolism of three molecular species, namely, ammonia, nitrate, and nitrogen, dominates the oxic–anoxic integrated biological nitrogen cycle. In the oxic biotechnology, nitrification plays a major role, while denitrification/fixation governs anoxibiosis in the process biotechnology. A striking balance between oxic–anoxic bioprocessing and water pollution is necessary to achieve the objective of wastewater purification.

2.2.5.2.4 Oxic–anoxic design scheme

In order to overcome the problem of eutrophication in wastewater purification, nitrogen in water must be removed; otherwise, it will cause BOD increase. The following are the operational characteristics of three schemes for designs of nitrification and denitrification (Figure 2.1):

- **Scheme 1:** This scheme depends upon the endogenous respiration of the activated sludge to achieve denitrification.
- **Scheme 2:** Here, a portion of the influent wastewater is passed to the denitrification tank to provide food for facultative organisms, thereby increasing the respiration rate and hence the denitrification rate.
- **Scheme 3:** It uses the influent wastewater that is nitrogen deficient as a food source for denitrifying organisms. The wastewater should contain a readily available carbon source.

In the Scheme 1, while the bioprocessing achieves a low nitrogen effluent, the slow rate of denitrification under endogenous respiration conditions results in a large denitrification tank. Scheme 2, although somewhat reduces the required size of the denitrification tank, has the disadvantage of increasing the unoxidized nitrogen in the treated effluent and in most cases increasing the effluent BOD. The application of Scheme 2 will not contribute nitrogen to the effluent; however, careful controlled operation is required to avoid increase of the effluent BOD. Experimental results and experience have indicated that the economic use of this process scheme increases the respiration rate, resultantly, increasing denitrification rate in the denitrifying unit. It is also double of that required for only carbonaceous BOD removal. If higher O_2 level is maintained in the aeration tank for maximum nitrification, the power requirement will be about two and half times that required for the conventional activated sludge process (ASP) operation.

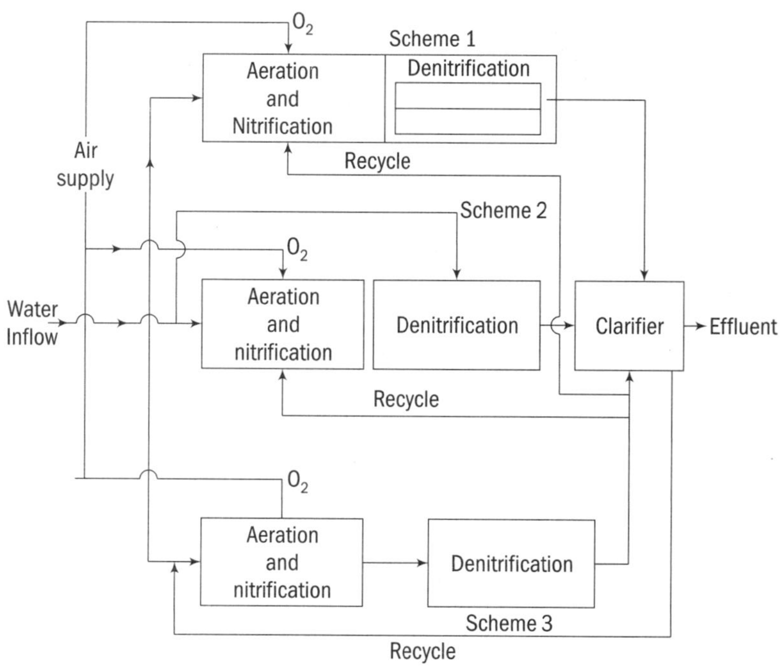

Figure 2.1 Operational process in designs for nitrification and denitrification

2.2.5.2.4 Acceleration of denitrification

Biodenitrification for removal of nitrate and nitrite could be accelerated by adding an appropriate amount of methanol. Based on stoichiometric equations of biodenitrificaiton, it could be computed that 1 mol NO is equivalent to $\frac{5}{6}$ mol by adding an extra amount of methanol. From these considerations, the total amount of methanol (C_m) requirement could be computed by the following relation:

$$C_m = 2.47N_o + 1.53N_i + 0.87DO \quad ...(2.1)$$

Here, N_o is the initial nitrate, N_i is the initial nitrite, and DO is the initial dissolved oxygen concentrations.

2.2.5.3 Chlorine disinfection

Purification of drinking water by chlorine disinfection is an age-old process. It is very effective in killing hazardous microorganisms in potable water. However, biomolecular design engineering concerns of water chlorination hazard have been raised in more recent years. Over-chlorination of water is hazardous to human health. The relationship between the concentration of available chlorine and the time taken to kill organisms in water is exponential:

$$c^n t = \text{constant} \quad ...(2.2)$$

Table 2.2 Water uniqueness and resources

Uniqueness	Water resource categories
Prime natural resource	Rainwater
Essential for survival	Surface water
In water/hydrological cycle	Stored water
It can be neither increased nor diminished	Groundwater
Essential for process industry	Waste or polluted water

Table 2.3 Comparison between India and the world regarding freshwater withdrawal

Country/region	Approximate annual withdrawal (%)		Sector (%)		
	Percentage of water resources per capita (m^3)		D	I	A
India	18–20	612	3–5	4–8	93–87
Worldwide	8	660	8	23	69
North and Central America	10	1692	9	42	49

D– domestic; I– industry; A– agriculture

Table 2.4 Important factors specifying pollution and their approximate usability specifications in water for different purposes

Factor	Approximate usability specifications of water		
	Drinking	Process industry	Agriculture
Physical appearance	Clear, colourless	Clear, colourless	Preferably clear
Odour	Nil	Nil	Nil
pH	7.0	6.9	5.5–9.0
Chemicals (ppm)	500	–	2000
Chloride as Cl^-	200	–	600
Sulphate	200	150	1000
Lead	Nil	Nil	Nil
Arsenic	0.005	–	–
Oil and grease	0.3	1.0	1.0
BOD (for 5 days at 20°C) (ppm)	4.5–5.0	4.5–5.0	6–7
COD (ppm)	7–8.5	–	–

BOD – biological oxygen demand; COD – chemical oxygen demand; ppm – parts per million

Grand Canyon part of the Colorado River has been suffering from severe pollution. Contamination has occurred near older cities, including Hack Canyon and Orphan Mines, where erosion and problems with containment have caused uranium to seep into waterways such as Horn Creek (Fedarko 2016).

2.2.5.4 Chemical processing and associated problems

Water treatment is a significant and persistent problem for scientists/engineers. Water treatment is necessary both before and after water is put to use. Pretreatment of water depends on its usage. In water softening plants, the proper amount of lime dosage is essential for controlling the precipitation of carbonate. After the water has been used, it has to be treated to remove dissolved organics and other nutrients, such as phosphorus and nitrogen.

Many examples are available that illustrate the application of reaction kinetics and equilibria in water treatment. A spectrum of problems need solution for dealing with pretreatment and post-treatment. The pretreatment of water is essential. Most of the reactions involved are physical, chemo-physical, and biochemical. In the example of phosphorus removal by land-based disposal of wastewater, kinetics of the physical process of absorption are illustrated. Biochemical kinetics of nitrification and denitrification are explored in the example of nitrogen removal from wastewater. The water pretreatment involves lime and soda water softening. The control of alkalinity and specifically the maintenance of maximum carbonate concentration are critical.

2.2.5.5 Optimum calcium removal in lime softening

It is common practice to use the results of alkalinity titrations in the operation of lime–soda water softening plants. Operating instructions for such plants frequently include guidelines based upon the values of total alkalinity (T) and phenolphthalein alkalinity ($\overline{P}$). When calcium reduction in hard water is the only objective, it can be achieved by operating at the lime dosage that results in maximum carbonate concentration, that is, maximum precipitation of $CaCO_3$, which occurs when $T/\overline{P} = 2$ (Shah, Novak, Schircrholz, *et al.* 1975).

The basic defining equations for total alkalinity and phenolphthalein alkalinity are as follows:

$$T = HCO_3^- + CO_3^{2-} + OH^- - H^+$$

$$\overline{P} = \frac{1}{2}\, CO_3^{2-} + OH^-$$

where the unit is mg/L.

The equilibrium relations are

$$CaCO_3(s) \rightleftarrows Ca^{2+} + CO_3^{2-}$$

$$H_2O \rightleftarrows H^+ + OH^-$$

$$HCO_3^- \rightleftarrows H^+ + CO_3^{2-}$$

The corrected dissociation constants are defined as

$$K'_{sp} = [Ca^{2+}][CO_3^{2-}]$$

$$K'_W = [H^+][OH^-]$$

$$K'_2 = \frac{[H+][CO_3^{2-}]}{[HCO_3^-]}$$

These constants are functions of temperature, dissolved solids, and normal dissociation. The brackets denote molar concentrations. The above equations can be used to derive the following expressions for the concentration of each ion (Shah, Coulman, Novak, *et al.* 1975):

$$[HCO_3^-] = \frac{T/50{,}000 + [H^+] - K'_W/[H^+]}{1 + 2K'_2/[H^+]}$$

$$[HCO_3^-] = \frac{T/50{,}000 + [H^+] - K'_W/[H^+]}{2(1 + [H^+]/2K'_2)}$$

$$[Ca^{2+}] = \frac{2K'_{sp}\,(1 + [H+]/2K'_2)}{T/50{,}000 + [H^+] - K'_W/[H^+]}$$

Problem 1

By using the above relations, prove that the maximum carbonate concentration occurs when $T/\overline{P} = 2$ for pH > 8.5. Prove this by showing that $[H^+]$, when $T/\overline{P} = 2$, is identical to the $[H^+]$ found when carbonate concentration is maximum.

Solution

As $[H^+]$ is negligible for pH > 8.5 compared to other alkalinity terms, the following expression results on equating $T/\overline{P} = 2$:

$$\frac{T}{\overline{P}} = \frac{HCO_3^- + CO_3^{2-} + OH^-}{CO_3^{2-}/2 + OH^-} = 2$$

This reduces to $HCO_3^- = OH^-$, which indicates that when $T/\overline{P}$ is 2, HCO_3^- must equal OH^-. Since and $[OH^-] = [HCO_3^-]$ can be substituted into the expression for $[HCO_3^-]$, this results in

$$\frac{K'_W}{[H^+]} = \frac{T/50{,}000 - K'_W/[H^+]}{1 + 2K'_2/[H^+]}$$

This expression can be rearranged and solved for $[H^+]$ using the quadratic formula with the following results:

$$[H^+] = \frac{K'_W \times 50{,}000}{T}\left(1 + \sqrt{1 + \frac{2K'_2 T}{50{,}000K'_W}}\right)$$

Next $[H^+]$ is determined for the condition where $[CO_3^{2-}]$ is maximum. This is found by differentiating the expression for $[CO_3]$ with respect to $[H^+]$ and equating to zero (again), $[H^+]$ $T/50{,}000 - K'_W/[H^+]$), that is,

$$\frac{1}{K'_2}\frac{d[CO_3^{2-}]}{d[H^+]} = \frac{2K'_2K'_W/[H^+]^2 + 2K'_2/[H^+] - T/50{,}000}{([H^+] + 2K'_2\,)^2} = 0$$

This expression can be zero only when the numerator is zero. Thus

$$[H^+]^2 - \frac{2K'_2 \times 50{,}000}{T}[H^+] - \frac{2 \times K'_2 \times 2K'_W \times 50{,}000}{T} = 0$$

which, when solved for $[H^+]$, yields

$$[H^+] = \frac{K'_W \times 50{,}000}{T}\left(1 + \sqrt{1 + \frac{2K'_2 \times T}{50{,}000\,K'_W}}\right) \tag{P.1}$$

Examination of the second derivative confirms that a maximum is determined.

As hydrogen concentration is the same for both the cases, the hypothesis that carbonate concentration is maximum when $T/\overline{P} = 2$ is proven. The following section discusses the subject in more detail.

2.2.5.6 Phosphate adsorption kinetics

The interaction of phosphates with soils has become a topic of interest to engineers. Recent environmental concerns and legislations have made this topic more significant as soils are increasingly considered for the treatment of agricultural, domestic, and industrial wastewaters and materials.

Phosphates are present in agricultural, domestic, and certain industrial wastewaters, and soils have only a finite capacity for removal of phosphates from wastewaters. This phosphate removal process in the initial stage involves adsorption of orthophosphates on soil particles. To quantitatively predict the movement of orthophosphates in soils, one must have equilibrium data on the orthophosphate–water–soil system and kinetic data on orthophosphate adsorption. With these data, one can formulate kinetic models that, when coupled with a material balance, will allow the prediction of orthophosphate movement in soil.

Problem 2

The equilibrium data for the orthophosphate–water–soil system have been found to be represented by the Langmuir equilibrium relationship:

$$C^* = \frac{(1/Kb)M}{1 - (1/b)M} \tag{P.2}$$

From the following information, find the values of "K" and "b".

One gram of dry soil was placed with 50 mL of orthophosphate solution into an Erlenmeyer flask. The flask was agitated and the amount of orthophosphate remaining in solution was determined periodically. After 3 h, adsorption equilibrium was reached. The amount of adsorbed orthophosphate – P on the dry soil charged to the flask was determined to be 3.67 ppm.

Data time (hour)	C orthophosphate – P in solution (ppm)
0	3.08
3	1.30
0	6.28
3	3.38
0	11.01
3	8.64

ppm – parts per million

Solution

The amount of absorbed P on the soil was determined from the following material balance on P:

$$M = M_i + V_L/W_S\,(C_i - C^*) \tag{P.3}$$

where M_i is 3.67 ppm and V_L/W_S is 50 mL/g. Sample calculation is

$$M = 3.67 + 50(3.08 - 1.39) = 92.7 \text{ ppm}$$

The Langmuir "K" and "b" values can be determined from a plot of C^*/M versus C^* as Eq. (P.2) can be rearranged to

$$C^*/M = \left(\frac{1}{b}\right)C^* + \frac{1}{Kb} \tag{P.4}$$

When $C^* = 1.30$, $C^*/M = 1.30/92.7 = 1.402 \times 10^{-2}$. From the slope and intercept of the plot of C^*/M versus C^*, we find that $K = 1.08$ and $b = 196$ ppm.

Problem 3

Assuming that Langmuir kinetic mechanism represents the rate of orthophosphate – P absorption by soils, determine the kinetic coefficient *"ka"* from the following in formation. Batch experiments were run using agitated Erlenmeyer flasks. One gram of dry soil was placed with 50 mL of orthophosphate solution into the flask and the orthophosphate remaining in solution was determined as adsorption took place.

Time (hour)	C orthophosphate – P in solution (ppm)
0.0	3.06
0.5	1.62
1.0	1.43
3.0	1.30
0.0	6.28
0.5	4.03
1.0	3.77
3.0	3.38
0.0	11.81
0.5	8.96
1.0	8.96
3.0	8.64

ppm – parts per million

Solution

For the batch experiment, the material balance and Langmuir kinetics model combine to give

$$-\frac{dc}{dt} = ka\left[C(b - M) - \frac{M}{L}\right] \tag{P.5}$$

and

$$M = M_i + \frac{V_L}{W_S}(C_i - C) \tag{P.6}$$

where M_i is 3.67 ppm, V_L/W_S is 50 mL/g, K is 1.08, and b is 196.

Equations (P.5) and (P.6) can be combined and integrated to give

$$\frac{1}{\xi}\text{In}\left(\frac{2\alpha C + \beta - \xi}{2\alpha C + \beta + \xi}\right)c(t)\ c(t = 0) = -\ ka(t - 0) \tag{P.7}$$

where

$$\alpha = \frac{V_L}{W_S}$$

$$\beta = b + \frac{a}{K} - aC_i - M_i$$

$$\gamma = \frac{aC_i}{K} + \frac{M_i}{K}$$

$$\xi = + \sqrt{4\alpha\gamma + \beta^2}$$

The kinetic coefficient "*ka*" can be determined by plotting the left-hand side of Eq. (P.7) versus time or by plotting *C* versus *t* for various values of *ka* until a desirable fit is obtained. The latter method is illustrated in Figure 2.2.

Nitrogen removal in continuous systems

Traditional activated sludge plants can be modified to remove nitrogen and reduce biological oxidation demand as shown by many researchers. The nitrogen removal is accomplished via nitrification and denitrification reactions. A modified process will consist of three stages. In the first stage, much of the carbon is removed by microbial oxidation. Nitrification is carried out in the second stage in which ammonium nitrogen is oxidized to nitrate nitrogen. Nitrate nitrogen is used as an electron acceptor in an anaerobic environment in presence of an organic carbon source and is converted to atmospheric nitrogen. This dentrification reaction is carried out in the third stage. All the three reactions are carried out in continuous stirred tank reactors.

Until now, all the microbial reaction kinetics have been assumed to follow Michaelis–Menten expression with respect to the growth-limiting substrate. This is applicable only if all other substrates are in excess. In nitrification and denitrification, many times situations could arise where more than one substrate becomes growth limiting. The following problems illustrate the application of equation of continuity coupled with multiple substrate limited microbial growth kinetics.

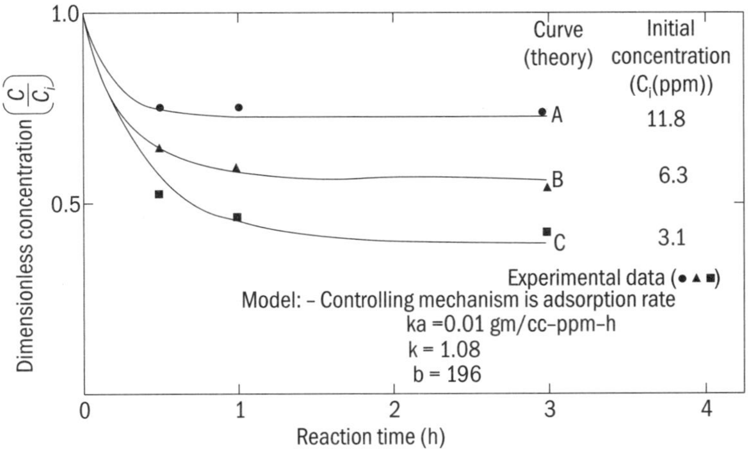

Figure 2.2 Plot of Eq. (P.7)

ppm – parts per million

Problem 4

Determine the concentrations of organic carbon and nitrate nitrogen remaining in the denitrifier operated as the continuous stirred tank reactor.

Dilution rate = 1 day^{-1}

Assume growth kinetics to be

$$\mu = \mu_m \left(\frac{S_1}{K_1 + S_1}\right)\left(\frac{S_2}{K_2 + S_2}\right)$$

where

$$S_1 \equiv NO_3^- -N$$
$$S_2 \equiv \text{Org. C.}$$
$$\mu_m = 5 \text{ day}^{-1}$$
$$K_1 = 25 \text{ mg/L}$$
$$K_2 = 75 \text{ mg/L}$$

Incoming nitrate nitrogen = 40 mg/L

Incoming organic carbon = 120 mg/L

Y_1(yield coefficient for NO_3^- –N) = 0.68

Y_2(yield coefficient for organic carbon) = 0.17

Solution

For a continuous system, material balance of substrates and cell mass results in the following equations:

$$X: \quad \frac{dx}{dt} = (\mu - D)x$$

$$S_1: \quad \frac{dS_1}{dt} = D(S_{1i} - S_1) - (\mu x/Y_1) \tag{P.8}$$

$$S_2: \quad \frac{dS_2}{dt} = D(S_{2i} - S_2) - (\mu x/Y_2)$$

Here, $\mu = \mu_m \left(\frac{S_1}{K_1 + S_1}\right)\left(\frac{S_2}{K_2 + S_2}\right)$

x = concentration of microbes, mg/L

Equation (P.8), at steady state, reduces to

$$\mu = D = \mu_m \left(\frac{S_1}{K_1 + S_1}\right)\left(\frac{S_2}{K_2 + S_2}\right)$$

$$S_{1i} - S_1 = x | y_1 \tag{P.9}$$
$$S_{2i} - S_2 = x | y_2$$

Solving Eq. (P.9) would give S_1, S_2, and x in the reactor effluent as follows:

$$S_2 = \left(\frac{-\beta \pm \sqrt{\beta^2 - 4\alpha\gamma}}{2\alpha}\right)$$

$$X = Y_2(S_{2i} - S_2) \tag{P.10}$$
$$S_1 = S_{1i} - X/Y_1$$

where

$$\alpha = \frac{D}{\mu_m}\frac{Y_2}{Y_1} - \frac{Y_2}{Y_1}$$

$$\beta = \frac{D}{\mu_m}\left(K_1 + S_{1i} - \frac{Y_2}{Y_1}S_{2i} + \frac{Y_2}{Y_1}K_2\right) - S_{1i} + \frac{Y_2}{Y_1}S_{2i}$$

$$\gamma = \frac{D}{\mu_m}K_2\left(K_1 + S_{1i} - \frac{Y_2}{Y_1}(S_{2i})\right)$$

Putting proper values, one obtains S_1 = 23.286 mg/L and S_2 = 53.144051. In the given example, $Y_1/Y_2 = 4$; that is, four parts of organic carbon are needed for every part of reduced. As the original supply ratio is only 3, complete removal cannot be accomplished even at very high residence times. If S_{2i} = 160 mg/L in place of 120 mg/L, the effluent substrate concentrations would be given by S_1 = 17.597 mg/L and S_2 = 70.388 mg/L. The degree of denitrification achieved is increased by providing proper amounts of the energy-yielding substrate. If a recycle is carried out, as is customary in activated sludge plants, the degree of reduction would increase.

Steady-state equations for this system are as follows:

$$\mu X_1 = D[(1 + r)X_1 - rX_3]$$

$$D(S_{1i} - S_1) = \frac{\mu X_2}{Y_1} \tag{P.11}$$

$$D(S_{2i} - S_2) = \frac{\mu X_2}{Y_2}$$

where

$$\mu = \mu_m\left(\frac{S_1}{K_1 + S_1}\right)\left(\frac{S_2}{K_2 + S_2}\right)$$

If r = 0.5 and $X_3 = 2X_1$, one gets $D = 2\mu$. Solving Eq. (P.11) gives S_1 = 10.046 mg/L and S_2 = 40.183 mg/L. Here, removal efficiency is 74.89% against 56% for no recycle system.

Problem 5

Perform similar calculations for the nitrification reactor.

Dilution rate = 0.25 h^{-1}

Assume $\mu = \mu_m\left(\frac{S_1}{K_1 + S_1}\right)\left(\frac{S_2}{K_2 + S_2}\right)$

where

$S_1 = NH_4^+ - N$

S_2 = dissolved oxygen

S_{1i} = 30 mg/L

Let $K_La = 60h^{-1}$(Sinclair and Ryder 1975), μ_m = 0.8 h, Y_2 = 1.4, Y_1 = 0.7, K_2 = 2 mg/L, and K_1 = 5 mg/L.

Solution

One of the limiting substrates for nitrification is dissolved oxygen concentration. Its material balance will have to include transfer of oxygen from the bubbling phase to the liquid phase. Steady-state equations for this case are

$$\mu = D = 0.8\left(\frac{S_1}{5 + S_1}\right)\left(\frac{S_2}{2 + S_2}\right)$$

$$D(S_{2i} - S_2) + K_L\, a(S_2^* - S_2) - \frac{\mu m}{Y_2} = 0 \qquad \text{(P.12)}$$

$$(S_{1i} - S_1)\, Y_1 = X$$

Here, S_2^* is the equilibrium concentration of oxygen in liquid with the bubbling air. Let $S_{2i} = S_2^*$ = 8 mg/L. Solving Eq. (P.12) gives

$$S_2 = \left(\frac{-\beta \pm R\beta^2 - 4\alpha\gamma}{2\alpha}\right)$$

$$X = \frac{y_2}{D}(8 - S_2)(60 + D) \qquad \text{(P.13)}$$

$$S_1 = S_{1i} - \frac{X}{Y_1}$$

where

$$\alpha = \frac{Y_2}{Y_1D}(60 + D)\left(\frac{D}{\mu m} - 1\right)$$

$$\beta = \frac{D}{\mu_m}\left[\frac{Y_2}{Y_1D}(60 + D)K_2 + K_1 + S_{1i}\frac{8Y_2}{Y_1D}(60 + D)\right] - S_{1i} + \frac{8Y}{Y_1D}(60 + D)$$

$$\gamma = \frac{D}{\mu m}K_2\left[K_1 - S_{1i} - S_{1i}\frac{8Y_2}{Y_1D}(60 + D)\right]$$

Substituting numerical values of all the constants in Eq. (P.13) gives S_2 = 7.94 mg/L and S_1 = 3.21 mg/L. As the value of S_2 is near the equilibrium value and is much higher than K_2, growth kinetics can be reduced to a single limiting substrate model. The form of expression for μ reduces very conveniently to Monod equation if one of the substrates is in excess. It has been shown experimentally that nitrification should be carried out in a separate stage from carbon oxidation because the growth rate of nitrifiers is much smaller and would require higher residence times. Obviously, the values of the constants μ_m, K_1, and K_2 chosen in the above example are not correct because they predict almost complete nitrification even at 4 h residence time.

Nomenclature

a	liquid–solid interfacial area per volume of solution, cm^{-1}
[A]	molar concentration of species A
b	Langmuir adsorption maximum, ppm
C	liquid-phase P concentration, ppm
C^*	liquid-phase P concentration in equilibrium with the concentration of phosphate, ppm
D	dilution rate, $time^{-1}$
K	ratio of forward to the reverse Langmuir kinetic coefficient, ppm^{-1}
k	kinetic rate coefficient
k_1, k_2	Michaelis–Menten constants (concentration of substrates at which $\mu = \frac{1}{2}\mu_m$)
K'_W, K'_2, K'_{SP}	dissociation constants
K_La	mass transfer coefficient, h^{-1}
M	solid-phase P concentration, ppm
$\overline{P}$	phenolphthalein alkalinity
r	fraction of the effluent recycled
S_j	limiting substrate concentration, mg/L
S_2^*	concentration of dissolved oxygen in equilibrium with gas phase, mg/L
t	time, h
T	total alkalinity
V_L	liquid volume, cm^3
W_S	weight of soil, g
X	concentration of microbes, mg/L
y	yield coefficient, mg microbes produced per mg substrate used
μ	growth rate of microbes, $time^{-1}$
α	constant
β	constant
γ	constant

Physical treatments

With regard to polluted water, gross particulate removal (GPR) is crucial and it is referred to as the primary physical treatment. The process includes screening, gravity settling or sedimentation, and flotation. In screening, the cross-sectional area of a screen channel is so designed that the velocity of flow through the screen is not less than 2 ft/s and greater than 3 ft/s. The cross-sectional area(s) is calculated using the following equation:

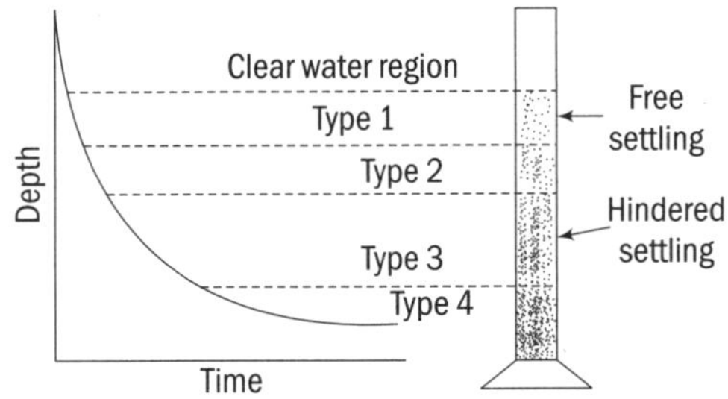

Figure 2.3 Suspended solid settling phenomenon

$$S = \frac{Q}{ev} \tag{2.3}$$

Here, S is in ft^2, Q is volumetric flow rate in ft^3/s, e is the ratio of the area of screen openings to the cross-sectional area of channel, and $v$is average velocity through screen (ft/s). Sedimentation plays a key role in GPR. Sedimentation may take place by free settling or hindered settling. Settling type/phenomenon, depending on the concentration of suspended solids, may be depicted with the help of a time versus depth relation in a cylinder containing suspended solids (Figure 2.3).

2.3 ESSENTIALITY OF WATER

Essentiality of water is associated with its structure and related characteristic properties as discussed in the following sections.

2.3.1 Structure of Water

In order to understand the function of water, first one needs to be thoroughly acquainted with its structure. Its state (solid, liquid, or gas) when present in a material, particularly a biological system, needs to be identified in order to understand its consequent behaviour. Structurally, the six valence electrons of oxygen in a water molecule are hybridized to four sp^3 orbitals. These orbitals are elongated to the corners of a tetrahedron. Of the four hybrid orbitals, two form O–H covalent bonds with a bond angle of 105° for H–O–H. The O–H covalent bonds, due to the highly electronegative oxygen, have a partial (40%) ionic character. Figures 2.4–2.6 show the molecular structure of water.

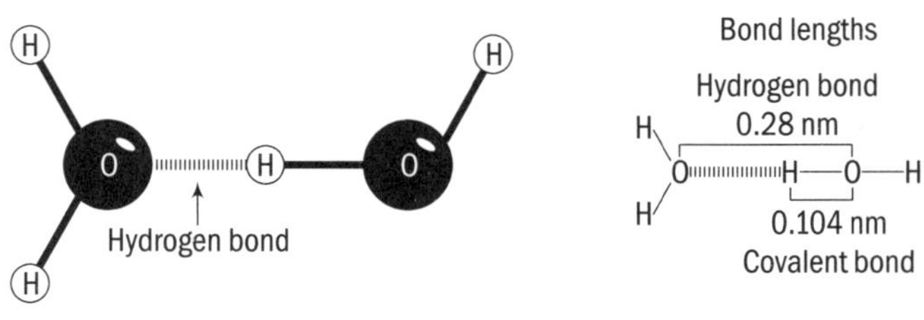

Figure 2.4 Feature of hydrogen bond formed between two water molecules

Two atoms of different elements connected by a covalent bond exert different forces of attraction on electrons. In such cases, the bond is dipolar, with one end slightly negatively charged (δ^-) and the other end slightly positively charged (δ^+). A covalent bond in which both donor and acceptor atoms are of the same element is called non-polar. The donor and acceptor properties of water relate to the following features:

(i) Tetrahedral structure

(ii) Aggregation property

(iii) Structural geometry (*C* is the centre of gravity)

(iv) Orbital model

The electronegative atom that participates in the formation of the polar covalent bond is called the hydrogen donor, while the electronegative atom that takes part in the formation of the hydrogen bond is called the hydrogen acceptor. In the given example, oxygen is the hydrogen donor and nitrogen is the hydrogen acceptor. A hydrogen bond can also be thought of as an intermediate type of acid–base reaction. It is to be noted, however, that the electronegative bond is an electrostatic one; no electrons are shared or exchanged between atoms in such a bond.

Hydrogen bonds are highly directional (note the arrows in Figure 2.4) and are strongest when all three atoms are collinear. The bond energy of a hydrogen bond is in the range of about 3–7 kcal/mol. This bond energy is intermediate between the energy of a covalent bond and that of a van der Waals bond. The energy of the bond is strongest when the hydrogen acceptor is fluorine, oxygen, or nitrogen. Hydrogen bonds formed with any of the three halogens are significant.

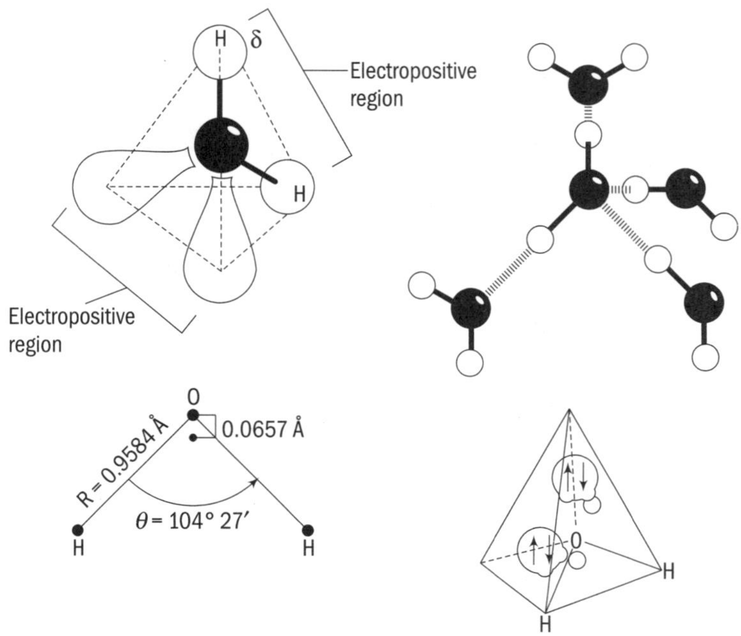

Figure 2.5 Structural bonding of a water molecule

Hydrogen bonds are responsible for the structure of water and its special properties as a biological solvent. There is extensive hydrogen bonding between water molecules, forming the water matrix. This structure has significant effects on the freezing and boiling points of water and its solubility properties. Any molecule capable of forming a hydrogen bond usually does so when mixed with water. Thus, a number of compounds dissociate and are soluble in water.

Hydrogen bonds are responsible for maintaining the shape of proteins. As shape is crucial to their function (both enzymes and structural components), this bonding is extremely important. For example, hydrogen bonds maintain the helical shape of keratin and collagen molecules and give them their characteristic strength and flexibility.

More important, hydrogen bonds hold together the two helices of a deoxyribonucleic acid (DNA). The bonding occurs between base pairs: adenine (A) with thymine (T) and guanine (G) with cytosine (C). The intermediate bond strength of the hydrogen bond is ideal for the function of DNA; it is strong enough to give the molecule stability, yet weak enough to be broken with sufficient ease for replication and ribonucleic acid (RNA) synthesis. The formation of a hydrogen bond between two adjacent water molecules is shown in Figure 2.5. The figure also indicates the lengths of the hydrogen and covalent bonds.

Although a water molecule has an overall neutral charge, the electrons within it are asymmetrically distributed, which makes the molecule polar. The nucleus of the oxygen draws electrons away from the nuclei of hydrogen atoms, leaving these nuclei with a small net positive charge. Thus, an excess of electron density on the oxygen atom creates weak negative regions at the other two corners of an imaginary tetrahedron (Figure 2.6).

Molecules of water join together transitionally in a hydrogen-bonded lattice (Figure 2.5). The property of aggregation of water, $(H_2O)_n$, molecules has great atmospheric implications.

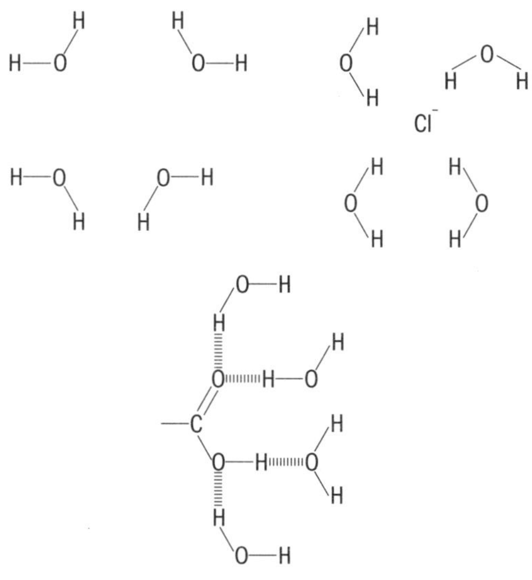

Figure 2.6 Cluster of water molecules around ions

Even at 37°C, 15% of water molecules are joined by four others in a short-lived assembly known as a flickering cluster. The cohesive nature of water is responsible for many of its unusual properties, such as high surface tension, specific heat, and heat of vaporization.

2.3.2 Properties of Water

2.3.2.1 General

The chemistry of life on this planet is essentially the chemistry of water and carbon. Water is the most abundant molecule both in a living cell and on earth. In fact, it makes up 70%–90% of the weight of most life forms.

The properties of water shaped the chemistry of living organisms when life began in oceans. Life easily developed in the liquid phase as reactions in solutions are more rapid than reactions between solids. In addition, water is an excellent solvent for living systems. A major advantage of this property is that water can stay in liquid state over a wide range of temperature variations. Almost all chemicals present in the living matter are soluble in water. The polar nature of water molecules allows them to remain as associated molecules. Water molecules cluster around ions and other polar molecules (Figure 2.6). Similar to ions shown in Figure 2.6, molecules that can be accommodated in the hydrogen-bonded structure of water are hydrophilic and relatively water soluble.

Scientifically, water exists in monomeric, dimeric, trimeric, and tetrameric forms in nature. The presence of water in natural hypercycles is crucial. For both biological and non-biological activities, water is the most important substance. Recent discovery of liquid water beneath the surface of Mars has further confirmed that it is a universal liquid. This discovery may shed some light on the existence of extraterrestrial life. Water has both chemo-physical and biogenic properties.

2.3.2.2 Chemo-physical

It is important to consider that water acts as a strong hydrogen bonding agent for biological systems. Intramolecular bonds may be formed even in strong polar solvents such as water. For better understanding of the role of water in biology, knowledge of physical biochemistry, computational chemistry, and biochemistry in relation to aqueous solvent systems is very important. This is because water contains significant numbers of aggregates.

Water in liquid form has a unique structure that is responsible for its chemo-physical properties in relation to temperature. The increase in density of water that occurs on warming from 0°C to 4°C may be attributed to a decrease in the concentration of aggregates and increase in the number of complexes with other denser structures. The structural features of liquid water reveal a crystal structure with a hydrogen-bonded framework of 46 water molecules per cubic unit, with eight water molecules per cubic unit at the centre of the polyhydra formed without hydrogen bonds. The number of hydrogen bonds in liquid form of water is 85% more than that in ice as indicated by the enthalpy of fusion. The density of a water crystal is 1 g/cm^3, which is the same as that of liquid water. Furthermore, it is found on analysis that the pentagonal dodecahedra structure may be arranged relative to one another in a number of ways such that random structures for liquid water may be based upon aggregates of water molecules bonded to one another in different ways. The radial distribution function for water as calculated from X-ray diffraction pattern, the dispersion of dielectric constant, and some otherchemo-physical properties of water is found to be compatible with this structure. An advantage the structure provides water is the availability of spaces between intermolecular spaces in associated/aggregated liquid water molecules, which is responsible for its solvent property. The hydrogen atoms, or protons, in water have a tendency to "wrap" themselves with electrons available from whatever sources. In aqueous solution, protons combine with water and form H_3O^+ (hydronium) ion complex.

$$H^+ + H_2O \longleftrightarrow H_2O^+ \tag{2.4}$$

If water is used as a solvent, the above ionization shifts to the right such that strengths of strong acids in aqueous solutions cannot be estimated. On the contrary, the strength of weak acids such as acetic, propionic, and phosphoric acids can be estimated as they do not ionize easily in aqueous solutions.

Water is a weak base, but its conjugate acid, H_3O^+, is a strong base. The influence of water as a solvent on the ionization and apparent strength of bases is also similar to that of acids. The strength of weak bases can be estimated in aqueous solutions because water is a weak proton acceptor. In biological systems and many industrial processes, constancy of pH is an essential prerequisite. Different types of solvents are known to keep the pH of solutions at a constant level. Such solutions are known as buffer solutions. The amount of base added to change the pH of a buffer solution by unity is called its buffer capacity.

$$\beta = \frac{db}{d(pH)} \tag{2.5}$$

where β is the buffer capacity and b is the amount of base added.
Based upon electroneutrality, Eq.(2.5) may be written as

$$b + C_H^+ = C_{OH-} \tag{2.6}$$

From Eqs (2.5) and (2.6), it is possible to arrive at an expression of buffer capacity of water.

$$\beta = 2.303\ [C_H^+ + C_{OH}^-] \tag{2.7}$$

The calculated value of β for water (Eq. (2.7)) shows an extremely small value when water is neutral (pH 7). This value rises to a certain extent when the solution is acidic or alkaline (below 2.4 and above 11.6 as experimentally determined in biosystems and non-biosystems).

In general, water in cells and cellular functions exhibits properties similar to those of water in bulk aqueous solutions. However, alternative biogenic descriptions of properties of cell water with regard to various cellular activities are also available. Functions such as water-powered bio-pumping in cellular activities have been studied. Ions can make and break water molecules *in vitro*. Water pricing based upon chemo-physical property has industrial implications. Table 2.5 lists the different chemo-physical properties of water with regard to bio-aqueous solvent systems.

Table 2.5 Chemo-physical properties of water

Properties	Value	Unit
Colour/odour (ideal)	Colourless/odourless	
Density (at 4°C)	1.00	g/cm^3
Density (at boiling point)	0.958	g/cm^3
Boiling point	100.0	°C
Freezing point (at 1 atm pressure)	0.0023	°C
Triple point (at 4.58 Hg)	+ 0.0098	°C
pH	7.00	–
Ion product (at 25°C) constant	1.008×10^{-14}	mol^2/dm^6
Viscosity (at 20°C)	0.001	Pa·s
Permittivity (at 18°C)	81.7	–
Specific conductance (at 18°C)	4×10^{-6}	$ohm^{-1}\ cm^{-1}$
Thermal conductivity	0.537	$kcal/h/m^2/°C$
Heat of fusion	1.434	kcal/mol
Heat capacity (liquid)	0.9986	cal/g/°C
Heat capacity (vapour)	0.484	cal/g/°C
Heat of formation (20°C)	68.35	kcal/mol
Heat of vaporization	9.719	kcal/mol
Latent heat of vaporization	576.6	cal/g
Critical temperature	374.1	°C
Surface tension (at boiling point)	58.8	dyn/cm
Electrolytic dissociation constant (at 18°C)		
K_1 ($H^2O \rightarrow H+ + OH^-$)	1.07×10^{16}	–
K_2 ($OH^- \rightarrow H+ + O^{2-}$)	$<10^{-36}$	
Dipole moment of vapour	1.85	cgs units $\times 10^{48}$
Strength of a hydrogen bond	5.0	kcal/mol
Hbond length in water	0.26 – 0.31	nm
H–O–H valence angle	104.5	degree
Redox potential (E_h)	2.13	V (volt)

2.3.2.3 *Biogenic*

The existence of life forms might not have been possible if water did not possess biogenic properties. Factors affecting the biogenic characteristics of water are listed in Table 2.6. Moreover, water has atmospheric weather implications for sustenance of life forms on earth.

Table 2.6 Factors affecting the biogenic characteristics of water

Items	Per cent (%)
Quantum of active radiation consumed in photosynthesis with respect to total incident radiation on the earth's surface	43
Maximum absorption of active radiation for consumption in photosynthesis (20% lost)	80
Conversion efficiency of absorbed light	23–30
Loss of biomass by respiration at night	20–40

Note It is clear that the maximum overall efficiency $E = 0.43 \times 0.8 \times (0.23 - 0.30) \times (0.60 - 0.80) = 5\% - 8\%$. In India, $E = (0.05\text{–}0.08) \times 4950\ \text{kcal/m}^2\text{/day} \approx 250 - 400\ \text{kcal/m}^2\text{/day}$.

For such sustenance, it is well known that the earth's temperature, weather, and climate are determined by absorption of solar radiation, with water playing a dominant role in the process.

Scientists have developed an "etch canal" method to assess the distinction between chemical and biogenic weathering of micrometeorites. Materials composed of olivine and interstitial glass bars have been found etched out in deep-sea spherules due to chemical etching in water. However, in spherules extracted from Greenland sediments (made of cocoons of siderobacteria), olivine is etched out through biogenic weathering.

The densities and partial pressures of hydrogen-bonded water complexes $-(H_2O)_n-$ indicate that they absorb a part of solar radiation. Their spectra shift and broaden from those of water vapour monomers. Therefore, the role of $(H_2O)_n$, when n is greater than 2, requires a study of the kinetics of $nH_2O \rightarrow (H_2O)_n$ thoroughly. Recent studies inferred that dimers and trimers of water influence the amount of solar radiation striking the earth's surface. However, higher oligomers of mole fractions less than 10^{-11} are unlikely to absorb appreciable amounts of radiation. This finding relates to the fact that the variation in the amount of solar radiation striking the earth's surface affects global biomass production potential. In addition, the amount of solar radiation reaching the earth's surface depends upon climatic conditions and varies with the geographical location. The level of dependence of the corresponding country on agricultural production will thus change accordingly. Further, several physical and biological constraints are involved in it.

The different biogenic properties of water with regard to bio-aqueous solvent systems are as follows:

- It is non-toxic to living beings.
- It is required as an aqueous solvent in a human body bioreactor.
- It is essential in biogenic weathering.
- It is multiphasically heterogeneous.
- It supports ionic interactions between chemicals of life in aqueous media.
- It aids in developing multiphase systems separated by membranes.
- It is required in water-powered bio-pump operations.
- By virtue of the biogenic weathering property of water, ice crystal nucleation or formation control can be stimulated by bacteria.

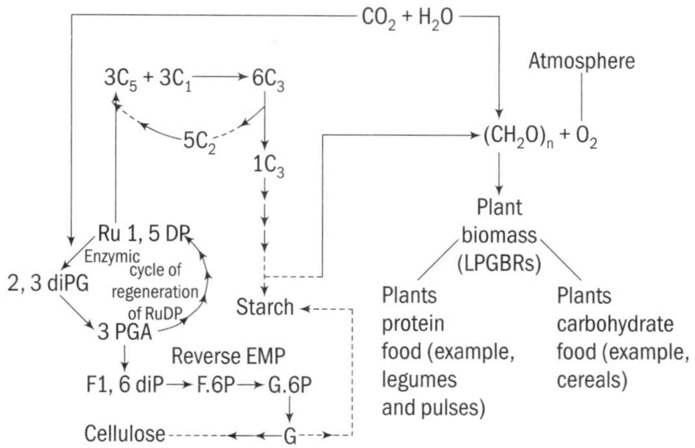

Figure 2.7 Atmospheric cycles in plant biomass synthesis/living plant genomic bioreactors (LPGBRs)

- Biogenic ions can make and break water molecules *in vivo.*
- Ions can make and break water molecules *in vivo.*
- In protein folding and unfolding in cells, entropy release of water plays a critical role.

Water pricing based upon biogenic property has cultural values and value judgement implications as it is a basic human and animal need. This amount of solar energy incidence, in combination with temperate weather, has made India an agricultural country, that is, most suitable for outdoor photosynthetic food production. Ribulose 1,5-diP is the key compound of photosynthesis in plants. It fixes CO_2 through reverse Embden–Meyerhof–Parnas (EMP) pathway in which participation of water and water quality are crucial (Figures 2.7 and 2.8).

The various properties of water assist in assuming the particular state in which it exists. Depending upon the pressure and temperature conditions, the transitional states are water (liquid), ice (solid), and vapour (gas) (Figure 2.9). The variation in the states of water has made it a multipurpose prime resource on earth and it plays a significant role in ecological balance. This property of water also makes it a vital material for industrial growth, especially in the beverage industry and in the preservation of food (Mukhopadhyay 1995; Mukhopadhyay and Baisya 1997). Thus, in ecological engineering, the biogenic character of water serves as an essential component of living matter and agricultural produce as it shapes the chemistry of all living organisms. Because of the biogenic properties, water dissolves waste products of metabolism in body reactors and assists in their removal from organisms to the environment as excreta/pollutants (Figure 2.10). Water functions as an insulator in body heat regulations. Owing to its high specific heat capacity, water has a good capacity of absorbing heat with only minimal changes in its own temperature.

Excess heat energy in the body is dissipated by breaking the hydrogen bonds of water molecules. In this way, living organisms are protected against sudden thermal changes.

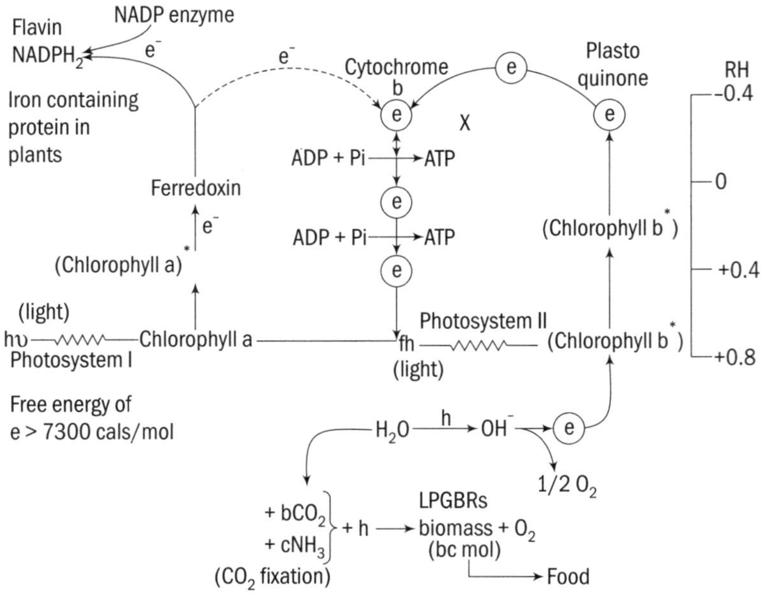

Figure 2.8 Biological redox systems in biomass production

Notes (i) * Indicates excited state after absorption of photons and electrons

(ii) LPGBRs - living plant genomic bioreactors

(iii) RH - relative humidity

Further, the conductivity property of water makes it possible for heat to be distributed evenly throughout bodily tissues. Water is also present in bodily fluids that serve as lubricants whenever one organ or bone rubs against another. There are many other significant functions of water that make it a unique fluid in biological systems.

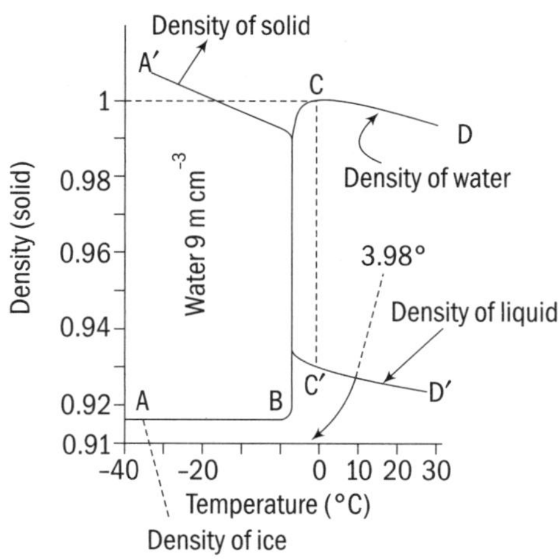

Figure 2.9 Transitional states of water depending upon pressure and temperature conditions

Notes (i) Density of solid state of water (ice) is less than its liquid phase (A, B, C, D).

(ii) Density of most substances is greater in solid than in the liquid state (A′, B′, C′, D′).

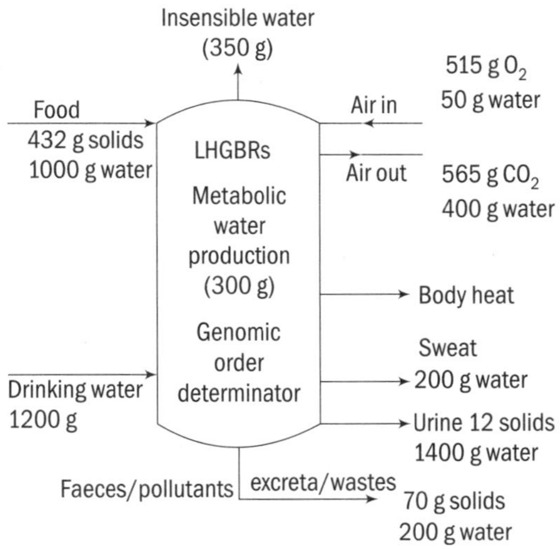

Figure 2.10 Fate of drinking water via food metabolism in human body bioreactor

Notes (i) LHGBRs - living human genomic body/bioreactors

(ii) Correlations of physiological variables in human body bioreactor and steady state mass balances on the body for an average person at rest per day (neglecting solid content in sweat less than 2 g).

A general mass balance can be written as

$$\sum m_{in} - \sum m_{our} = \frac{d_m t}{dt} \quad \text{(zero at steady state)} \tag{2.8}$$

where m is the mass flow rate and m_t is the total body mass at any time t (Figure 2.10).

2.3.2.4 *Driver of life forms*

All living beings daily use water from natural resources such as rainwater, surface water (for example, oceans, seas, rivers, and lakes), stored water (for example, ponds), and groundwater for various purposes. This ultimately leads to pollution in these waterbodies and generates another type of water known as wastewater. Human activities for domestic, municipal, or industrial purposes mostly produce this wastewater. Nearly 97% of water on earth is surface water that is extensively used by humans as aquatic transport avenues or waterways. In such uses, the presence of dissolved or suspended foreign matters (for example, silt, chemicals, faeces, metallic elements, organic materials/nutrients or oil spills) in concentrations higher than normal may cause pollution in natural waterways by various phenomena, such as eutrophication, dissolved oxygen depletion, followed by death of aquatic life, and other so on. Water resources receive microbial cells, organic nitrogen, and other polluting factors from soil, air, sewerage, municipal wastes, dead plants and animals, and process industry wastes (Rich 1973). This makes water resources polluted and the need to define purity of water standard arises. Different water standard specifications have been used for different purposes. These specifications indicate the need for the development of strategies to break the transmission chain and further requirement of ecological approaches to reduce the burden of water-borne diseases.

2.3.3 Ecology Concerns

All living matters try to maintain a dynamic coexistence on earth. It is helpful towards keeping equilibrium between biological synthesis and cleavage reactions (Eqs (2.9) and (2.11)) in water ecology. If none of these ecosystem processes is disturbed, a peaceful earth will be established for living systems with their environments. This is called ecology. Thus, in biospheric and abiospheric worlds, ecology plays an important role. The balance of these two worlds is essential. However, ecological imbalance is caused by pollution in water, soil, and air. Among them, soil and water are the major storehouses of various types of micro- and macro-organisms that are the principal agents for redesigning one biological form to another and for maintaining their food cycle.

Biocellular enzymes

$$C_6H_{12}O_6 + NH_2 + O_2 \rightarrow (C_5H_7NO_2)_n + 4H_2O + CO_2 \tag{2.9}$$

Biodegradations

$$C_5H_7NO_2 + 12H^+ \xrightarrow{\text{Enzyme (anoxic)}} 4CH_4 + CO_2 + NH_3 \tag{2.10}$$

$$C_5H_7NO_2 + O_2 \xrightarrow{\text{Enzyme (oxic)}} 5CO_2 + 2H_2O + NH_3 \tag{2.11}$$

2.3.4 Treatment, Management, and Pollution Abatement Concerns

Abatement of water pollution has now become an essential mission of all nations. For pollution abatement/management, water is treated by various methods (Table 2.7). These methods are broadly divided into four categories: (i) biological treatment, (ii) chemical treatment, (iii) electrochemical treatment, and (iv) physical separation/purification. Biological treatment is still the most widely used procedure for managing polluted water. As it cannot cope with many specific industrial pollutants in water, many other methods of treating such wastewaters have been adopted. Solvent treatment, electrodialysis, membrane separation (for example, reverse osmosis and ultrafiltration), and chemical oxidation are some of the methods used. Biological treatment/management makes use of microbial biomass to break down/consume organic matter, resulting in stabilization of pollutants.

The methodology of water purification or treatment/management may be considered for drinking water and waste/effluent water. In the management of drinking water purification, filter media systems have now been designed for high filtering speed with the greatest possible reduction in pollutants. They can also been customized corresponding to specific requirements. The selection of filter media has been oriented by the volume of water to be purified and the types and quantity of impurities. The filter media vary to a large extent

Table 2.7 Ecology balancing water pollution management by treatment methods

Treatment method	Specification
(a) Physical/chemical	
Aeration	Waterfall aeration
Bubble aeration	
Sedimentation/flocculation	Discrete particle
Flocculent	
Hindered	
Compression	
Filtration	Carbon
Sand	
Membrane	
Disinfection	Chlorine
Demineralization	Ion exchange
(b) Biological	
Aerobic	
Activated sludge process (ASP)	Conventional
Step aeration	High rate
Contact stabilization	Trickling filter
Extended aeration	
Aerated lagoons	
Anaerobic	
Anoxic digestion	Mesophilic
Biotrophicmethanation	Thermophilic

depending upon the application properties of drinking water. These filter media have been graded accordingly. For example, a filter medium of grade 1579 is a pure cellulose paper filter without adsorbents and is the best filter for the purpose of eliminating chlorine from treated water. Also, the presence of activated carbon is absolutely necessary for reducing pollutants like atrazine (a pesticide used in Europe and the United States) and similar compounds (for example, lindane, a pesticide used in the United States) in drinking water. Potable water must conform to biological and tolerable standards. For biological standards of potable water, the levels of *Escherichia coli* count in any 10 mL sample, coliform count in 95% of 100 mL samples in any year, and coliform organisms in a second sample should be zero or coliform organisms in any 100 mL sample should be at a maximum of 10. In case of the total coliform count, 1/100 mL is the acceptable standard. Tolerable standards (highest limit) for toxic chemical components to be taken care for ecological acceptability are as follows: As: 0.05, Cd: 0.005, NO_3^-: 45.00 (all expressed in parts per million), polynuclear aromatic

Table 2.8 Components affecting the ecological acceptability of water

Components	Highest limit
Colouring (on platinum and cobalt scale)	5–50 units
Odour	Non-objectionable
Taste	Non-objectionable
Turbidity	5 in NTU
pH	7–7.5
TS	500 mg/L
Total hardness	2 meq/L
Fe	0.1 mg/L
Mn	0.25 mg/L
Cu	0.05 mg/L
Zn	5.00 mg/L
Ca	75.00 mg/L
Mg	30.00 mg/L
SO_4	200.00 mg/L
Cl	200.00 mg/L
Phenolics	0.001 mg/L
Mineral oil	0.01–0.30 mg/L
Anionic detergents (as MABs)	0.2–1.0 mg/L

MAB - monoclonal antibody; NTU - number of turbidity units

hydrocarbons: 0.20 mg/L, radioactive substance: gross activity 3.0 pCi/L. The standard set of components affecting the ecological acceptability of water is given in Table 2.8.

In the management of drinking water technology, the most conventional method of treatment of water is chlorination. Disinfective chlorine demand (Figure 2.11) indicates that its overdose may pose biohazard risk (Figure 2.12) and, therefore, influence ecology (Mukhopadhyay and Baisya 1997). In addition to chlorination of water for rendering it safe, the industrial processor often further chlorinates water in the plant (such as food processing plants) for disinfecting products and machinery. As wastewaters may contain a wide variety of materials, their compositions and pollution loads vary greatly. Nevertheless, wastewaters can be broadly classified according to the nature of their impurities and pollution potentials (as stated earlier). These will determine the methodology suitable for taking care of ecological balance. Primary methods of physical nature (such as screening out gross particulates) and secondary methods of chemical and biological nature have been widely used. Many wastewaters usually require neutralization by addition of acids and alkalis to a pH range of 6–9 before being discharged to sewage treatment plants (STPs) or natural waters. Secondary biological management methods include trickling filtration (TF), activated sludge processing (ASP), anaerobic digestion (AD), and aerobic lagooning. Therefore, environment and environmental biotechnology play a vital role in balancing

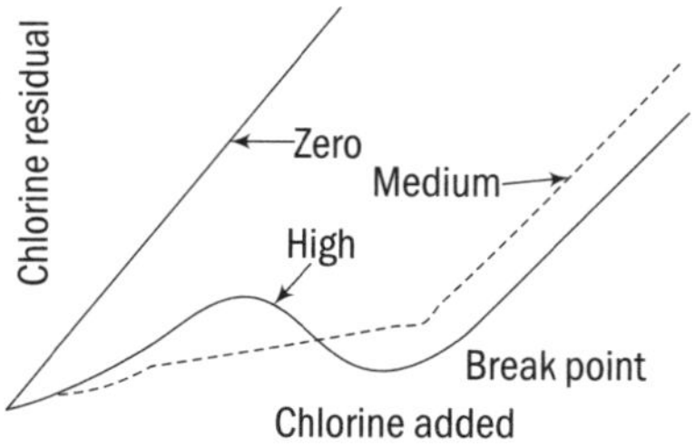

Figure 2.11 Disinfective chlorine demand characteristics of water

domains of ecological engineering (Mukhopadhyay and Mukhopadhyay 2000). A tertiary method of demineralization by ion exchange is also in use. It is obvious that handling, care, and protection of environment for targeting balanced ecology through implementation of ecological engineering are vital for improving human lives.

2.3.5 Essentiality in Relation to Chemo-physical and Biogenic Properties

Cellular membranes are highly permeable to water and impermeable to protons, which has been a dilemma in biogenic physiology. Bioscientists reported that this is due to the presence of aquaporin-1, which is the water channel found in mammalian red blood cells. Its molecular features include an aqueous pathway line with hydrophobic residues, allowing rapid water transport, and a pore spanning just 3 Å, which is narrow enough to let water through but not protons.

Water, being nature's biological solvent, is highly influential in determining survival after exposure to liquid nitrogen. In its cryogenic state, water has served extensively in

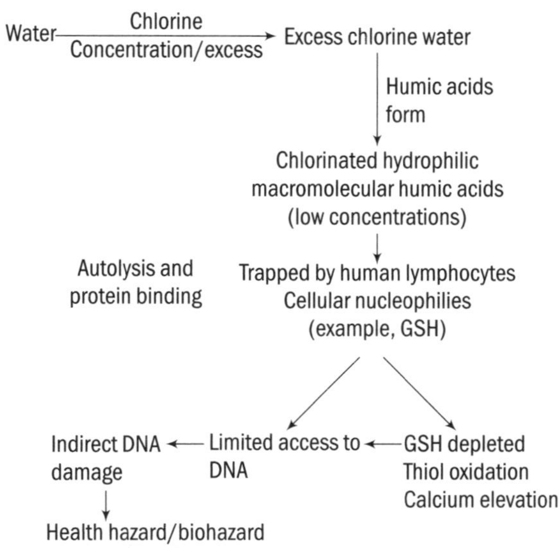

Figure 2.12 Overdose chlorination of water on biohazard possibility

cryopreservation of life forms. Biogenic water properties are essential for the formation of calcified matrix that confers lacuno-canalicular system characteristics in bone through its viscoelastic properties. Further, in mineralized bone tissue formation, the biogenic property of water is essential. Bone water occurs at various locations and in different binding states. It is associated with mineral phase bound to organic phase (collagen and cementing substance), and a large fraction occurs in more or less free form (bulk water). The most tightly bound water occupies the calcium ion coordination sites in the appetite-like crystaler (about 35 mg of water per gram mineral). This water cannot be replaced by simple drying at 100°C. A significantly less tightly bound reaction is the association of water with collagen fibrils. In the unmineralized state, as the bone matrix is laid down by osteoblasts, the collagen fibres produced contain a large fraction of water (about 60%), which is essentially needed.

2.4 WASTEWATER RESOURCES

2.4.1 Domestic/Household

Domestic wastewaters contain discarded or biomass derived materials from LHGBRs in the form of solid waste foods and used water and excreta. These wastewater, solid biomass, and organic wastes may be converted successfully by biotransformations/bioconversions in anaerobic digesters or by aerobic bioprocessing. Treatment of household wastewater starts with the crude septic tank in houses. A number of improved process designs, including "standard rate" digestion, stage digestion, "high rate digestion," and anaerobic contact process, have evolved during nearly 100 years of anaerobic sludge digestion applications to sewage sludge stabilization. Household/domestic (also called sanitary) wastewaters are discharged from residences as well as commercial, institutional, and similar facilities/ sources.

In computing the availability of waste water, knowledge of the rate of public water supplies is very important. Moreover, it is also important to determine the amount of water withdrawn or produced (often called consumed) and the amount of water actually used. The difference between both the parameters is the amount of water lost or unaccounted for in the distribution system plus the amount used for various public services, such as fire fighting, street washing, municipal park, and nursery needs.

2.4.2 Municipal

Municipal water supply becomes domestic wastewater after its use. Whenever it is possible, the use of municipal wastewater treatment facilities should be given preference over individual household systems. However, in cases where municipal facilities are unavailable, the individual system—the septic tank system—can often be used with satisfactory results, provided the soil conditions are favourable and the system is designed properly. In many rural villages in India, such systems are in operation.

The septic tank in the municipal wastewater facility consists of two components, each with its own treatment objective and special requirements. The wastewater from the originating household is discharged into a septic tank, a concrete chamber of rectangular design with an average hydraulic retention time (HRT) of nearly 2 days (Figure 2.13). The sludge particulates settle at the bottom, while the flotage (oily and greasy materials) separates as scum and rises to the surface. Scum is kept away from the discharge liquid of the tank by baffles as shown in the figure. The organic solids at the bottom and scum at the top of the chamber undergo anaerobic digestion (AD). This results in the conversion of organic matter into CO_2 and CH_4 gases. The accumulated solids in the chamber are removed periodically. This solid with a good N:P:K ratio may be used or sold as a biomanure after anti-pathogenic treatments.

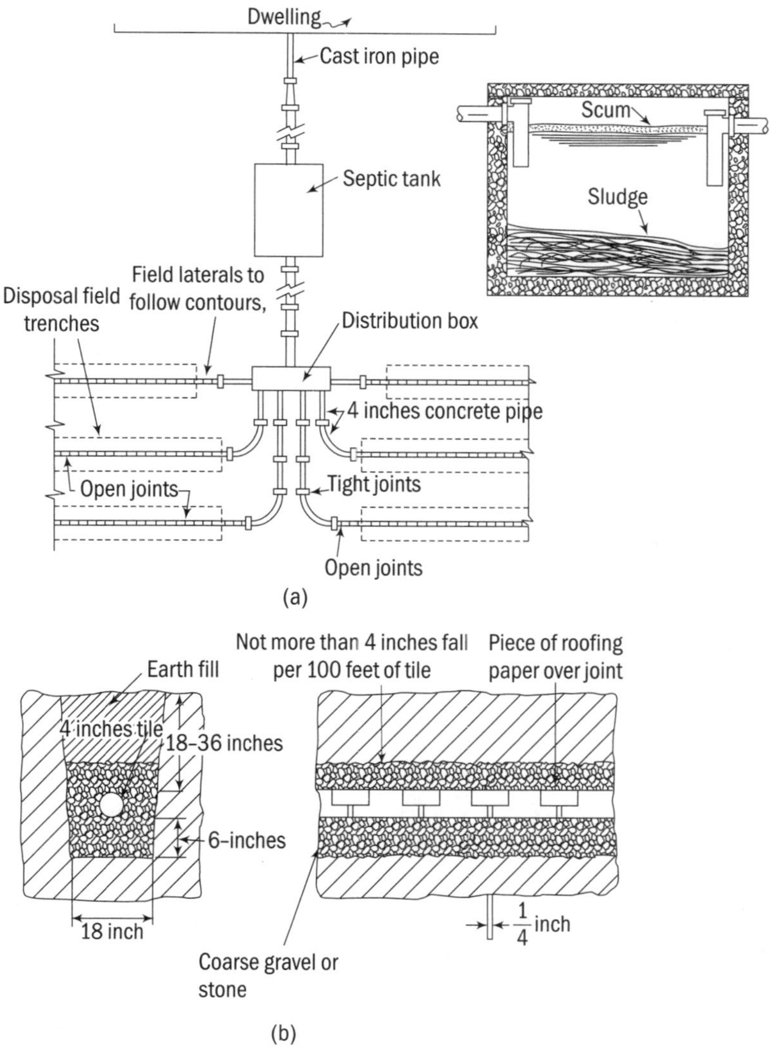

Figure 2.13 Septic tank system: (a) layout of disposal field and (b) details of the absorption field tile line

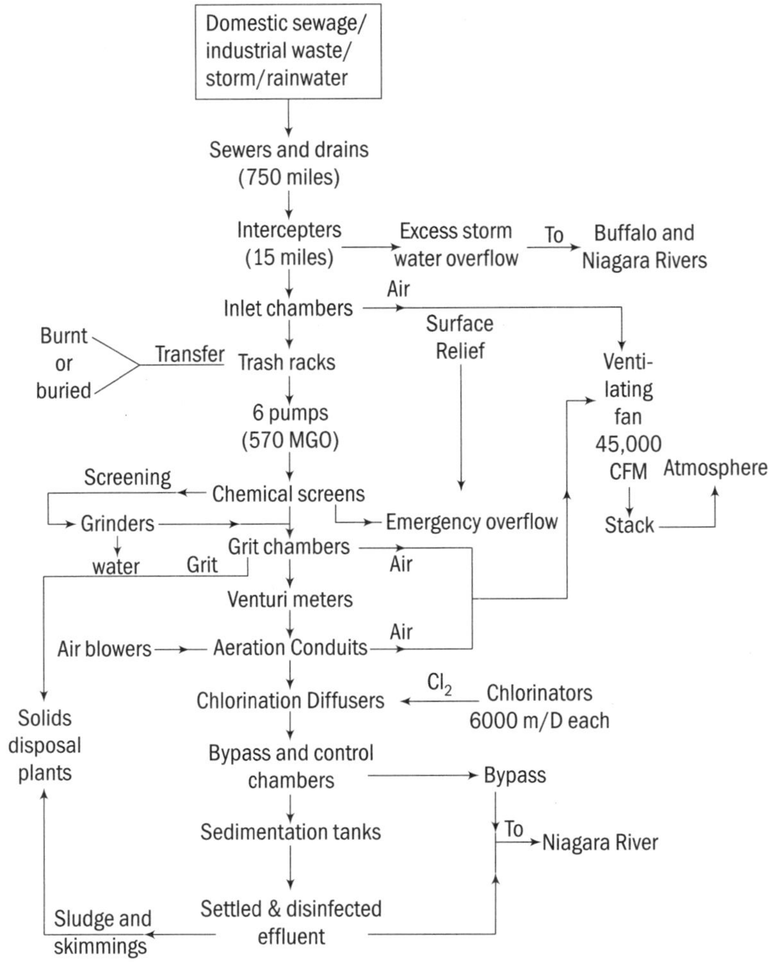

Figure 2.14 Operation of sewage treatment plant at Buffalo, USA—a flow sheet

The effluent from the septic tank flows to a tile system. As a result, the wastewater is spread uniformly at a subsurface level over a soil absorption field. By this distribution, the organics remaining in the effluent water are degraded aerobically (Figure 2.13). The volume of the septic chamber is generally based on the number of persons involved in the disposal system. In this disposal, the maximum rate of application of the absorption field may be calculated using the following relationship:

$$q = \frac{195}{t^{1/2}} \tag{2.12}$$

Here, q is the maximum rate of distribution ($m^3/(m^2\ d)$) and t is the rate of subsidence (min/cm). In deciding the location of the septic chamber, the focus should be on the topography of effluent water supply. A typical municipal sewage treatment plant at Buffalo, USA is shown in Figure 2.14.

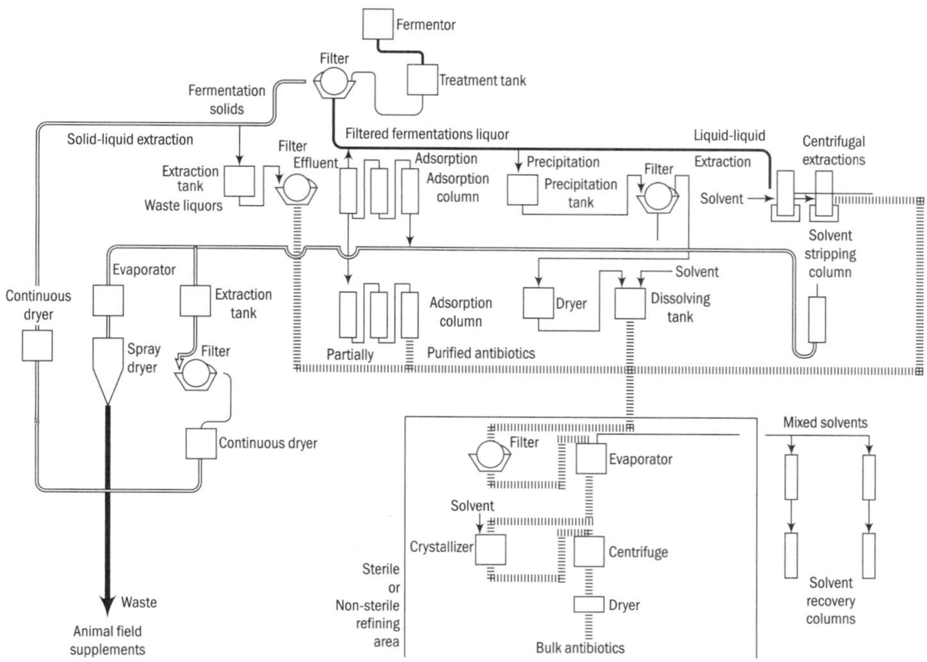

Figure 2.15 Basic flow sheet for recovery of antibiotics with waste generation and sources of wastewaters in industrial penicillin recovery

2.4.3 Industrial

2.4.3.1 Antibiotic

Antibiotic industries (for example, penicillin industry) generate wastewaters with characteristics as given in Table 2.9. The sources of wastes and wastewaters are shown in Figure 2.15. The following streams come out during the processes:

- Liquid and solid wastes from fermentation processes (strong)
- Liquid and solid wastes from extraction and purification processes (strong)
- Liquid and solid wastes from recovery processes (strong)
- Floor and equipment washings (varied)

Table 2.9 Antibiotic wastewater characteristics (penicillin industry)

Resource	BOD (mg/L)	Total solids (mg/L)	Total suspended solids (mg/L)	pH	Colour (mg/L)	Optimum temperature (°C)
Liquid wastewater	15,000–35,000	From 2×10^3 to 50×10^3	From 10×10^3 to 40×10^3	3–10.5	From 20×10^3 to 30×10^3	95–100
Mixed liquid and floor washings	2,000–8,000	3,000–15,000	2,000–9,000	2–11.5	1,000–5,000	58–60

BOD - biological oxygen demand

- Waste cooling water (uncontaminated)
- Laboratory waters as well assanitary and miscellaneous wastes/wastewater (varied)

The following are the difficulties encountered in handling this wastewater:

- The hydraulic load is irregular from time to time as the production activity may change daily.
- The daily waste composition is greatly varied as the productivity is changed.
- The production waste has unusual characteristics: high undissolved colloidal solids, varying pH, high BOD, high temperature, toxicity, colour, odour, and so on.

The treatment process involves the following three stages:

(i) Liquid waste/wastewater is pasteurized with lime or NaOH and pH of 8–8.5 is maintained prior to its treatment in plant.

(ii) Pasteurized liquids are discharged to a 15,000 gal flow equalization holding tank and then pumped to two high-rate aeration tanks, each with a net volume of 140,000 gal. Settled solids in the equalization holding tank are pumped to sludge digesters. A predetermined amount of weak liquid waste is added constantly to the aeration tank for dilution purpose. Both diffused air and turbulent aeration are used in aeration tanks. The settled solid concentration in the aeration system is maintained at 6000–8000 mg/L level. Per pound (lb) of raw $BOD_{(20°C)}$ received approximately 175 ft^3/min air. pH of the treatment liquid is kept at 7.5–8.5. Temperature in aeration tank is maintained at 95–100°F. Aeration time is approximately 12 h.

(iii) The mixed liquor now flows to a clarifier—thickener from which sludge is returned to the two-stage aeration tanks at two different points. The volume of sludge recycled is 60%–100% of the total antibiotic wastewater received by both aeration tanks.

2.4.3.2 Dirty wastewater/aquatic seasonal food provider

Water chestnuts, called *singhara* in Hindi, are farmed in Bijnor district of Uttar Pradesh. They are also grown in Yamuna River. Figure 2.16 shows the harvesting of caltrop or water chestnut in the Yamuna River.

Figure 2.16 Harvesting of water chestnut in Yamuna River fed wetland in Delhi

Table 2.10 A few important sources of food wastes in north India

Source	Computed daily availability (kg)	Energy equivalent (kcal/kg waste)
Fruit and vegetable processing wastes	7.5×10^4	1768
Dairy and meat processing shed wastes (dung)	4.0×10^8	1128
Fisheries	3.0×10^5	1344
Rice mill (husk)	30.0×10^6	975
Sugar mills (bagasse)	15.0×10^6	784
Distilleries	15.0×10^6	720

The harvested aquatic food fetches ₹40 per kilogram. It is also sold in wholesale market. Yamuna has turned out to be very lucrative for the people engaged in this cultivation. They come to the city twice a year to sell this seasonal fruit. These people have been farming water chestnut in a private wetland fed by polluted Yamuna water for two decades.

2.4.3.3 *Food processing wastes*

Food processing industries generate a substantial amount of waste (Table 2.10). It is digested in two steps by the use of mixed microbial flora.

In the first liquefaction and acidification step, acidogenic bacteria like acetic acid bacteria play an important role for nearly 7–10 days. Liquefied acidic materials are acted upon by methane forming organisms in the second step and digestion is completed in another 5–8 days. Improvement has been made in this basic process biotechnology of biogas generation from food wastes. This has resulted in the reduction in the digestion time and increase in the methane productivity to a great extent. A list of computed daily availability of biogas from food waste and its energy equivalent is given in Table 2.10. These important waste resources have not been exploited for small-scale bioconversion into methane.

In the wastewater generated from food processing industries like dairy, the principle of permeability of membranes is employed in whey separation and treatments. Also, for purification and fractionation, as well as in waste disposal and by-product recovery from waste liquid effluents, membrane processing is very useful. Membrane treatments primarily provide major advantage in retaining natural aroma and fresh flavour. Therefore, in treatment of fruit juices and other beverages, membranes have been favoured extensively. In beer and alcohol distilleries, clarification and removal of yeast by microfiltration (MF) are now in practice.

In the dairy, bakery, and confectionary food industries, canned food liquids and wastes are generated as effluents in large amounts. These must be treated before disposal. For example, in the production of fruit yogurt, only fruits are used and the syrup is discarded from canned fruits. A certain dairy plant in Japan discharges 2 MT of syrup per day, containing approximately 200 g sugar per litre. It is a serious problem for industry that wastewater treatment plants cannot treat effluents with such a high sugar concentration.

Table 2.11 Sugar composition of the discarded waste syrup

Constituents	Amount (g/L)
Total sugar	205
Glucose	35
Fructose	33
Sucrose	137

Source Ueno, Ozawa, Ishikawa, *et al.* (2003)

Therefore, factories must transfer the syrup to waste disposer for final treatment. Sugar composition of the discarded waste syrup is given in Table 2.11.

A solution to treat such a high sugar effluent lies in producing lactic acid from canned pineapple syrup through bioconversion. However, in this approach, sucrose cannot be metabolized by *Lactococcus lactis* because the bacteria preferentially use glucose and not fructose. *L. lactis* is one of the more widely used bacteria for industrial production of lactic acid and polylactic acid.

Molasses is produced in abundance from sugar industry. Molasses, which is a waste liquid, has been used in more recent years to produce the aroma compound 2-phenylethanol (2PE) by yeast fermentation. 2PE aromatic alcohol imparts a rose-like odour and is an important raw material in the fragrance industry as well as to some extent in food industry. In the food industry, it is used to enhance the taste of soft drinks, candy, cookies and some other foods.

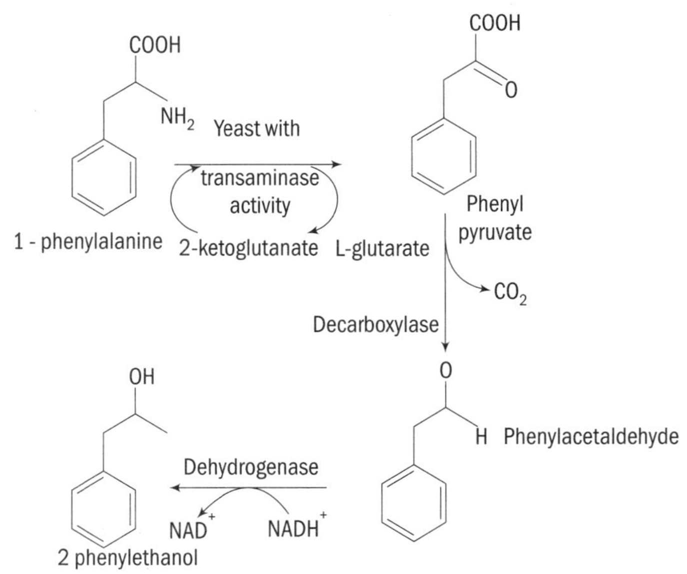

Figure 2.17 Waste utilization pathway in food industry

World's annual production of 2PE from molasses waste is about 700–800 tonnes. This food additive production from food waste molasses is rising because this natural flavouring is found to have no side effects or carcinogenic effects as observed with chemically produced materials. As 2PE is a metabolite of microbial fermentation by yeasts, its production from food waste or natural biowastes is obvious. *Kluyveromycesmarxianus* CBS 600 and CBS 397 have been used. The different pathways leading to 2PE production in yeasts have been published, of which Ehrlich pathway, shown in Figure 2.17, is well known.

2.5 WASTEWATER TREATMENT PROCESSES

Process biotechnology theory and practice for biological wastes and wastewater treatments relate to biological oxidation (BO) principle. It may be carried out in mesophilic or thermophilic temperature ranges depending on the predominant microorganisms involved. Both can be oxic or anoxic mode of operation. Oxic thermophilic digestion (OTD) principle is shown in Figure 2.18.

2.5.1 Biological (Aerobic/Oxic) Oxidation

In this degradation type of reaction, cells derive energy in the form of ATP from the controlled oxidation reaction of hydrogen with oxygen to form water.

Thermophilic process designs

Oxic thermophilic digestion (OTD) and anoxic thermophilic digestion (ATD) strategies have been developed with the aim of meeting the following objectives:

- Thermophilic process biotechnological treatments of WF (waste food) abating pollution
- Retrieval of value-added products (VAPs) in the form of biomanure and bioenergy
- Earning cash from trash

Oxic thermophilic digestion system

It consists of an insulated 2.5 L glass digester for carrying out oxic thermophilic microbial reactions. The digester body is well insulated by a half inch glass layer of industrial glass wool, followed by a one-fourth inch layer of thermocol. It is then covered by aluminium foil. Flexible silicone tubings are used to connect various units and they are insulated in a

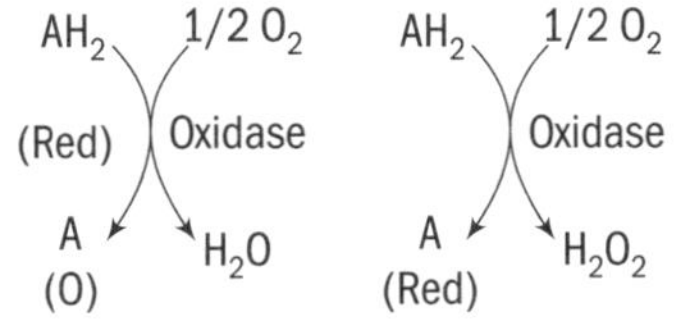

Figure 2.18 Fundamental principle of OTD of BO

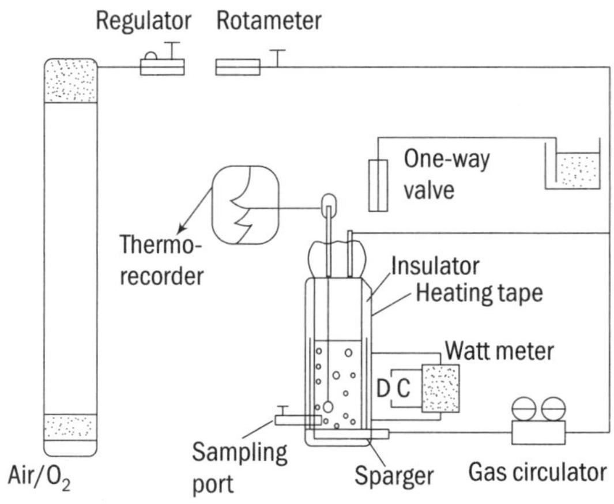

Figure 2.19 Schematic of bench-scale batch thermophilic aerobic digester

similar manner (Figure 2.19). This is done to ensure that the heat lost from digester body and tubings is as low as possible. A cylindrical aluminum pipe with an internal cross-section of 0.1 cm^2 is installed as an air sparging device at the bottom end of the digester. The sparging hole diameter is 1.2 mm.

Operation

The biological sludge obtained from the clarifier bottom (recycle line in aeration tanks) of Delhi Jal Board (DJB) containing total solids (TS) of 8.0 g/L, total suspended solids (TSS) of 7.4 g/Land volatile suspended solids (VSS) of 6.0 g/L can be placed in the digester. Air or oxygen is sparged into the content of the digester through holes positioned downwards to facilitate mixing, avoiding settling of solids when oxic digestion of sludge is carried out.

2.5.2 Activated Sludge Process

It is defined as a system in which biological grow this continuously contacted with wastewater in presence of air/oxygen.

2.5.2.1 Principle

It is usually carried out in the mesophilic temperature range. When thermophilic organisms are predominant, ASP is performed by OTD. The primary aim of mesophilic and thermophilic digestion processes is to reduce BOD to a safe limit. This process results in biologically active cell biomass that is able to absorb/convert organic/biodegradable pollutants from wastewaters.

2.5.2.2 Important factors of activated sludge process

After absorption, organic matters are converted by oxidative enzyme systems to simple products such as CO_2, H_2O, NO_3, and SO_2. The main units of ASP are aeration tank (AT) and settling tank (ST). The process involves an oxic step, followed by a solid–liquid separation step. The separated sludge is recycled back for and added with the waste. In ASP, the most important factor is the food/microorganism ratio (F/M).

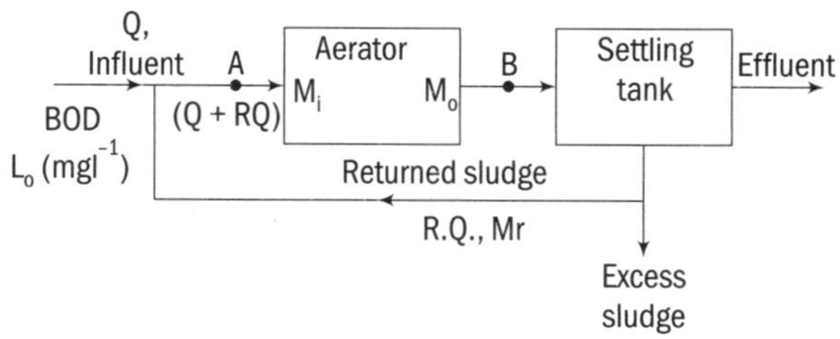

Figure 2.20 Layout of AT operation in an ASP

F/*M* = lbs BOD/day/lbs MLSS in AT = lbs BOD/day/lb VSS in AT

After a definite contact period of sludge with waste, depending on the nature of the sludge and wastewater, the mixed liquor is transferred from the aeration tank to the settler or clarifier (ST). In the AT, O_2 is usually supplied from air bubbles injected into sludge-liquid mass under turbulence. The aeration in AT involves the following three steps:

1. A rapid absorption of waste substrate by the active sludge
2. Progressive oxidation and synthesis of absorbed organics and inorganics concurrently removed from solution
3. Further/extended aeration results in oxidation and dispersion of sludge particles

2.5.2.3 *In settler or clarifier*

Here, activated sludge (AS) settles at the bottom. There is always an excess production of biological cell mass or sludge in the aerobic treatment process. The required activated sludge is returned to AT, while the excess sludge is removed from the system as the process effluent. The clear liquid above the settled sludge in the ST leaves the treatment plant as the process effluent.

2.5.2.4 *Typical layout of activated sludge process*

Both theoretically and experimentally, it is found that the greater the *F*/*M* value, the lesser is the BOD removal, whereas the lesser the *F*/*M* value (that is, greater the microbial population present), the greater is the BOD removal (Figure 2.20).

2.5.2.5 *Microorganisms used*

Mixed aerobic bacteria are employed. Bacteria are highly effective than fungi in this process biotechnology.

The process influent, that is, untreated wastewater containing organic pollution load, enters the AT that contains microorganisms. Here, organic materials, both unsettleable and dissolved, are utilized as food by microorganisms for their growth and energy, which is called bacterial metabolism. This actively growing microbial population is collectively called "activated sludge". The activated sludge mixed with influent wastewater is called "mixed liquor suspended solids" (MLSS).

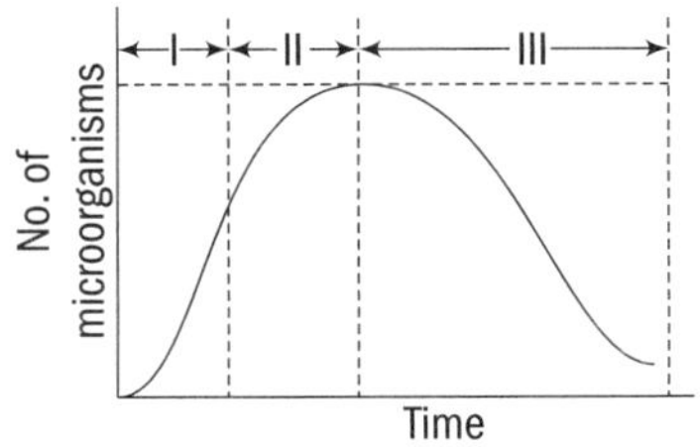

Figure 2.21 Biological growth in ASP

In ASP, the rate of removal of organic pollutants from wastewaters depends on the phase of biological growth. The growth curve is shown in Figure 2.21. In this figure, I is the log phase, II is the declining phase, and III is the endogenous respiration phase. BOD theory assumes a first-order reaction of pollutant removal. At high organic content, the reaction is likely to be zero order with the constant rate of removal of pollutants per unit cell mass/weight. Thus,

$$\frac{a}{S_a} \cdot \frac{dL}{dt} = K \text{ (zero order)} \tag{2.13}$$

Here, K is constant, a = mass of VSS, L = UBOD. Solving Eq. (2.13), we get

$$\frac{aL_r}{S_{at}} = K \tag{2.14}$$

where L_r = BOD removed.

When the organic concentration is reduced to a limiting value, the rate of removal of BOD becomes dependent on the concentration of the organic matter/pollutant, that is, the first-order rate. Thus,

$$\frac{a}{S_a} \cdot \frac{dL}{dt} = KL \text{ (first order)} \tag{2.15}$$

$$\frac{a}{S_a} \cdot \frac{dL}{dt} = KL^2 \text{ (second order)} \tag{2.16}$$

About one-third of COD of a waste is used for energy and the remaining two-thirds is used to synthesize new cells. The sludge production is increased as VSS ranges from 0.2 to 0.8 kg per kg COD removed depending on the wastewater and the time of aeration.

2.5.2.5.1 Operation parameters

1. **Sludge production**: VSS accumulation is given by

$$S_t = C(S_i + aL_r) - bS \tag{2.17}$$

where S_t is the mass of VSS accumulated per unit time, S_i is the mass of VSS in influent, a is the synthesis constant (lb MLVSS produced per lb substrate removed), b is the endogenous

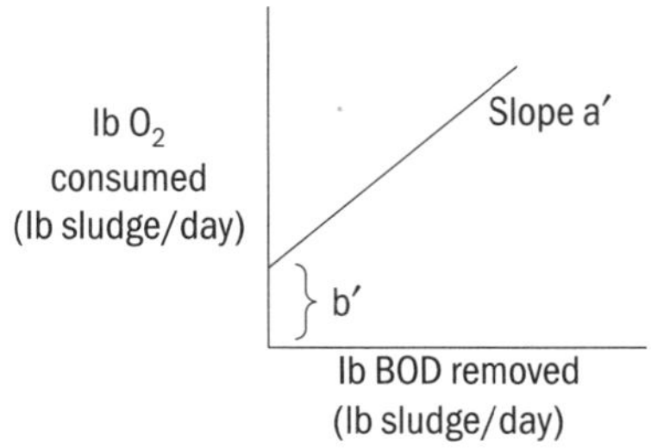

Figure 2.22 Oxygen requirement as a function of mixed liquid volatile suspended solid (MLVSS)

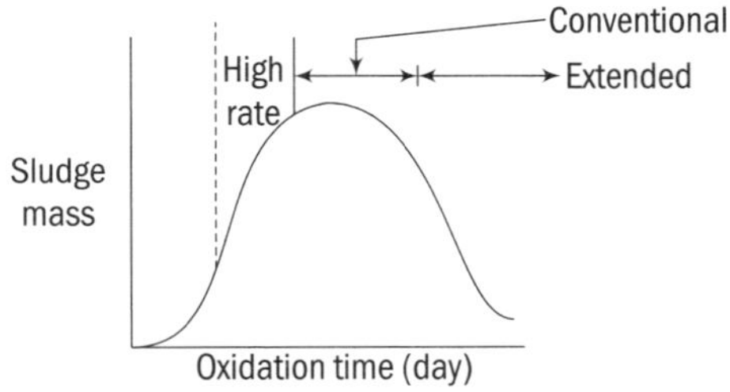

Figure 2.23 Operating zones of biological oxidation process

respiration constant, and C is the fraction of non-biodegradable VSS in influent. For aerobic/oxic oxidation, typical values of a and b would be 0.55 and 0.15, respectively.

2. Oxygen requirement: Oxygen must be supplied to provide energy for synthesis and to permit endogenous respiration (Figures 2.22 and 23). The oxygen requirement per unit time is given by

$$O_2/\text{day} = a'\text{lb BOD/day} + b'\text{lb MLVSS}$$

$$O_2 = 0.39L_r + b'S \tag{2.18}$$

Oxygenation capacity (OC) is given by

$$OC = k_La_{(20°)}\cdot C_SV \text{ (lb/h)} \tag{2.19}$$

$$K_La_{(T)} = K_La_{20°} \times 1.024^{(T-20)} \tag{2.20}$$

When BOD removal is the second-order kinetics,

$$\frac{a}{S_a} \cdot \frac{dL}{dt} = KL^2 \tag{2.21}$$

On integration, this becomes

$$\int_{L_o}^{L_e} \frac{dL}{L2} = \frac{S_a}{a} \cdot K \int_0^t dt \tag{2.22}$$

or

$$\frac{L_e}{L_o} = \frac{a}{1 + KS_at} \tag{2.23}$$

From Monod's equation,

$$\frac{ds}{dt} = \mu_m \frac{L}{K_S + L} \cdot S \tag{2.24}$$

and

$$a \cdot \frac{dL}{dt} = \frac{ds}{dt} \tag{2.25}$$

Here, s = VSS and L = BOD.

Therefore,

$$a \frac{dL}{dt} = \mu_m \frac{L}{K_S + L} \cdot S \tag{2.26}$$

From this equation, the following two cases may arise:

(a) **Case I**: For high organic concentration,

$L \gg$ Ka′ that is, KS + L ≈ L

Therefore, from Eq. (2.26)

$$\frac{dL}{dt} = \frac{\mu_m}{a} \cdot S \tag{2.27}$$

(b) **Case II:** For low organic concentration,

$K_S + L \approx K_S$

Therefore,

$$\frac{dL}{dt} = \frac{\mu_m}{aK_S} \cdot L \cdot S = K \cdot L \cdot S \tag{2.28}$$

3. Temperature of treatment: Fraction of BOD removal, p, is dependent on temperature of operation of the treatment plant. It is given by

$$p = 1 - 10^{-K_{20}tM} = 1 - f \tag{2.29}$$

where k_{20} is the BOD coefficient in common log at 20°C with the unit day^{-1}, t is the time in days, and M is the substitution factor. Here

$$M = C^{(T - 20)}$$

T is the temperature in °C and

$$C = \left(\frac{t_1 \log f_2}{t_2 \log f_1} \right)^{\beta} \tag{2.30}$$

$$\beta = \frac{1}{T_2 - T_1}$$

The effect of temperature on biological population is such that the number of types of microorganisms is lowest in winter and highest in summer, but the number of individuals decreases in summer and increases in winter. The temperature–respiration relationship for AS in the range 0–25°C is given by

$$\frac{R_{T_1}}{R_{T_2}} = e^{\beta(T1 - T2)} \tag{2.31}$$

where R_{T_1} and R_{T_2} are respiration rates at temperatures T_1 and T_2, respectively, and β is a constant.

4. Loading: Loading is considered in two terms: process loading rate (PLR) and BOD loading rate (BLR). PLR is defined as

$$\text{PLR} = \frac{\text{Ib BOD applied}}{\text{day/Ib MLVSS in AT}}$$

$$= \frac{QL_S}{(6.24/10^6)VM_i} \tag{2.32}$$

$$= \frac{24L_a}{S_{av}\ (l + R)} \tag{2.33}$$

Here, L_a = ppm (mg/L) BOD applied
Q = loading of influent, MG (million gallons)
S_{av} = average MLVSS in AT
V = AT volume, MG
M_i = MLVSS at inlet of AT, mg/L
R = recycle ratio
t = aeration retention time, day

BLR determines the mode of ASP operation. It is defined as

$$\text{BLR} = \frac{\text{Ib BOD applied}}{\text{day/AT volume}}$$

$$= \frac{QL_a \times 10}{V/1000}$$

$$= \frac{QL_a \times 10^4}{\left(\frac{V \times 6.24}{10^6}\right)M_i \times 20}$$

$$= \frac{QL_a}{\left(\frac{V \times 6.24}{10^6} \times \text{Mi}\right)} \times 10^3 = \text{PLR} \times 10^3$$

So, $$\text{BLR} = \text{PLR} \times 10^3 \tag{2.34}$$

5. Recycle ratio (R): From material balance across AT,

$$QO + RQM_r = Q(1 + R)M_i \tag{2.35}$$

M_r is the MLVSS of return sludge. From Eq. (2.35),

$$M_i = \frac{R}{(1 + R)} M_r$$

or

$$R = \frac{M_i}{M_r - M_i} \tag{2.36}$$

Another way of representing sludge recycle ratio per cent for ASP (except extended aeration) is

$$R(\%) = \frac{100}{120\psi S_a - 1} \tag{2.37}$$

where

$$\psi \text{ (Mohlman index)} = \frac{V}{X_v}$$

S_a = AT solid concentration (%)

V = AT volume

X_v = MLVSS (%)

6. Total suspended solids (TSS): Suspended solid control is necessary. For this,

$$S_R = \frac{(1 + R_s)(S_a - \text{IS})}{R_s} \tag{2.38}$$

where I is the volume of incoming flow to AT and S_s is the raw waste solid concentration entering aeration tank.

$$R = \frac{100 R_s}{I} = \frac{(S_a - S_s) \times 100}{S_R - S_a} \tag{2.39}$$

$$S_x = \frac{R_s S_R + I_x S_x}{R_s + I_x} \tag{2.40}$$

Here, R is in % and R_s, I, and I_x are in million gallons per day (MGD). Other quantities are in mg/L. I_x is the total primary effluent up to and including any point x. S_x is the suspended solid to the section of channel between two adjacent points of entrance of return sludge.

7. H^+ concentration: Based on laboratory experiments, the best ASP is obtained at pH values from 7.0 to 7.5 although the removal of oxygen consumed, suspended solids, and bacteria is high at pH values ranging from 6 to 9.

8. Efficiency and BOD removal: BOD removal from a completely mixed tank/basin for a relatively low soluble effluent is given by

$$KS_e = \frac{L_i - L_e}{X_a t} \tag{2.41}$$

Here, K is the removal rate coefficient of BOD, L_e is the soluble effluent BOD, X_a is the average MLSS, L_i is the influent BOD, and t is the aeration time. So, per cent efficiency (E) of O_2 transfer is

$$E = \frac{100\,W_o}{W_i} \tag{2.42}$$

where W_o is the weight of O_2 dissolved and W_i is the weight of O_2 fed to AT. Now per cent efficiency of BOD removal is given by

$$e = \frac{100 \times (L_i - L_e)}{L_i}$$

$$= 100 \times \left[1 - \frac{L_e}{L_i}\right] \tag{2.43}$$

9. Residence time (RT) in AT: RT is calculated from the following two criteria to determine which one controls the design:

(i) Effluent quality: Depending on the ubstrate removal rate, it is

$$t = \frac{S_o - S_e}{KV_{vs}(S_e - S_n)} \tag{2.44}$$

(ii) Organic loading: It is evaluated from F/M ratio for optimum flocculation and settling of sludge. As

$$\frac{F}{M} = \frac{\text{lb of sunstrate in influent}}{\text{(lb MLVSS in AT) (day)}}$$

$$F = Q \cdot S_o \times 8.34 \text{ lb/day}$$

$$M = X_{va} \times 8.34 \text{ lb}$$

Q in MGD and so X_{va} in mg/L,

$$\frac{F}{M} = \frac{QS_o}{X_{va} \times V}$$

$$= \frac{1}{t} \times \frac{S_o}{X_{vs}}$$

As $$\frac{V}{Q} = RT \text{ day}^{-1}$$

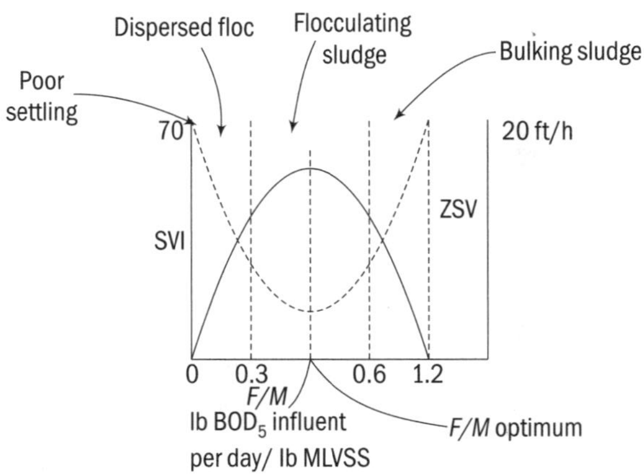

Figure 2.24 Variation of parameters as a function of F/M ratio in a typical ASP

SVI - sludge volume index; ZSV - zone settling velocity

or
$$t = \frac{S_o}{X_{va}{}^{(F/M)}\ \text{OPT}} \tag{2.45}$$

Of Eqs. (2.44) and (2.45) of *t*, the larger value is used for design. The relation of *F*/*M* as a function of other parameters of ASP is shown in Figure 2.24.

10. Nutrient requirement: Usually for N_2 and P deficient wastewaters, the nutrients are to be added in traces. The requirements are determined by N_2 and P lost by wastage of MLVSS.

$N_2 - 0.12\ \Delta X_v$ lb/day

Here, ΔX_v is the wasted MLSS (assuming it contains 2% P and 12% N_2).

Amount lost in effluent is

N_2: $Q_F \times 10.5 \times 8.34$ lb/day

Q_F is the effluent flow in MGD.

P: $Q_F \times 0.5 \times 8.34$ lb/day

So, the total is

$N_2 = 0.12\ \Delta X_v + Q_F \times 1.0 \times 8.34$ lb/day

$P\ = 0.02\ \Delta X_v + Q_F \times 0.5 \times 8.34$ lb/day

11. Other parameters

(a) Sludge volume index (SVI): For maintaining good quality sludge in ASP, the following indices are used to assist control.

$$\text{SVI} = \frac{\%\ \text{settled volume of sludge in 30 min.}}{\text{MLSS (\%)}}$$

SVI ranges from 40 to 100 for good sludge, but it is more than 200 for poor sludge with tendency to bulking.

(b) Sludge density index (SDI)

$$\text{SDI} = \frac{\text{MLSS (\%)} \times 100}{\%\ \text{settled volume of sludge in 30 min}}$$

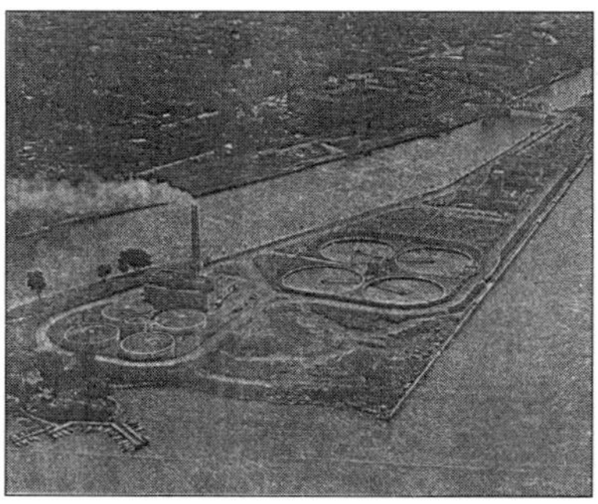

Figure 2.25 Arrangement of a typical ASP plant (with four sedimentation tanks, four sludge digestion tanks, and an incinerator nearby)

Table 2.12 N:P:K comparison of different field manures

Manure type	Nitrogen (%)	Phosphorus [as P_2O_5 (%)]	Potassium [as K_2O (%)]
Farm yard manure	0.40	0.20	0.40
Rural compost/urban compost	0.60	0.50	0.60
Green manure	0.600	0.100	1.25
Biomanure (50% dry)	1.802	1.001	0.600

(c) Sludge age (*G*)

$$G = \frac{\text{Aeration zone volume (m}^3\text{)} \times \text{MLSS (mg/L)}}{\text{Flow (m}^3\text{/day)} \times \text{SS in feed (mg/L)}}$$

A *G* value of 4–6 days is considered desirable in a conventional plant.

Arrangement in a typical ASP is shown in Figure 2.25.

2.5.2.5.2 Oxic thermophilic digestion products

After oxic digestion of waste food (WF) for a period of 8 days in the thermophilic condition (60°C) in the batch autothermal digester that followed the above described kinetics, the solid digested biomass obtained was dewatered and sun-dried. Yield of 120 kg wet weight digested solids per m^3 digestion volume could be measured. Presence of pathogens in this digested solid was tested using a pathogen indicator organism. As no pathogenic indication could be detected, it was assumed to be pasteurized under thermic (60°C) digestion condition. Next this solid mass was analysed for detecting its N:P:K content and its strength was compared with different field manures. The results shown in Table 2.12 indicate its suitability as a biomanure.

Although this digested solid biomanure is being sold in Delhi for horticulture/nursery purposes at a price of ₹5 per kg, its economics and cost–benefit estimates need to be worked out. This is primarily to assess the usefulness of this biomanure, which is found very suitable as a value-added product to enrich soil with agronutrients to improve the crop yield.

2.5.2.5.3 Oxic thermophilic digestion analysis

The biokinetic pattern of OTD may be expressed in terms of the increment rate of mixed liquor suspended solids with the removal of volatile suspended solids (VSS). For this, one may write

$$\frac{dX}{dt} = \mu X \tag{2.46}$$

where X is MLSS concentration during digestion (g/(L day)) and μ is the specific MLSS increment coefficient. Thus

$$\mu = \frac{\mu_m S}{K_s + S}$$

or

$$\frac{1}{\mu} = \frac{\mu_m}{K_s + S} + \frac{1}{\mu_m} \tag{2.47}$$

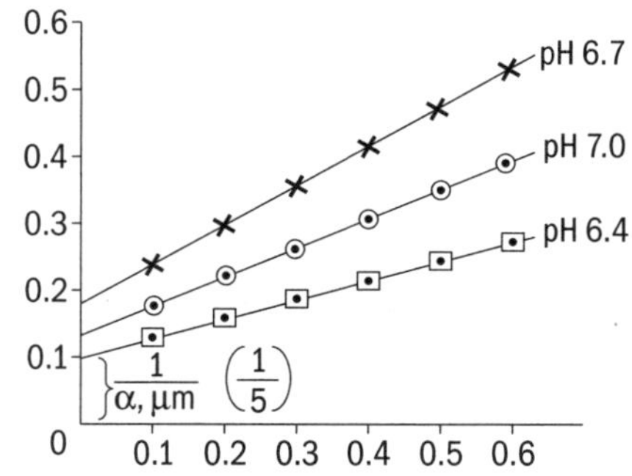

Figure 2.26 Double-reciprocal plot of $\left(\frac{dp}{dt}\right)^{-1}$ versus S^{-1}

So, plot of $1/\mu$ versus $1/s$ will provide values of μ_m and K_s. In Eq. (2.47), μ_m is the maximum value of μ (day^{-1}), S is limiting VSS substrate concentration (gVSS/(L day)), and K_s is the VSS affinity constant. As

$$\mu \propto q \text{ (specific VSS removal rate)} \tag{2.48}$$

$$\mu = Yq \tag{2.49}$$

where Y is the MLSS yield coefficient.

So

$$\frac{ds}{dt} = qX = \frac{\mu}{Y}X = \frac{\mu_m SX}{Y(K_s + S)} \tag{2.50}$$

In the equation, μ_m/Y is the maximum value of quantity of VSS removed per unit quantity of MLSS increased (lb VSS/1b MLSS or lb/VSS/1b BOD). As it is an MLSS increase associated nitrogenous biofertilizer/biomanure (product P) formation system (measured in terms of sludge protein content [SPC], g/L), one may write

$$\frac{dp}{dt} = \alpha\frac{dx}{dt} \tag{2.51}$$

or

$$\frac{dp}{dt} = \alpha\left(\frac{\mu_m S}{K_s + S}\right) \tag{2.52}$$

Taking double reciprocal,

$$\frac{1}{dp/dt} = \frac{K_s}{\alpha\mu_m} \times \frac{1}{S} + \frac{1}{\alpha\mu_m} \tag{2.53}$$

It shows that a double-reciprocal plot of Eq. (2.53) with pH as a control parameter would result in a linear profile as shown in Figure 2.26.

Values of kinetics constants are listed in Table 2.13.

2.5.3 Trickling Filter Beds/Biofilters

A typical attached growth/biofilm process is seen in the conventional trickling filter operation. In principle, it is a packed bed trickle bed bioreactor. The packed bed bioreactor

Table 2.13 Values of kinetics constants/coefficients as a function of pH of oxic thermophilic digestion

Constant/coefficient	Unit	Values at pH		
		6.4	6.7	7.0
μ_m	Day^{-1}	0.21	0.42	0.31
K_s	mg/L	68	120	83
q	gBOD/gVSS/day	12	18	14
Y	gVSS/gMLSS or gVSS/g	0.12	0.56	0.23
SRT	BOD	6	6	6
$SVI = V_{30}/V_0X$	Days	40	54	38
α	mL/gMLSS (g/L)	24	15	36

consists of some medium covered with a biological slime film, over which wastewater is passed. Oxygen and organic matter diffuses into the slime film, where biological oxidation occurs (Figure 2.27).

A packing material maybe ascribed to be ideal if it facilitates the following functions:

(i) It offers a surface on which the biological slime can grow successfully.

(ii) It permits the waste liquid to flow over the bios complex in as thin a film as possible to allow for rapid O_2 transfer from air.

(iii) It has adequate void space so that there is a steady supply of air to the film.

(iv) It has adequate void space to allow free passage of treated effluent and free removal of bios film scoured from the packing.

(v) It is inert and chemically stable.

(vi) It takes loadings up to 1 lbBOD/yd^3 of packing.

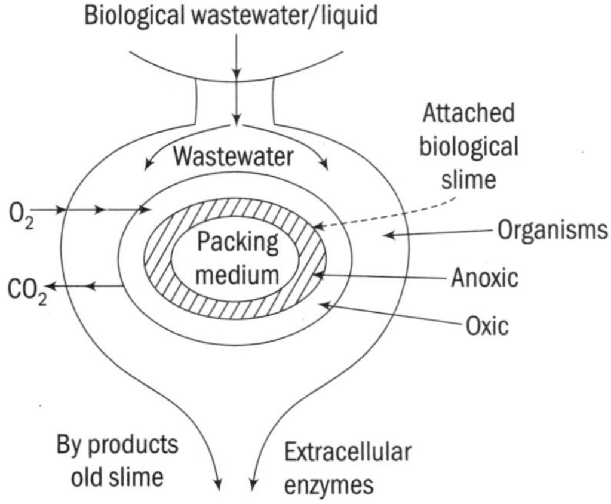

Figure 2.27 Principle of trickle bed bioreactor

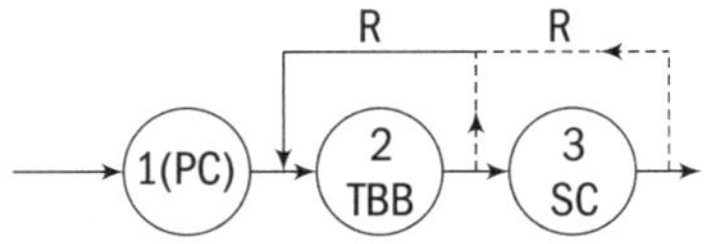

Figure 2.28 Operation of single-stage trickle bed bioreactor (TBB). 1: primary clarifier (PC), 2: TBB, and 3: secondary clarifier (SC)

The principle of operation of a single-stage trickle bed bioreactor is shown in Figure 2.28. Its operation has two aspects: general and biological.

1. **General**: TBBs are used to process a large volume of sewage or organic industrial wastewaters like that of food processing biotechnology (for example, whey). TBBs remove soluble organics by bio-oxidation as shown in Figure 2.27. TBBs are beds of crushed stones/other porous media (1.5″–3″ size) through which industrial/sewage wastewater is percolated/trickled. It is a fixed bed over which sewage or wastewater is intermittently or continuously discharged and contacted with biological slime film on the filter/support media.

A single-stage TBB arrangement consists of units as shown in Figure 2.28. It has a primary clarifier to remove insoluble solids that settle rapidly. Non-removal of these solids clogs the TBB. From PC (primary clarifier), the effluent passes to a secondary clarifier (SC). Recirculation (*R*) may be directly from the exit of TBB or from SC back into the influent to the TBB. Aerobic/oxic condition is maintained by the flow of air through filter bed induced by the difference in weight of the atmosphere inside and outside the bed. In some cases, forced draft ventilation is used. Periodic sloughing off slime/filter film is discharged to SC unit. Sewage or wastewater may be introduced into the filter either at a low rate or at a high rate. Accordingly, characteristics of TBB operation include those given in Table 2.14.

2. **Biological**
   (i) The primary removal of organic matter is not accomplished through a filtering or straining action. It occurs by an adsorption process that takes place at the surface of biological slime covering the filter media.
   (ii) Subsequent to their adsorption, the waste organic matters are utilized by the biomass slime for growth and energy.
   (iii) Microbes that take part in the slime are predominantly fungi. They are more efficient than bacteria in TBB for removing organic pollutant load from the waste. Algae may also take part in the action.

The schematic representation of a TBB filter action is shown in Figure 2.29. The action involves the following parts:

Table 2.14 TBB characteristics

Flow rate	Loading		Depth (ft)	Dosing
	Hydraulic (mg/acre/day)	Organic (lb BOD/acre ft/day)		
Low	2–6	1500–1700	6	Intermittent
High	10–30	3000–5000	4	With recirculation

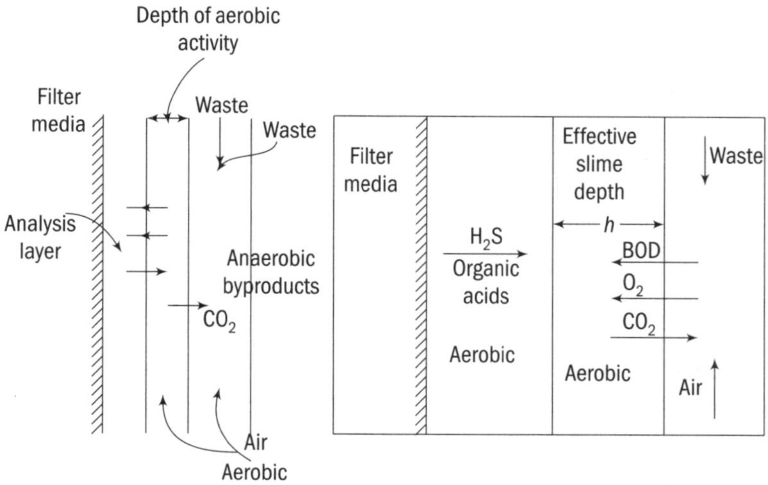

Figure 2.29 Schematic representation of TBB filter action

1. **Aerobic**
   (i) As liquid waste passes around the filter media, it is brought into contact with the biological slime layer.
   (ii) Waste is adsorbed on the slime, where it is oxidized and the produced end products are discharged.
   (iii) Oxygen/air diffuses from the atmosphere into the outer layer of the slime to support aerobic/oxic growth.
   (vi) The biological growth is most active at the outer layer of the slime, where the concentration of the organic matter and O_2 is the highest.
   (v) As the slime layer increases in thickness, the amount of diffused O_2 becomes limited and anaerobic/anoxic activity is established that aids in the sloughing off of old slime by scouring caused by influent flow.
2. **Anaerobic/anoxic**
   (i) When slime layer becomes very thick, anaerobic/anoxic action dominates, starting from a thin anoxic layer. Anoxic slime layer builds up as the slime coat thickens.
   (ii) At a certain condition of hydraulic pressure, slimes lose their zoogleal properties near supporting media.
   (iii) The slime layer sloughs off under hydraulic pressure as stated earlier.

2.5.3.1 Mechanism of biological oxygen demand removal in trickle bed bioreactor

At the low BOD concentration, usually applied to a TBB, the fraction of BOD remaining is directly related to contact time and is given by the following equation:

$$\frac{L_e}{L_o} = e^{-kt} \tag{2.54}$$

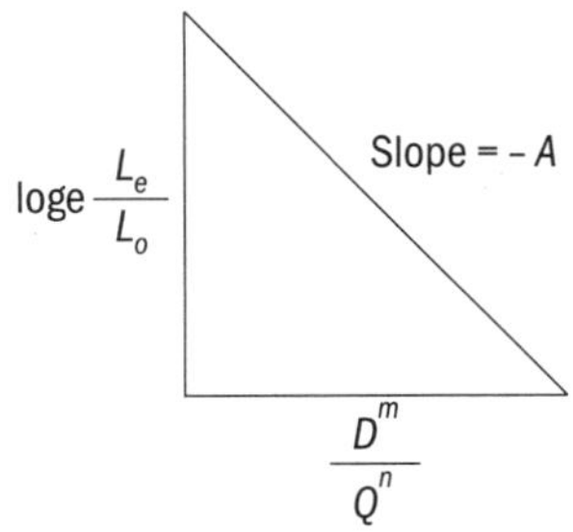

Figure 2.30 Plot of Eq. (2.57)

Here, L_e is the BOD remaining in the effluent, L_o is the BOD applied to TBB after mixing with recycle, k is the rate constant and a function of the active film/unit volume, and t is the contact time that is related to three factors (that is, depth of the bed, hydraulic loading, and physical characteristic of the filter media).

The mean contact time t can be expressed as

$$t = c \times \frac{D^m}{Q^n} \tag{2.55}$$

where c is constant, D is filter bed depth, Q is hydraulic loading, and m and n are exponents. The constant c and the exponents m and n incorporate surface and viscosity effects and will vary with the type of filter media. Combining the above two equations gives

$$\frac{L_e}{L_o} = e^{-ck(D^m/Q^n)} = e^{-A(D^m/Q^n)} \tag{2.56}$$

where the constant $A = ck$.

So,
$$\log\left(\frac{L_e}{L_o}\right) = -A \times \frac{D^m}{Q^n} \tag{2.57}$$

The plot of Eq. (2.57) gives the value of A (Figure 2.30).

When film is uniformly distributed throughout the filter depth, $m = 1$, and so in that case

$$\frac{L_e}{L_o} = e^{-A\frac{D}{Q^n}} \tag{2.58}$$

Theoretical analysis indicates the following:

- For turbulent flow, $n \to 1/3$

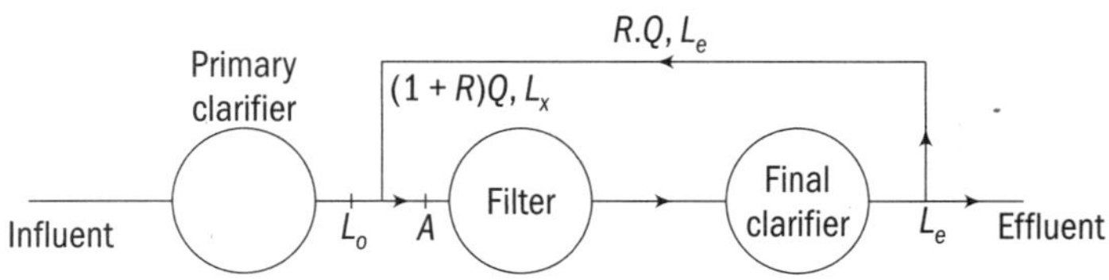

Figure 2.31 A schematic of the recirculation system

- For laminar flow, $n \rightarrow 2/3$

2.5.3.2 Efficiency of treatment

Efficiency of treatment= portion of BOD removed

$$= \left[1 - \frac{L_e}{L_o}\right] = 1 - e^{-A(Dm/Qn)}$$

Therefore, per cent BOD removal $= \left(1 - \frac{L_e}{L_o}\right) \times 100$

or $$\text{Per cent treatment} = [1 - e^{-A(Dm/Qn)}] \times 100 \tag{2.59}$$

It signifies that for the same organic loading and depth, if Q, that is, hydraulic loading, increases, the efficiency decreases.

2.5.3.3 Need of recirculation for a trickling filter

Let us consider a system with recirculation (Figure 2.31). Initially, when there is no circulation,

$$L_e = L_o \cdot e^{-kt}$$

Therefore, Per cent BOD removal $= \left[1 - \frac{L_e}{L_o}\right] \times 100$

$$= [1 - e^{-kt}] \times 100 \tag{2.60}$$

When there is circulation, the general equation for performance of tricking filter with recirculation can be written as

$$\frac{L_e}{L_o} = \frac{e^{-k'(D/Q^n)}}{[(1 + n) - Ne^{-k'(D/Q^n)}]} \tag{2.61}$$

where N is the recirculation ratio.

Hydraulic loading at the point A is $(Q + QR)$ in Figure 2.31 and R is the recycle ratio. So, BOD at A is

$$L_x = \frac{(QL_o + QRL_e}{Q(1 + R}\,)$$

or $$L_x = \frac{L_o + RL_e}{1 + R} \tag{2.62}$$

Thus, the per cent BOD removal is given by

$$\left[1 + \frac{L_e}{L_x}\right] \times 100$$

So, it may happen that

$$\left[1 - \frac{L_e}{L_x}\right] \times 100 \lesseqgtr \left(1 - \frac{L_e}{L_o}\right) \tag{2.63}$$

$$\text{or}\quad \left(1 - \frac{L_e}{L_x}\right) \lesseqgtr \left(1 - \frac{L_e}{L_o}\right) \tag{2.64}$$

Now for checking whether

$$\left(1 - \frac{L_e}{L_x}\right) > \left(1 - \frac{L_e}{L_o}\right) \tag{2.65}$$

one needs to show that $L_o > L_x$ and in that case the per cent BOD removal with recycle will be always greater than per cent removal with no recycle, and this will prove the essentiality of recycling.

From Eq. (2.65), one has

$$L_x = \frac{L_o + RL_e}{1 + R} \tag{2.66}$$

As L_e is equal to a fraction of L_o, say, $L_e = pL_o$, where p is the factor less than 1,

$$L_x = \frac{L_o + RpL_o}{1 + R}$$

$$= \frac{L_o(1 + Rp)}{1 + R} \tag{2.67}$$

As $(1 + Rp) < (1 + R)$ [$\because p < 1$],

$L_x < L_o$ or $L_o > L_x$

This proves that recycling is essential in TBB.

2.5.3.4 Tricking filter design

Three basic empirical design approaches are used for trickling filters (Blakebrough1967). These are National Research Council (NRC) formula (1946), Eckenfelder formula (1963), and Galler and Gotaas formula (1966). The design engineer should be acquainted with strengths and weaknesses of each of these approaches and understand the fact that more formal mechanistic models are forthcoming. Also, an effort should be made to examine the original papers cited here to access figures and monographs for making estimates.

The first good trickling filter formula and one still widely used is the NRC formula. It is based upon data and evaluations made at units installed at military establishments.

$$E = \frac{1}{1 + C\left(\frac{W}{VF}\right)^{0.5}} \tag{2.68}$$

where E is the fraction of BOD removed, W is the influent organic load (lb/day), V is the filter volume (1000 ft^3), F is the recirculation factor, and C is the constant (0.0561). The organic loading W is actually a product of the flow rate Q, the influent BOD (L_i expressed in mg/L), and a constant 8.34. The recirculation factor F is defined as

$$F = \frac{1 + R}{1 + (1 - p)R^2} \tag{2.69}$$

where R is the recirculation ratio and p is the weighting factor (≈0.9). The recirculation ratio is the ratio of recycle flow to plant influent flow. In cases where second-stage filters are used, there is a noticeable "decrease in treat ability" of the effluent waste from the first filter. The magnitude of this factor is dependent upon the amount of removal of BOD in the first stage. This removal effect is handled by introducing a factor to retard organic loading to $1/(1 - E_1)$, where E_1 is the efficiency of the first stage. Thus, the equation for the second stage is

$$E_2 = \frac{1}{1 + 0.0561\left(\dfrac{W_2}{VF(1 - E_1)^2}\right)^{0.5}} \tag{2.70}$$

where E_2 is the efficiency of the second stage and W_2 is the organic loading of the influent to the second stage. These equations are based on the efficiency of the trickling filter and the settling tanks following the trickling filter. The equation for the second-stage filter is based on the presence of a settler between first-stage and second-stage filters.

Velz (1948) proposed a design formula for BOD removal based on the principle that the rate of extraction of organic matter per interval of filter depth is proportional to the remaining concentration of organic matter as measured in terms of its removability.

$$\frac{L_D}{L} = 10^{-KD} \tag{2.71}$$

where L is the total removable fraction of BOD and L_D is the corresponding BOD at depth D.

Schulze and Lipe (1964) modified this equation somewhat to account for hydraulic loading and a temperature correction factor and for recirculation. The time, X, required for a fluid to travel through a trickling filter was estimated by using the following expression:

$$X = \frac{(1.035)^{(T-20)} D}{(Q/A)^{2/s}}(1 + R)^{1/3} \tag{2.72}$$

where X is the time of travel, T is the temperature in °C, and A is the surface area of the bed. In 1960, Schulze developed a similar expression:

$$\frac{L_e}{L_i} = 10\frac{KD}{(Q/A)^{2/s}} \tag{2.73}$$

Here, L_e is the BOD in filter effluent (mg/L), L_i is the initial BOD (mg/L), D is the filter depth (ft), Q is the hydraulic load (10^6 gal/acre/day), and K is a constant.

The temperature correction factor

$$\frac{L_e}{L_i} = 10^{-b} K_{20}\left(\frac{D}{Q^{2/s}}\right) \tag{2.74}$$

$$b = 1.035^{(T-20)}$$

The recirculation of the combined hydraulic load Q_2 is

$$Q_2 = Q_1(1 + 2) = 3Q_1 \tag{2.75}$$

Equation (2.75) was expanded by Eckenfelder and O' Connor (1961) and Eckenfelder (1963) to cover the effect of the decreasing removal of BOD per unit depth with increasing filter depth, that is, residence time.

$$\frac{L_e}{L_o} = e^{-(KD1-m)}/(Q/A)^n \tag{2.76}$$

where L_o is the BOD of influent, D is the depth, $1 - m$ is the exponent of $D = 1.0$ for a filter where biological slime is uniform with depth. If a waste is so complex that there is a decrease in the ratio of removal of BOD with time, the equation becomes

$$\frac{L_e}{L_o} = \frac{1}{1 + \dfrac{CD(1 - m)}{(Q/A)^n}} \tag{2.77}$$

Certain values are established on filter performance $C = 2.5$, $1 - m = 0.67$, $n = 0.5$ for A in acres. D is the depth in feet and Q is the feed rate in 10^6 gal/day. Recirculation is treated strictly as a diluent of the influent feed L_i and expressed as

$$L_o = \frac{L_i + RL_e}{1 + R} \tag{2.78}$$

Gallerand Gotaas (1964) used multiple regression analysis on data from existing plants and gave the following expression:

$$L_e = \frac{0.464\,(L_o)^{1.19}(1 + R)^{0.28}(Q/A)^{0.13}}{(1 + D)^{0.67} \times 7^{0.15}} \tag{2.79}$$

The recirculation variable of Eq. (2.79) should be substituted for the applied BOD, that is, L_o for filters using recirculation. To obtain volumes by means of these three primary design approaches, the following formulas are used:

First stage

$$V = 0.0263\, QL_i \frac{(1 + 0.1R)^2}{1 + R}\left(\frac{E}{1 - E}\right)^2 \tag{2.80}$$

Second stage

$$V = 0.0263 \times \frac{QL_i}{F}\left(\frac{E_2}{(1 - E_1)\,(1 - E_2)}\right) \tag{2.81}$$

Eckenfelder: First stage

$$V = 7.0 \times \frac{Q}{D^{0.33}}\left(\frac{E/(1 - E)}{1 + R}\right)^2 \tag{2.82}$$

Second stage

$$V = 7.0 \times \frac{Q}{D^{0.55}}\left(\frac{E_2}{(1 - E_2)(1 - E_2)(1 + R}\right) \tag{2.83}$$

Galler–Gotaas: First stage

$$V = 0.1355D\left(\frac{Q^{0.15}L_i^{0.19}[(1+R(1-E)]^{0.19}}{7^{0.15}(1+D)^{0.67}(1-E)(1+R)^{0.75}}\right)\frac{1}{(1-E_1)^{0.5}} \quad (2.84)$$

Second stage

$$V = 0.1355D\left(\frac{Q^{0.15}L_i^{0.19}[1+R(1-E)]^{0.19}}{7^{0.15}(1+D)^{0.67}(1-E)(1+R)^{0.73}(1-E_1)^{0.5}}\right)^8 \quad (2.85)$$

The trickling filter volume, when compared by the NRC and Eckenfelder formulas, will vary directly with both flow rate and efficiency function $E/(1-E)^2$. Volume will decrease with increase in recirculation, reaching practical limiting recirculation of 4 or 5. Volume varies with influent BOD in the NRC formula but is not a parameter in the Eckenfelder formula. Also, in the Eckenfelder formula, volume varies with the depth function $1/D^{0.33}$, which causes volume to decrease with depth. In the Galler–Gotaas approach, volume increases with the flow rate, influent BOD, and required efficiency. It decreases with increases in temperature and depth. Recirculation and efficiency are independent parameters.

A volume decrease is observed only with higher efficiencies at practical limiting ratios of 4 or 5. Two-stage trickling filters can be designed by all the three formulas. The optimal volume can be approximated by designing stages of equal volume with the NRC and the Eckenfelder formula and using a ratio of 1:2 in the Galler–Gotaas formula. A two-stage unit requires less volume than a single-stage unit.

2.5.3.5 Design cost

Optimization analysis for the design of trickling filter is presented in considerable detailsin Galler and Gotaas (1966). A linear programming procedure was developed for a trickling filter in which radius, depth, recirculation, feed rate, and BOD levels were considered. Primary and secondary settlers of the rectangular design were included. Both design and cost calculations were dependent on the BOD in the filter influent and the desired BOD reduction, and they were independent of the plant influent volumetric rate. The hydraulic rate through a filter should be set at the maximum level permitted until the circulation ratio is approximately 4:1. Calculations show that the hydraulic rate should be decreased after this ratio is achieved. In cases where pumping of influent is not a factor, trickling filter should be designed as deep as possible while maintaining adequate air circulation. Deep filters with forced air circulation are more economical than shallow filters in reducing BOD. However, they lose their advantage if pumping is required.

2.5.3.6 Recycling

Although recycling is commonly practised in conventional systems, there is no single reason for justifying its use. Some of the factors involved are as follows:

1. Valuable nutrients are brought to the site of maximum growth again.
2. It allows the recycle effluent to pick up activities of extracellular enzymes.

3. It dilutes the feed, lowering the level of any inhibiting compound present in the influent.
4. It often helps increase the hydraulic flow around the packing; therefore, adequate scouring takes place.
5. It continually re-modulates the system with a balanced set of microbial flora.

2.5.3.7 Oxygen transfer and utilization in trickle bed bioreactor

Oxygen is transferred from air passing through the filter to the films of liquid/water passing over the slime as shown in Figure 2.27. Experimentation with various types of TBBs has shown that the rate of O_2 absorption by a liquid passing over slimes follows the following relationship:

$$-\frac{dc}{dD} = k(C_s - C_e) \tag{2.86}$$

where C is the concentration of dissolved oxygen, D is the depth of the filter bed, C_s is the saturation concentration of O_2 in the film, C_e is the DO concentration at equilibrium, and $(C_s - C_e)$ is the oxygen deficit. Integration of above equation gives

$$\int_1^2 \frac{dC}{C_s - C_e} = \int_{D_1}^{D_2} K dD \tag{2.87}$$

So

$$\frac{(C_s - C_e)_1}{C_s - C_e)_2} = e^{k(D_2 - D_1)} \tag{2.88}$$

in which k is the transfer rate coefficient related to the hydraulic loading Q to the TBB and is given by

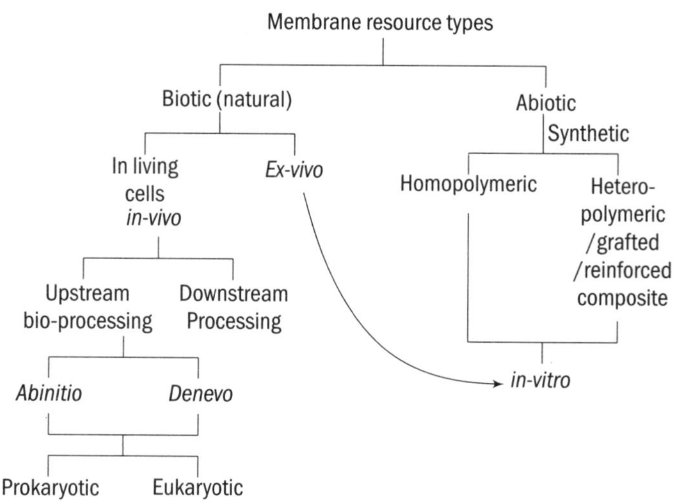

Figure 2.32 Membrane types in wastewater reuse

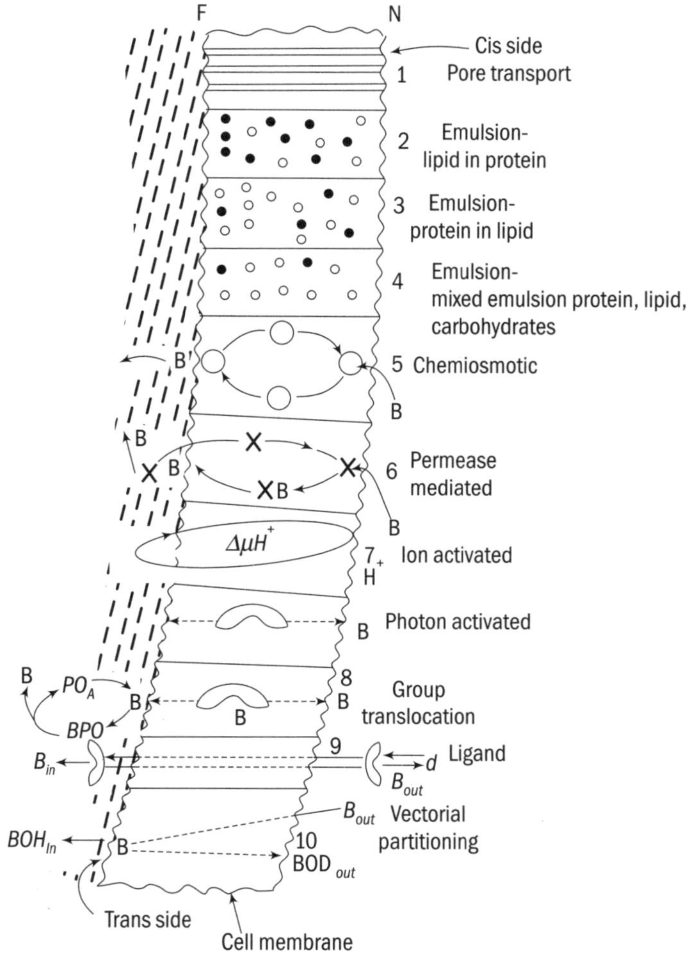

Figure 2.33 Schematic of a microbial *in vivo* cell membrane active in pollutant removal mechanisms

$$k = \frac{c}{Q^n} \tag{2.89}$$

Here, n depends on filter medium and C is a coefficient that also depends on the filter media. Total depth (h) of aerobic/oxic film is estimated from

$$h = \sqrt{\frac{2D_L C_L}{K_r \rho}} \tag{2.90}$$

Here, h is the depth of film to which O_2 penetrates, D_L is the diffusivity of O_2 through filter film, C_L is the concentration of dissolved O_2 in the liquid passing the film surface, ρ is the density of filter film, and K_r is the unit O_2 utilization rate of the film. Oxygen transferred to a typical polygrid filter is given by

$$N = 5.0 \times 10^{-4}\ \mathrm{K}\ \frac{(C_s - C_e)}{Q} \tag{2.91}$$

in which N is in lb $O_2/h/ft^2$, C_s is O_2 saturation in mg/L, C_e is equilibrium DO concentration in mg/L, and Q is hydraulic loading in gal/min/ft^2.

2.5.3.8 Temperature effect

Performance of TBB varies with the temperature of filter films and the liquid passing over the films. The change of temperature causes changes in respiration rate, O_2 transfer rate, and O_2 saturation. The efficiency of TBB (E_t) and temperature (T) can be expressed by the following equation:

$$E_T = E_{20^\circ} \times 1.935^{(T-20)} \tag{2.92}$$

2.5.4 Membrane Bioreactor Processing

2.5.4.1 Membrane types and criteria

For reuse of wastewaters *in vivo* and *in vitro*, membranes have played crucial roles. The efficient use of water in production processes, depending on membrane resources and their participation, helps create a greener image. Figure 2.32 shows the membrane resource types.

2.5.4.2 In vivo membrane involvement

For efficient wastewater treatments, the participation of microbial cells is well known. Trickling filter and immobilized cell reactors are the examples. In microbial cell participation in wastewater treatment *in vivo*, membrane involvement may become prominent. *In vivo* membrane function mechanisms playa crucial role in removing pollutants from waste waters (Figure 2.33). Because of these removal advantages in *in vivo* membranes, environmental engineers did try to mimic operation of synthetic abiotic membrane in bio-and non-bio separation processes. For separation of pollutants, the membrane in general need to meet the following features:

- It should be semi-permeable.
- It should be high enough in rates of transport to be economical.
- It should chemically and mechanically be able to withstand excursions in pH and temperature.
- It should be stable in different solvents and in relatively high pressures.
- It should be non-biodegradable.
- It should be easy to regenerate.
- It should be economical to cost.

Water reuse is gaining significance due to the overall water shortage in many countries around the world. It has become almost mandatory to think of ways to reuse or recycle water in certain parts of the world. Sustainability is the new keyword and membrane bioreactors (MBRs) help to achieve just that: a relatively simple way to treat wastewater and make it available for other uses (for example, tempering). However, the increased popularity of MBRs cannot be attributed only to countries with severe water shortage. Even in countries like Germany MBRs are becoming increasingly popular as a way to address complex wastewater treatment problems.

Both conventional wastewater treatment plants (WWTPs) and MBRs use activated sludge to clean wastewater. Subsequently, the activated sludge and the clean water need to be separated once the activated sludge has completed its task. In a conventional WWTP, the activated sludge settles in the secondary clarifier and the clean water proceeds to the drainage system. This separation process is quite difficult to control and, therefore, the separation is never perfect. Certain particles, bacteria, and other components will enter the discharge system with the clean water and eventually leave WWTP. This makes reuse of the clean water somewhat challenging since any bacterial contamination poses a major problem when the wateris recycled as process water. In turn, this is one of the biggest advantages of an MBR system. The membranes will be submerged directly into the activated sludge or in a separate filtration chamber and ensure that no particles or bacteria will enter the discharge system along with the clean water.

Submerged modules for MBRs exist in a variety of different formats. The differentiating criteria are the packing density of the membrane area, the hydraulic conditions that will influence modules for braiding or silting, and the possibility to backwash the system. Some modules are based on flat sheet membranes that have been converted into laminated sandwich-like structures that not only guarantee high packing density but can also be back washed with permeate.

What is common to all different makes of submerged modules is the use of cross-flow aeration. The cross-flow aeration cleans the membrane surface continuously. The use of fine bubble aerations is generally a good idea because then a portion of the air injected is available for the activated sludge process.

2.5.4.3 Technical know-how improves performance

Industrial areas are often limited in space and room for expansion. By implementing the membrane as a secure "shield" against all particles, the actual load of total solids in the activated sludge can be increased to 10–12 g/L (conventional plants operate with 3–5 g/L). This in turn results in a much smaller filtration chamber—often up to 2–3 times smaller than a conventional WWTP.

Besides being used as a space-saving technology, MBRs also represent an important feature of wastewater treatment plants when treating very complex wastewater, for example, the one that contain smicro-pollutants. The biological treatment process of such highly polluted wastewater does not remove all micro-pollutants (Figure 2.34). Further treatment steps are required. By applying an MBR and, therefore, providing particle-free wastewater, the treatment steps with ozone or activated carbon can be sized and operated much more efficiently. The treatment with ozone will break up the dissolved micro-pollutants. The activated carbon can absorb micro-pollutants. Both options become more economic when an MBR is included. The chemical and pharmaceutical industry, as well as the farming industry, will benefit comparably from such systems.

In order to further improve these systems, the cleaning strategy for the membrane and energy efficiency of an MBR are of importance. A clean membrane allows for a higher flux, which in turn results in lower energy consumption for the same amount of wastewater.

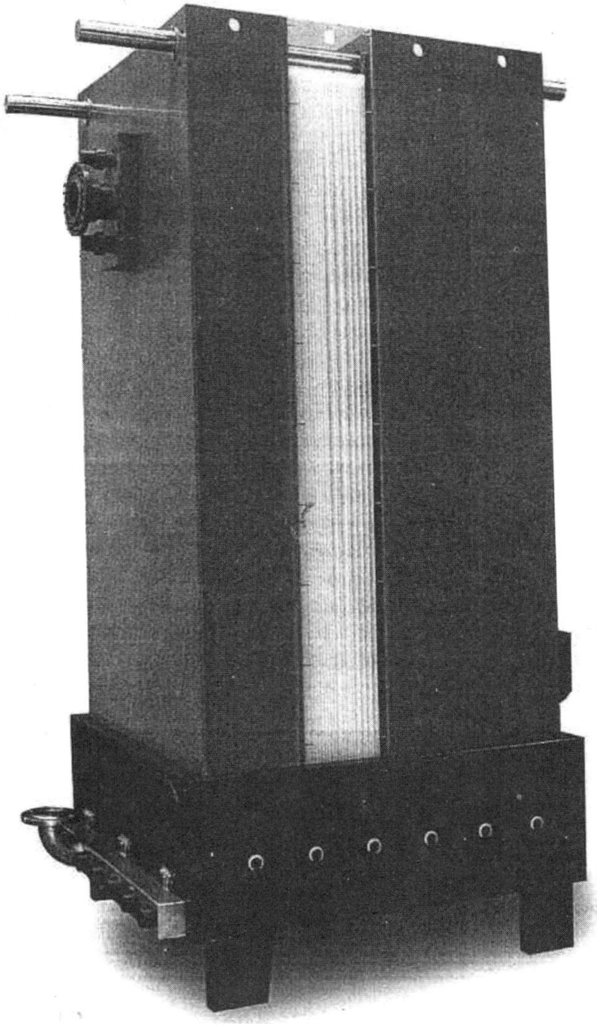

Figure 2.34 Microdyn-Nadir has developed the MBR bio-cel for an efficient and high-quality wastewater treatment at low-space requirement

The recently developed mechanical cleaning process (MCP) for submerged membranes is the latest innovation in the wastewater treatment sector. It was specially developed for the above-mentioned modules that use laminated membranes that can be backwashed. When applying MCP, polymeric granules are added to wastewater to mechanically clean the membrane surface. The aforementioned cross-flow aeration will move beads along the membrane surface. The contact of beads with the membrane surface will break up the layer on the membrane and the cross-flow aeration will move it away from the membrane. The result is a very clean membrane throughout the entire operating cycle.

By applying MCP to the MBR process, a multitude of advantages can be observed. Not only is the need for chemical cleaning decreased to almost zero, but the average flux is also increased by approximately 25%. This results in the less membrane area needed and therefore small filtration chambers. If less membrane area is needed for the same amount of wastewater, the overall energy consumption is reduced.

In summary, MBR plants overall have a lot to offer to the operator. The capacity of existing WWTPs can be increased. Small ecological footprint allows for efficient wastewater treatment even if space is limited. The high quality of the effluent allows for a range of potential reuse applications. In combination with other treatment processes, even very complex wastewater streams can be treated efficiently (Hoffman and Lyko 2013).

2.5.5 Rotating Biological Contactors

2.5.5.1 *Internal film controlled*

Internal film controlled (IFC) form of rotating biological contactors (RBC) consists of a revolving drum equipped with internal baffles and buckets or concentric horizontal rotary cylinders with spargers. These are internal thin liquid film formed on the rotating surface of the contactors. The rotational speed of these drum/cylinders is limited to the low range of 15–25 rpm because of the economic reason and for reduction of foaming when air is sparged. These are also called hinterland contactors. A typical IFC form of RBC is shown in Figure 2.35. In case of high OTR requirement for producing chemicals/biochemicals from waste liquids using a pure microbial culture aerobic system, such types of contactors can

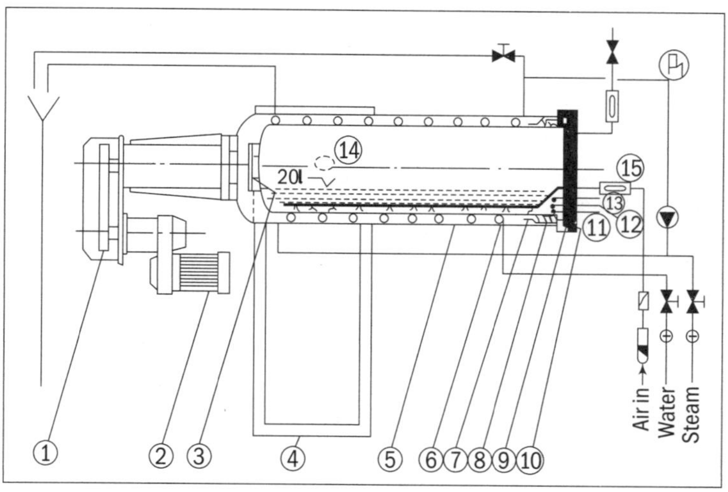

Figure 2.35 (a) IFC (60 L capacity) and (b) labelled IFC: 1. chain drive, 2. variable speed drive, 3. mechanical shaft seal, 4. switch and speed control board, 5. chamber for condensate, 6. cooling tube, 7. sliding bearing, 8. packing rings, 9. gland, 10. cover, 11. temperature probe, 12. DO probe, 13. pH probe, 14. spindle shaft, 15. air filters

be very useful. Production of citric or gluconicacid by pure culture of *Aspergillus niger* or *Pseudomonas ovalis* using molasses, which is a waste of sugar industry, as the substrate is a typical example.

From standard OTR equation, one has

$$C_L = \frac{1}{K_L a_o}\left[\frac{dC_L}{dt} + rx\right] + C^* \tag{2.93}$$

In horizontal rotary bioreactor (HRB),

$$K_L a_0 = K_L a_f + K_L a_s \tag{2.94}$$

where

$K_L a_f$ = VOTC (volumetric oxygen transfer coefficient) in the film by sparging and rotation

$K_L a_s$ = VOTC in the liquid with no rotation

$K_L a_f = K_L a_0 + K_L a_s$

Liquid film thickness is determined from (Figures 2.36 and 2.37)

$$h_f = \frac{\sqrt[3]{(\mu_{am} \times V_p)^2}}{6a^{1/6}\sqrt{\mu a^5}} \tag{2.95}$$

and

$$K_L a_f = \frac{\sqrt{Ds}}{h_f} \tag{2.96}$$

where

$$s = \frac{QN}{360} \tag{2.97}$$

so

$$K_L a_f = K\frac{\sqrt{N}}{h_f} \tag{2.98}$$

where

$$K = \text{constant} = \sqrt{\frac{D\theta}{360}}$$

$$\frac{a'}{t_c} = \frac{2\pi rl}{V}\left(\frac{\theta}{360}\right)N$$

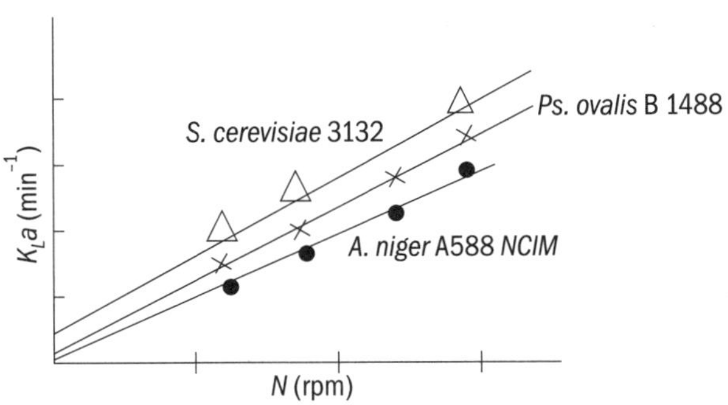

Figure 2.36 Variation of K_La as a function of N with different microorganisms in fermentation

Source Mukhopadhyay and Ghose (1978)

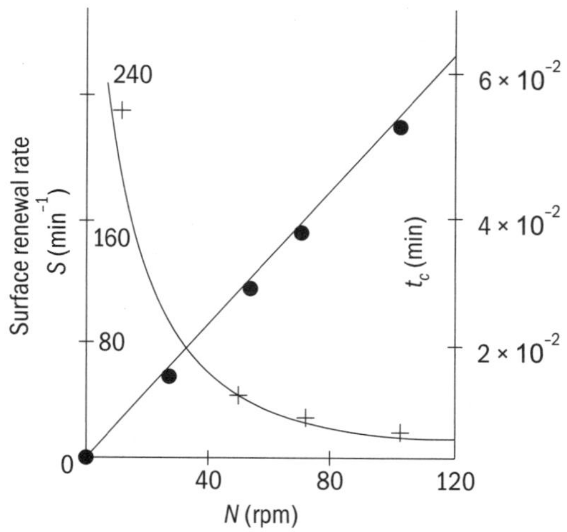

Figure 2.37 Variation of s and t_c as a function of N

As per surface renewal theory,

$$K_L = \sqrt{Ds} = \sqrt{D\left(\frac{\theta N}{360}\right)}$$

$$\frac{dC_L}{dt} = K_L a\,(C^* - C_L) + 9.7 \times 10^{-7} A_f Y$$

where A_f is the rate of surface formation or submergence in cm^2/h and Y is the volumetric or mol fraction in gaseous phase of gas being absorbed (Figure 2.38).

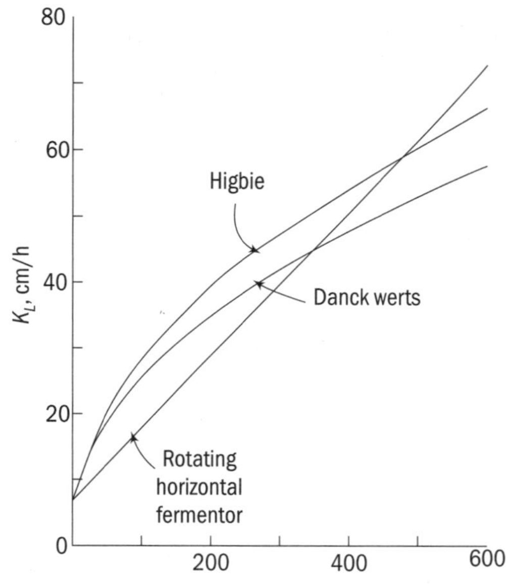

Figure 2.38 Comparison of values for K_L determined in the 3-inch diameter horizontal rotary thin film bioreactor (HRTFB) with those calculated from different theories

$$K_L = 3.3 \times 10^{-5} \times V_p + 6$$

2.5.5.2 External film controlled

External film controlled (EFC) form of RBC consists of a series of closely spaced circular disks of polystyrene or polyvinyl chloride (PVC) (Figures 2.39 and 40). These are partially submerged in wastewater and rotated slowly in it. During the rotation, biological growth results on the surface of disks that eventually forms a slime layer over the entire wetted surface area of disks. In this mode of wastewater treatment, the thickness of the slime film may be controlled by the rotation speed of the disks to regulate the oxygen mass transfer rate through the depth of the external microbial film thickness. Here, F/M ratio serves as a good parameter, indicating the sufficiency of the oxygen transfer rate for the aerobic microbial flora in the film.

Another design of EFC mode-wise operated RBC is a multi-rope system designed by CLRI, Chennai in India. In both modes of operation of RBC, because of high surface renewal for the oxygen mass transfer rate, BOD removal from wastewater occurs successfully.

Process description In operation, biological growths become attached to surfaces of disks and eventually form a slime layer over the entire wetted surface area of the disks. The rotation of disks alternately facilitates the contact of the biomass with the organic material in the wastewater and then with the atmosphere for adsorption of oxygen. The disk rotation affects oxygen transfer and maintains the biomass in an aerobic condition. The rotation is

Figure 2.39 Rotating biological contactor
Source Metcalf and Eddy, Inc. (1979)

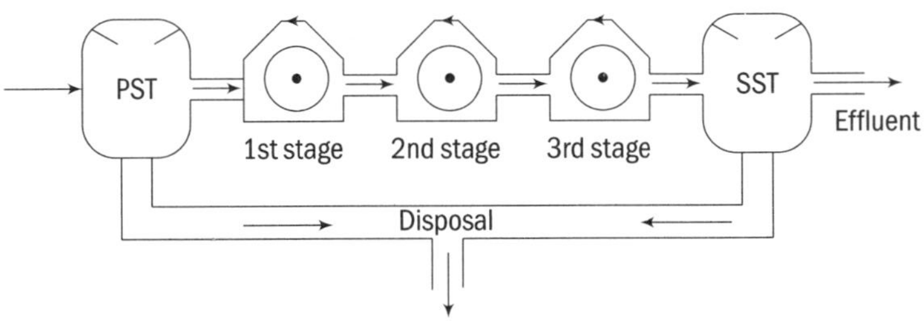

Figure 2.40 RBC treatment train

also the mechanism for removing excess solids from disks by the shearing forces it creates and maintaining the sloughed solids in suspension so that they can be carried from the unit to a clarifier. RBCs can be used for secondary treatment, and they can also be operated in the seasonal and continuous nitrification modes.

Process performance analysis: RBCs are usually designed on the basis of loading factors derived from pilot plant and full-scale installations, although, their performance can be analyzed using an approach similar to that for trickling filters. Both hydraulic and organic load in grate criteria are used in sizing units for secondary treatment. The loading rates for warm weather and year-round nitrification will be considerably lower than the corresponding rates for secondary treatment.

Rotating biological contactors generally are more reliable than other fixed-film processes because of the large amount of biological mass present (low operating *F/M*). This also permits them to withstand hydraulic and organic surges more effectively. The effect of staging in this plug-flow system eliminates short circuiting and damps shock loadings (Figure 2.41).

2.6 WASTE FEEDSTOCKS AND GASES USED FOR THE DEVELOPMENT OF SINGLE-CELL PROTEIN TECHNOLOGY BY IMPERIAL CHEMICAL INDUSTRY

Most people prefer to include meat, fish, eggs, and milk in their diets, given the opportunity to do so. This preference is reflected in the differing proportions of animal products in the total protein consumption in different parts of the world (Figure 2.42). For example, in North America, where there is an ample dietary opportunity, a very high proportion of protein is eaten as meat. The drive for more efficient production of meat has led to an increase in intensive animal production and specialized feedstuffs industries to supply effective diets. The ideal pig and poultry diets contain more crude protein than cereals can provide. Of

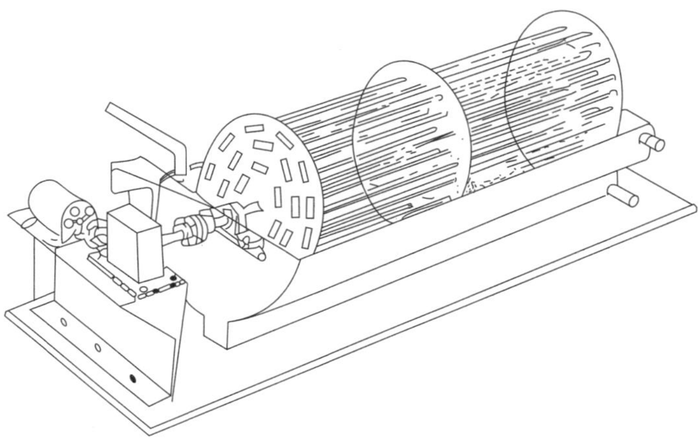

Figure 2.41 External film controlled RBC

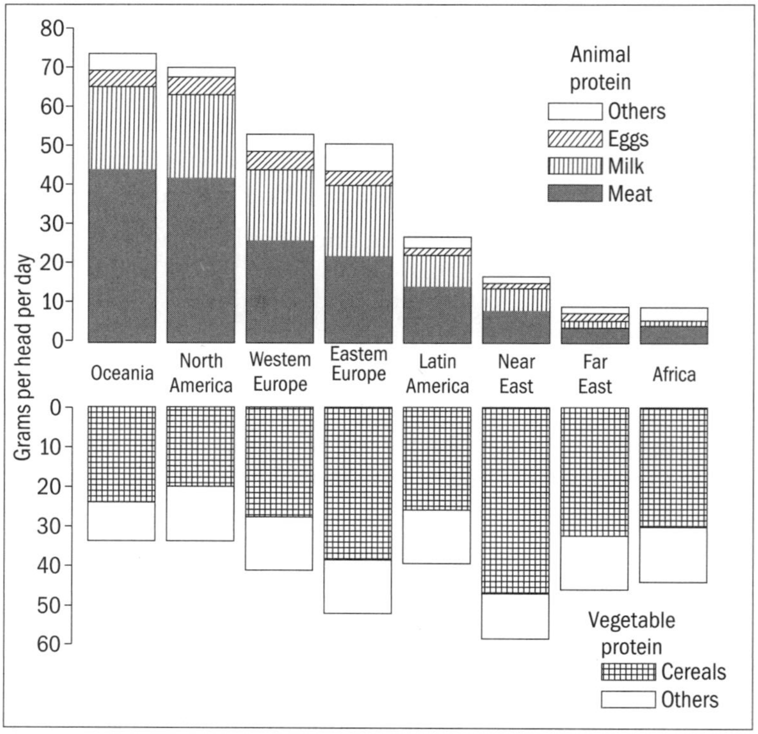

Figure 2.42 Global animal and protein consumption pattern

the high protein supplements used, soya meal is the dominant commodity. The EEC pays approximately $3000 million annually for its soyabean meal imports. It was upon this large market that the Agricultural Division of Imperial Chemical Industry (ICI) set its sights in 1968 when it began development of its single-cell protein (SCP) process. Many particle board mills generate 90% waste methanol in exhaust air containing VOCs.

The target was to produce what is essentially a feed grade product at bulk chemical costs by continuous fermentation.

2.6.1 Substrate and Organism Selection

The first problem was substrate selection. So, initial thinking of ICI was towards the natural gas methane as the substrate. Superficially methane is attractive, but an examination of the chemical and mechanical engineering problems revealed several difficulties. Methane is not very soluble in water and like oxygen it will pass through the fermenter unabsorbed. Recovery of this methane is difficult, and in addition there are intrinsic safety problems associated with the handling of oxygen and methane. Further, it was discovered that for microorganisms growing on methane, the first oxidation step of methane, which is to methanol, consumed energy, rather than making energy available for further reactions.

It was, therefore, concluded that this step was better left to industrial methanol synthesis technology. Therefore, methanol was selected as the substrate, followed by microorganism selection. The criteria for selection included the following:

(1) The culture should be stable.
(2) The organism should be non-pathogenic and non-toxic.
(3) It should have a high nutritional value.
(4) It should have a high yield on substrate and should grow at a temperature, as high as possible.
(5) In addition to methanol as the substrate, only simple inorganic salts should be added as the remainder of the diet.

In its search, ICI examined a large number of methanol-utilizing microorganisms. Those that appeared promising after first evaluation were subjected to preliminary screening for pathogenicity and toxicity. Laboratory fermentation studies were carried out in 5 L fermenters. The outcome of these studies was the selection of a single methanol obligate bacterium that was given the name *Methylophilus methylotrophus.*

Fermentation studies continued to 1 m^3 scale and trial quantities of product were produced. This provisional selection of the microorganism preceded commercial production by several years, during which feeding trials and toxicological testing of the product proceeded in parallel with process and engineering development.

2.6.2 Process Challenges of Large-scale Single-cell Protein Manufacture by Bacterial Fermentation of Methanol

The principal challenges of developing this technology were the following:

(1) Large-scale aerobic fermentation was required.
(2) It was necessary to carry out fermentation under sterile conditions.
(3) Continuous fermentation was necessary for the best process economics.
(4) It was necessary to develop a means of harvesting the bacteria of exceedingly small rods, approximately 1 μm long and 0.2 μm in diameter.
(5) In order to achieve water economy and reduce effluent problems, recycling was required.
(6) Establishment of appropriate drying techniques and formation of an effective product for the market were required.

2.6.3 Imperial Chemical Industry Pressure Cycle Fermenter

Imperial Chemical Industry invented its pressure cycle or airlift fermenter (Figure 2.43) in response to many challenges of large-scale, continuous, sterile, and aerobic fermentation. It was a bold move away from the fermentation industry norm of stirred vessels. The following were the major reasons for the initiative:

(1) There are no moving parts, thus reducing the risk of contamination by eliminating the moving shaft seal necessary for stirred fermenters.

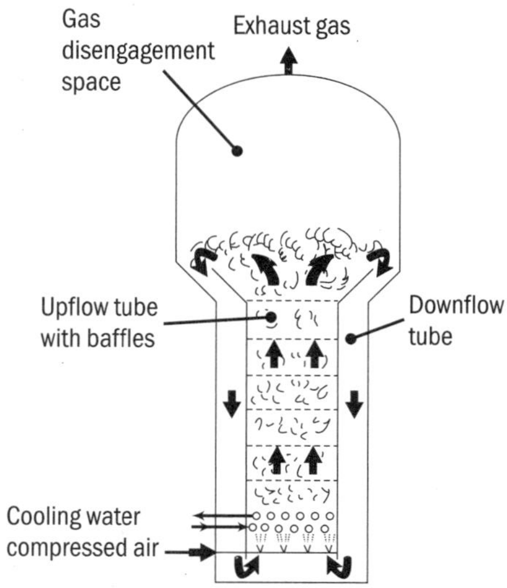

Figure 2.43 ICI pressure cycle/airlift fermenter

(2) A well-defined flow pattern is achieved.
(3) Large heat transfer duties can be accommodated within the fermenter.
(4) Much work is available to provide liquid mixing and surface renewal of bubbles.
(5) Scale-up is easier than that for stirred vessels.
(6) Scale-up is to the limit of mechanical transport or site erection limits.
(7) Pressure cycle fermenter can be designed for high oxygen utilization efficiency and transfer rate.
(8) Although air compressor power is high, there is a considerable saving in agitator power and sterile filtration costs are considerably lower because of the low volumetric rate at the relatively high delivery pressure.

The principle of operation of the airlift fermenter is as follows: The fermenter consists of a riser section, disengager section and downcomer. Air is injected at the base of the riser and effectively the culture is lifted up through coolers and baffle plates. The passage of culture through the baffle plates provides extremely good mixing and transfer of oxygen. At the top of the riser, the culture enters the disengagement section, where bubbles are desorbed. The culture liquid then flows down the downcomer. By carrying out an energy and pressure drop analysis, it is possible to design to have more than 50% of the input energy available for dissipation through coolers and baffles.

2.6.4 Pilot Plant

In 1972, ICI commissioned its 1000 tonnes/year pilot plant. It was built to test and further develop solutions to the process challenges previously described and also to provide tonnage quantities for nutritional and toxicological trials.

The plant incorporated an early version of the airlift fermenter with a working volume of approximately 40 m^3. Data such as circulation rate, air rate, dissolved oxygen concentration, and exit gas composition were logged as far as possible directly by computer and compared with a simulation computer program. This program became an invaluable design aid. The performance of the pilot airlift fermenter proved as good as predicted theoretically. It outperformed conventional fermenters in terms of available oxygen transfer intensity and oxygen transfer per unit of energy dissipated.

Some initial concerns arose because the achieved carbon conversion efficiency on the pilot plant was lower than that in laboratory fermenters. The installation of an appropriately designed methanol distribution system within the fermenter provided a solution to this problem. Experience was gained at this time of sterile operation. The achievement of extended sterile plant operation was probably the greatest engineering challenge faced by the project team. In order to achieve the process objective of growing the chosen microorganism in monoculture for months at a time, it was necessary to exclude other microorganisms from the fermenter system.

2.6.5 Sterile Engineering

Sterile operation should meet the following seven principal objectives:

(i) Achievement of initial sterility of the designated sterile area
(ii) Charging with sterile medium
(iii) Introduction of a pure in oculum in a sterile manner
(iv) Introduction of contaminant-free feed streams
(v) Removal of fluids from the sterile area without causing or allowing ingress of contaminants
(vi) Control of the process without contaminating it (instruments, valves, pump, and so on must be so designed to prevent their contamination)
(vii) Leakage paths from the sterile area must not occur

Failure to achieve the list items (i), (ii), and (iii) results in detection of contaminant after a grow up time that depends on fermenter size, while failure to achieve items (iv), (v), (vi), and (vii) will lead to the detection of a contaminant at the same time or later. Thus, it is necessary to ensure that the plant with its equipment is designed and operated in such a way that the likelihood of contamination is minimal. ICI has developed a sterile engineering code of practice for the design and operation of plants requiring sterile operation.

2.6.6 PRUTEEN

Following pilot plant development, the decision was taken in late 1976 to build the first commercial production plant. The plant was sanctioned as a £40 million project with a design capacity of 50,000 tonnes/year of granular product, up to 30,000 tonnes/year of which could be ground to a powered grade of product for sale into the higher value milk replacer market. ICI employed contractors John Brown (E&C) Ltd during the detailed engineering, design, and procurement phase of the project. The 600-tonne fermenter vessel in particular was a most impressive engineering feature. The vessel, which was 60 m tall, was fabricated at Dunkirk, France and transported in one piece to Billingham by seagoing barge and finally by a purpose-built road to the site.

Inoculation of the plant took place in December 1979. Many lessons were learned during the commissioning phase that followed. Problems encountered were particularly associated with output and efficiency. During 1980, several modifications were carried out, particularly in the fermenter area, in order to achieve the desired output and efficiency.

The process can be described by its mode of start-up and the subsequent operation. First, the empty fermenter and part of the harvesting system are sterilized. This is achieved by displacing air from the system with steam and raising the temperature to 140°C and holding for 2 h. The system is then depressurized and air is added through the sterile filtration system. Sterilized water is then added to the fermenter and when the level reaches the top of the downcomer, circulation is established. The temperature of the liquid is then reduced to 30°C by coolers and nutrients are added to support culture growth. The fermenter is then inoculated with a small quantity, 0.4 kg, of growing culture.

Initial growth is rapid, the cell mass doubling rapidly approximately every 2 h. After approximately 30 h, the cell concentration reaches 3%, there being 30 tonnes of living organisms in the fermenter—at this stage continuous harvesting commences. Culture is continuously withdrawn from the fermenter; it is heated to 80°C and then lowered in pH by addition of mineral acids. Methanol and other nutrients are then continuously added to the fermenter. This treatment of the culture by heat and acid causes flocs to be formed. The released carbon dioxide attaches to the flocs, causing them to rise. The flocs are continuously removed in a flotation stage at a solid concentration of 10%; the clear liquid from the flotation stage is recycled back to the fermenter. The harvesting rate and nutrient addition rate are maintainedat a constant cell concentration of 3% in the fermenter. The flocs are then centrifuged to 20% solids concentration before fed to driers. The concentrate is heat sterilized and returned to the fermenter.

Final dewatering is achieved in two flash driers, which produce a granular product. The drier loop comprises furnace, an airlift tube, cyclones for dried product collection, a granulator, and a gas scrubbing system. The cream from the centrifuge is neutralized to pH 7 and is then sprayed into the granulator. The wet granules fall into the base of the airlift tube and they are rapidly transported by the hot furnace gases from there to the top of the tube. Rapid moisture loss occurs and the granules are captured in the cyclones. Some of the granules are removed as the product and the remaining granules are returned to the granulator. The gases leaving the cyclones are scrubbed and the evaporated water condenses. The gases are recycled back to the furnace system to maintain a safe, low oxygen concentration. The product granules pass over a fluid bed cooler and are pneumatically conveyed to storage bunkers.

The plant has the capacity to grind a portion of the granular product into a fine powder. This product is used as a replacement for skimmed milk powder, particularly in the diets of young calves. The product from ICI's PRUTEEN plant has been widely accepted as an effective alternative to traditional protein sources such as soya bean meal and fish meal for inclusion in animal feed. The product is officially registered in 15 countries and has been sold in 25 countries. More than 120 million animals have been raised on PRUTEEN containing diets. PRUTEEN is now an established product in the animal feed industry.

2.6.7 Future of Single-cell Protein Technology

The SCP concept is 20 years old. During this time, several technologies have appeared, plants have been built, products sold, and hundreds of millions of animals fed SCP have been consumed. Yet it remains to many a technology of unfulfilled promise. This impression, as well as the very familiarity of the subject, leads to an undervaluing of the real and immediate contribution of this profound and proven resource transfer technology.

Industrial biomass production allows simple hydrocarbons—non-agricultural resources—to be introduced into the food chain. Most of the international food supply and food balance problems, which existed in the mid-1960s and which helped generate the enthusiasm and hope surrounding the early SCP days, remained in the middle 1980s. The majority of the world's population is still undernourished, receiving inadequate dietary protein and energy. The Eastern Hemisphere still depends heavily on the Western Hemisphere to meet its protein shortfall through imported soya beans; yet billions of dollars worth of gas continues to be flared each year and the world is experiencing massive overcapacity in methanol production. Hydrocarbon surpluses and protein deficiency are often with in the same country.

Despite the social, political, and strategic implications of an independent novel source of additional food and its positive reinforcing effects upon domestic agriculture, SCP production has not yet proliferated. Part of the explanation lies in the considerable changes in the relationship between protein and energy prices over the last decade, which have adversely affected SCP economics. Part of it lies in the fact that this technology does bridge two sectors of the economy—energy/petrochemical sector and agriculture—where investments of this scale and significance are judged completely differently.

With energy prices being the principal determinant of SCP production costs and product value being primarily influenced by the commodity protein market as represented by the price of soya bean meal, the simple soya–methanol price relationship provides a convenient measure of the recent changes in SCP economics.

In the late 1970s and early 1980s, just as proven methanol SCP technology was being made available, the second oil crisis, the ensuing recession, together with record traditional protein meal production construed to increase energy costs and decrease protein price at the same time with serious consequences on SCP economics. The impetus went out of the SCP projects, second-generation plants were not built, and the learning and benefits of further process development were slower. The economic situation is now recovering and projects are receiving renewed attention, but this experience has shed further light on some of the conflicts that resulted from such resource transfer technologies.

Methanol is consumed in SCP production. Methanol is a product of the petrochemical industry and derived from natural gas—normally a nationally owned primary fuel. It is typical then for investment in SCP production to be evaluated as a petrochemical project.

As such it is subjected to the rigorous investment analysis that may be applied to other petrochemical projects and requires to demonstrate worthwhile return on capital to be viable. However, the product is an agricultural commodity, an integral part of animal and food production. For most nations, securing a strong independent base in basic food production is of the highest priority and most branches of agriculture consequently benefit from substantial governmental support. The practice is no more evident in Europe. It is not a perfect system but serves to illustrate the fundamental importance attributed to food balance, food security, and the security of domestic agricultural production. It is a vital strategic consideration. The application of what may be simply called industrial investment appraisal to strategic food production will not always be appropriate.

The SCP production lies between these two cultures and has more often than not been viewed as a petrochemical project. Consequently, in recent times when the economics have not been able to show the required high rates of return needed to secure investment, there has been hesitancy and inaction. Viewed as an agricultural project, even the current modest rates of return on SCP plant should be viewed as a bonus over its effects on food security. SCP projects require the push from the energy sector and the pull from the agricultural sector to succeed, and for the strategic benefits to be fully appreciated, they must be viewed from the national political level.

Without SCP production, a nation could sell its abundant natural hydrocarbon resources (for example, methanol) to finance the importation of soya beans. This process is vulnerable to the vagaries of two world markets. With SCP production, methanol is converted directly to SCP, thus reducing the need for soya imports and bringing immediate significant foreign exchange savings. These savings alone could finance the SCP investment in a few years. Projects are being increasingly viewed in these terms and there is every reason to expect that SCPs will eventually play a significant role in food production in the near future.

A further major factor that will influence the future of SCP production is the growing evidence of the nutritional value, safety, and acceptability of protein isolates of SCP for direct human consumption. ICI has developed a process for extraction of nucleic acid from 'PRUTEEN'. It yields a protein concentrate that is not only highly nutritious but also possesses other desirable properties that could make it very attractive for incorporation into foods as a functional protein supplement.

The ICI 'PRUTEEN' plant is not just a means of producing food, but it is also the largest continuous monoculture fermentation and recovery plant for successful operation. The technology itself and the learning and skills associated with it represent a vast, largely untapped potential for diverse applications. Still, in the field of SCP, carbohydrates, which are in plentiful supply in a variety of forms, can be a very good source of substrate for SCP production. ICI's technology for the production of SCP from methanol can be readily adapted forthe production of SCP from carbohydrates. Microorganisms such as *Saccharomyces cerevisiae* and *Candida utilis* that readily grow on sugars are already approved for food use in most parts of the world. Development work on the continuous fermentation of both these microorganisms has been carried out to the extent that plants could readily be built to produce large tonnages of SCP by continuous fermentation. A variety of sugar-based substrates could be used and one such feedstock is sugar cane juice. Therefore, there is a potential for the conversion of sugar into protein in the country of origin. While cane is being cropped and crushed, a part of the resultant juice can be diverted to SCP manufacture.

An advantage of the process is that both invert sugar and sucrose in the cane can be utilized to produce SCP; this leads to the possibility of cropping cane throughout the year, provided land conditions permit. The SCP production plant envisaged comprises a fermentation stage with refrigerative cooling and air compressor and a drying stage. It is convenient to link the existing bagasse furnace steam supply to the SCP production plant, thus reducing the SCP production energy requirements.

There will be many wider applications of ICI technology. The concept of a biorefinery opens up an unexplored array of biological products that may supplement or substitute established synthesis routes in chemical production. One such example already explored by ICI is the production of a biodegradable, body compatible thermoplastic called Biopol.

Poly(3-hydroxybutyrate), PHB, is an optically active aliphatic polyester produced by microorganisms as a carbon and energy storage compound. Under carefully controlled fermentation conditions, selected strains of bacteria can be grown to produce up to 80% of their dry weight as PHB. A variety of carbon sources can be used to achieve this, including carbohydrates, methanol, acetic acid and carbon dioxide–hydrogen mixtures. All of these can be obtained from renewable sources, and so PHB production is not linked to petrochemical supply. Figure 2.44 shows an electron micrographic view of intact cells, containing granules of PHB. The polymer may be extracted from bacterial cells into a suitable solvent such as chloroform or methylene chloride. The resulting PHB is a true thermoplastic with melting point in the region of 180°C and may be processed using conventional plastic industry equipment.

In addition to producing PHB, some bacteria can be induced into manufacturing copolymers of 3-hydroxybutyrate (HB) and 3-hydroxypentanoate units. The latter monomer is usually referred to by its trivial name hydroxyvaleric acid (HV). The mechanical properties of these materials range from the very stiff and rather brittle HB homopolymer to tough,

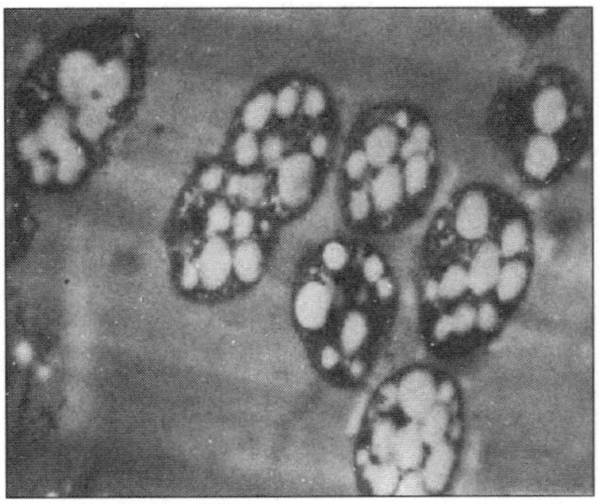

Figure 2.44 Electron microscopic view of yeast SCP

ductile copolymers containing up to 30 mol% HV units. Generally, these polymers are similar to the conventional commodity thermoplastics in terms of appearance, properties, and process ability. However, they retain a trace of their biological origin in that they are capable of being reassimilated by microorganisms. PHB and its copolymers are truly biodegradable as, under the suitable circumstance, they can be completely metabolized by fungi, bacteria, and algae. This property has opened up a new and exciting range of potential applications in disposable products, veterinary medicine, and horticulture.

Moreover, the degradation product of PHB is 3-hydroxybutyrate, which is a normal constituent of human tissue. The rate of biodegradation of PHB in the body is very slow due to the absence of degrading microorganisms, but it is believed to proceed by the action of non-specific enzymes, and several medical companies are evaluating the polymers in or thopaedic devices. The market development for this new and interesting range of materials is undertaken by a subsidiary of ICI called Marlborough Biopolymers Limited.

The development of new large-scale technologies such as SCP technology by ICI is time-consuming and expensive. The development principally suggests that products can be made continuously and reliably by biochemical means on a very large scale. The technology can be used to carry out transformation between products—both petrochemical and agricultural materials can be transformed. In many regions of the world, there are surpluses of such materials. If a consistent realistic pricing policy for such materials is allowed to prevail, large-scale production of a variety of products by fermentation means can be achieved.

2.7 NON-AQUEOUS LIQUID WASTES VOLATILE ORGANIC COMPOUNDS

2.7.1 Resource Industries

2.7.1.1 Transport

(i) On road/land: Transportation vehicles use petrol as the fuel. The menace of toxic emissions at petrol pumps, in cities, and in highways has come under the lens of National Green Tribunal (NGT). Ministry of Petroleum and Natural Gas (MoPNG), Central Pollution Control Board (CPCB), Indian Oil Corporation Limited (IOCL), and other organizations have shown concern on this alarming issue. Petroleum products at fuel stations contain traces of benzene, toluene, and xylene (BTX), all of which are also potentially toxic air pollutants among other volatile organic compounds (VOCs), and they have been subjected to significant investigations/research in almost all countries, especially in Western countries and India. BTX has been detected in both exhaust emissions from vehicles and evaporative emissions at fuel delivery outlets. It is pertinent to note that exposure to a high level of BTX causes neurotoxic health hazard symptoms. It is also claimed that persistent exposure to a high level of BTX may cause injury to human bone marrow. It will also cause DNA damage in mammalian cells. So BTX level in air must be always within permitted limits. In more recent years, it has prompted all state and central governments to take pollution mitigation measures. Energy and fuel industries contribute significantly to air pollution.

(ii) Aquatic avenues: Petroleum non-aqueous products in oceanic avenues of transportation leave oil spills through leakage of tankers. It may cause fire pollution hazards in such avenues. For oil spill pollution, fire hazard prevention by using the genetic engineering approach is now well known. GM superbug degrades oil spill when sprayed on it and prevents fire hazard.

(iii) Water avenues unfit even for bathing: India and many other countries have many rivers and beaches that people use for bathing and many other purposes. These waterbodies need regular cleaning. In India, rivers such as Ganges, Narmada, Jamuna, Teesta are laden with pollutants and phytoplanktons because of human activities. The polluting activities of LHGBRs in India have caused many rivers to remain only on maps and non-existent on the ground. In India, a dip in the river Ganga is believed to wash away all sins. But this dip in Ganga may also make people sick. The CPCB in its survey report in recent years has noted that water of Ganga is not fit for bathing. According to the official report of CPCB, Ganga water fails in meet almost all parameters of safety norms. CPCB conducted test of water along the river at 11 locations in Uttarakhand from Gangotri to Haridwar, a distance of 296 km. The four major water quality indicators, namely, temperature, dissolved oxygen (DO), biological oxygen demand (BOD), and coliform (bacterial), were assessed in Ganga River. Clinical tests of water samples showed high levels of BOD, coliform, and other toxic materials around Haridwar district.

Recently, efforts have been made to generate clean water from wastewater. Indian Institute of Technology Madras in collaboration with Chennai Petroleum Company Limited (CPCL) is developing a solar powered system for safe drinking water supply from wastewater for villages. In the northern India, recycling by domestic wastewater treatment system is gaining importance.

Bibliography

Biskner, C. D. and J. C. Young. 1977. *JWPCF* 49(2)

Blakebrough, N. (ed.). 1967. *Biochemical and Biological Engineering Science*, Vol. 1. New York: Academic Press

Chakraborty, A. M. and J. F. Brown, Jr. 1978.*Genetic Engineering*, edited by A. M. Chakraborty, pp. 185–193. CRC Press

Chipman, J. K. 1993. *Biotechnology News* 29: 6

Choudhury, N. 1994. Keynote lecture on Water Management. The National Concern, Challenges for Survival Organized by J. U. Alumni Association at India International Centre, July 30

Eckenfelder, W. W., Jr. Trickling filtration design and performance. 1963. *Trans. ASCE* 128: 371–384

Eckenfelder, W. W., Jr and D. J. O'Connor. 1961. *Biological Waste Treatment*. New York: Pergamon Press

Etschmann, M. M. W., W. Bluemke, D. Sell, and J. Schrader. 2002. Biotechnological production of 2-phenylethanol. *Appl. Microbiol. Biotechnol.* 59(1): 1–8

Etschmann, M. M. W., D. Sell, and J. Schrader. 2003. Screen of yeasts for the production of the aroma compound 2-phenylethanol in a molasses-based medium. *Biotechnol. Lett.* 25: 531–536

Fabre, C. E., P. J. Blanc and G. Goma. 1998. 2-Phenylethyl alcohol: an aroma profile. *Perfum. Flavor* 23: 43–45

Fedarko, K. 2016. Are We Losing the Grand Canyon? *National Geographic* September, p. 14

Galler, W. S. and J. B. Gotaas. 1966. Optimization analysis for biological filter design. *J. Sanit. Eng. Div.* 92: 163–182

Ghosh, S. and D. L. Klass. 1981. ACS Report, pp. 252–258

Hoffman, C. and M. Lyko. 2013. Efficient wastewater treatment. *Process Worldwide* 1: 40–41

Howe, R. H. L. 1962. Complete biological treatment of antibiotic production wastes. *Biotechnol. Bioeng.* 4(4): 161

Kim, I. 2004. The promising world of bio-oxidation. *CEP* January, pp. 8–11

Krishna, J. 1990. *Water Management: Perspectives, Problems and Policy Issues*. INAE Publication

Loehr, R. C. 1974. *Agricultural Waste Management: Problems, Processes and Approaches*. New York: Academic Press

Metcalf and Eddy, Inc. 1979. *Wastewater Engineering: treatment disposal reuse*, 2nd edn. New York: McGraw-Hill

Mukhopadhyay, S. N. 1987. The dimension of biotechnology in transportation system studies. In *Transport Systems Analysis and Policy Studies*, edited by P. S. Satsangi and A.L. Agarwal, pp. 350–354. New Delhi: Tata McGraw-Hill

Mukhopadhyay, S. N. 2008. Oxic and anoxic thermophilic biotechnological treatments of solid food waste for value added products retrieval: recent progress. Proceedings of the International Conference on Solid Waste Technology and Management, Philadelphia, 30 March to 2 April, pp. 760–771

Mukhopadhyay, S. N. 1995. Water resources, pollution and purification for environment and hazard protection. In *Proceedings of the Environmental Development of Process Industries*, pp. 189–200. New Delhi: Indian National Academy of Engineering

Mukhopadhyay, S. N. 1998. The dimension of biotechnology in transportation system studies. In *Proceedings of the International Conference on Transportation Systems Studies*, edited by P. S. Satsangi and A. L. Agarwal, pp. 350–355. New Delhi: TMH Publishing Co. Ltd

Mukhopadhyay, S. N. 2012. *Process Biotechnology: Theory and Practice*. New Delhi: TERI Press

Mukhopadhyay, S. N. 2016. *Food Engineering: Process and Technology*. New Delhi: Viva Books

Mukhopadhyay, S. N. 2018. *Process Biotechnology Fundamentals*, 4th edn. New Delhi: Viva Books

Mukhopadhyay, S. N. and R. K. Baisya. 1997. Role of water composition in relation to food biotechnology. Proceedings of APBioChEC'97 Beijing: Tsinghua University Press

Mukhopadhyay, S. N. and D. K. Das. 1994. *Oxygen Responses, Reactivities, and Measurements in Biosystems*, pp. 61–104. London: CRC Press

Mukhopadhyay, S. N. and T. K. Ghose. 1978. Oxygen transfer in laboratory horizontal rotary fermenter under varying rotation speeds. *J. Ferment. Technol.* 56: 558

Mukhopadhyay, S. N. and P. Mukhopadhyay. 2000. Water properties, pollution and management. In *Waste Recycling and Resource Management in Developing World, Ecological Engineering Approach*, edited by B. B. Jana, *et al.*, pp. 551–557. University of Kalyani, India and International Ecological Engineering Society, Switzerland

Mukhopadhyay, S. N. and N. Saha. 2000. Progress in thermophilicbiotropic *Methanobacterium* sp. Proceedings of Book of Abstracts, Vol. 4, Germany, pp. 74–76

Nosrati, M., T. R. Sreekrishnan, and S. N. Mukhopadhyay.2007. Energy audit, solids reduction, and pathogen inactivation in secondary sludges during batch thermophilic aerobic digestion process. *J. Environ. Eng.* 133(5): 477–484

Rich, L. G. 1973. *Environmental Systems Engineering*. Tokyo: McGraw-Hill

Schulze, K. L. and R. S. Lipe. 1964. Relationship between substrate concentration, growth rate and respiration rate of *Escherichia coli* in continuous culture. *Archiv.furMikrobiol.* 48: 1

Shah, D. B., G. A. Coulman, L. T. Novak, and B. G. Ellis. 1975. A mathematical model for phosphorus movement in soils. *J. Environ. Qual.* 5: 405

Shah, D. B., L. T. Novak, P. M. Schircrholz, and G. A. Columana. 1975. Water treatment: chemical engineering methods in action. *AIChE Symp. Ser.* 71(151): 360–366

Sinclair, C. G. and D. N. Ryder. 1975. *Biotechnol. Eng.* 17: 232

Smith, J. S. and Y. H. Hui (eds). 2014. *Food Processing: Principles and Applications*. Blackwell Publishing

Spellman, F. R. 2016. *Handbook of Environmental Engineering*. Boca Raton, FL: CRC Press

Times of India, New Delhi, 25 October 2016

Times of India, New Delhi, 25 June 2016, p. 23

Times of India, New Delhi, 5 February 2016, p. 21

Times of India, New Delhi, 4 April 2016, p. 7

Times of India, New Delhi, 18 May 2017, pp. 1–10

Ueno, T., Y. Ozawa, and M. Ishikawa, *et al.* 2003. Lactic acid production using two food processing wastes, canned pineapple syrup and grape in vertase, as substrate and enzyme. *Biotechnol. Lett.* 25(7): 573–577

van der Kooji, D. and P. W. J. J. van der Wielen (eds). 2014. *Microbial Growth in Drinking Water Supplies: Problems, Causes, Control and Research Needs.* London: IWA Publishing

Velz, C. J. 1948. A basic law for the performance of biological beds. *Sewage Works J.* 20(4): 607

World Health Organization. 1963. *International Standard for Drinking Water*, 2nd edn. Geneva: World Health Organization

3 Solid/Semisolid Waste Treatments/ Management for Business Developments

Time and again it has been realized that the cause of environmental deterioration is not the need but the greed of LHGBRs natural and unnatural desire to accumulate more and more.

3.1 INTRODUCTION

Everyday a large quantity of solid and semisolid wastes is generated by the major cosmopolitan cities of the world. For instance, Delhi, the capital of India, itself generates close to 12,300 tonne of solid waste daily. The habitable areas in these cities have sites where the solid waste is dumped. The waste composition may vary from city to city to some extent. In India, 51% of urban municipal solid waste (MSW) is organic, 17% is recyclable, and 31% is inert solid waste. If not cleared routinely, these solid wastes dumped for putrefaction create nuisance and filth in neighbouring areas and may cause epidemics, such as cholera, typhoid, and many others. To avoid such hazardous impacts, corporations usually transport the garbage to open landfill sites. At landfill sites, garbage is weighed before being dumped. Garbage is levelled and a layer of soil is spread over it. In the long term, it may pose health hazard risks. Solid wastes dumped in open landfills may cause groundwater contamination near the area. The removal of the accumulated garbage may be hampered by natural influences like rain as shown in Figure 3.1.

Solid organic wastes may be reused/recycled, depending on the composition, towards solid state fermentation (SSF) or composting such as vermiculture for soil fertility in agriculture. In developed countries such as the US, biomass and organic wastes, besides land filling, are also used to generate nearly 15% of the country's energy needs.

3.2 COLLECTION MODE AND THEIR SUCCESS PICTURE

In developed countries, segregating recyclable and non-recyclable wastes at the source has been a success. Householders dispose wastes in community bins that are separately coloured for recycled and non-recycled. Then the corporations collect and dispose them to approved locations. In Delhi, a similar collection strategy was introduced in 2005 by civic bodies with

Figure 3.1 Accumulation of garbage

blue and green bins for wet and dry wastes, respectively. However, complete noncompliance led to the strategy being roll backed in 2011. At the country level, auto tippers were engaged to collect waste from domestic households in over 200 wards in 2011. Segregation though was not done. The drive to segregate waste at source has been successful in some parts of India. Table 3.1 lists the successful models operating in many states.

Table 3.1 Locations/States of awareness and successful collection models in some states of India

Location/State	Awareness/success model
Pune/Maharashtra	Under 'Zero Waste Electoral Ward', about 7500 establishments in the Katraj Gaon ward give 9 tonne of waste to wastepickers daily. Nearly, 3 tonne of wet waste is sent to biogas/bioenergy plants. The dry waste is recycled.
Municipal Corporation of Delhi / Delhi	With the support of Chintan, a non-governmental organization, New Delhi Municipal Council (NDMC) and East Delhi Municipal Corporation work to include wastepickers in doorstep collection for 4000 households and 25,000 households, respectively. Chintan also runs a waste recovery centre.
Howrah Municipal Corporation/West Bengal	Residents are fully aware of recycling procedure. The corporation promoted waste pickers in door-to-door waste collection. Bally municipality takes care that the household wastes are dumped in landfill areas.
Uttarpara-Kotrung Municipality/West Bengal	A l63-year old municipality, it bet Auckland and Milan to win the global award on solid waste management improvement project held during 40 Mayors' Summit in Mexico City on December 1, 2016

The Centre for Science and Environment (CSE) has been working with cities to promote and implement source-segregation and decentralized model of waste management. On December 12, 2017, it launched the 'Forum of Cities that Segregate' which has 26 members. According to the 2017–18 assessment, Vengurla in Maharashtra received the Five Leaves Award, the highest award. Panchagani, another city with less than 0.1 million residents, got the Four Leaves Award along with Alappuzha and Thiruvananthapuram, two mid-level cities. In million plus cities, Indore and Mysuru got the Four Leaves Award.

It is well known that dumping wastes in landfills is not the solution for curbing pollution. Cities must learn to process waste. One of the best options has been followed in Kerala, where municipalities have withdrawn from the waste business. Householders segregate and compost wastes while informal recyclers collect and sell. This type of approach and awareness helps keep cities clean and work towards *Swachh Survekshan* to make 'Swachh Bharat'.

3.2.1 Impact

It is imperative that all living human genomic bioreactors (LHGBRs) and other life forms (LFs) be concerned about the environment. This is because LHGBRs 'live from', 'live in', and 'live with' the environment. LHGBRs live from the environment means that the latter provides livelihood to the former. The environment is the source of raw materials/resources for productive activities. LHGBRs mine resources from the environment, activate and harvest its fruits, and shape the contours of the land for human habitation and make roads and dredge rivers for transport. The term "live in it" means that the environment is not just a physical precondition for all living beings and their productive activities; it is here where the living beings lead their daily lives in the domain of birth and death. Events—social, aesthetic, or cultural—cannot take place in the absence of the environment. LHGBRs and LFs "live with it" mean that the environment existed prior to the arrival of living beings and shall outlast them. The presence of living beings can have a two-fold effect on the environment: beneficial and non-beneficial.

Beneficial impact: It relates to the concept of "existence value". It means the contributions' living beings make to the environment as long as they live. All living beings have inherent value. The environment is impacted as a result of loss of biodiversity, disappearance of particular habitats, and extinction of particular species of flora and fauna at the local and global levels. Beneficial impact is the idea of measuring the values of the environment in material terms or in terms of satisfaction of performances/preferences. However, social meaning, social relations, and evaluative commitments cannot exist in the absence of the environment and is difficult to assess in materialistic terms. Moreover, these values are intrinsic, rather than being extrinsic to human beings because they have spiritual bonding with them.

Non-beneficial impact: Time and again it has been realized that the cause of environmental deterioration is not the need but the greed of LHGBRs natural and unnatural desire to accumulate more and more. LHGBRs and LFs contribute pollution-causing materials to the environment that are s detrimental to their health and well-being too. These are the non-beneficial impacts imparted by the LHGHBRs and LFs.

In order to maintain balance between beneficial and non-beneficial impacts in the environment, it is imperative to revere traditional wisdom along with modern eco-friendly practices and learn to live in sync with the environment.

3.3 SOURCES AND CHARACTERISTICS

Solid wastes vary considerably in source and characteristics. Accordingly, collection activity and disposal vary dependent on the quantity and quality of solid waste generated and their sources. A huge quantity of cellulosic agro residues (straws or cobs) in the form of solid wastes is generated from agricultural fields in countries like India. Large amount of cellulosic wastes are also generated when milling industries process cereals, such as rice, wheat, corn, and barley. Figure 3.2 depicts paddy rice milling and its products as wastes while Figure 3.3 describes wet milling of corn, its products, and co-products including solid wastes.

3.4 RECYCLING THROUGH DIGESTER

Cellulosic solid wastes can be treated and recycled successfully as described in the ensuing paragraphs.

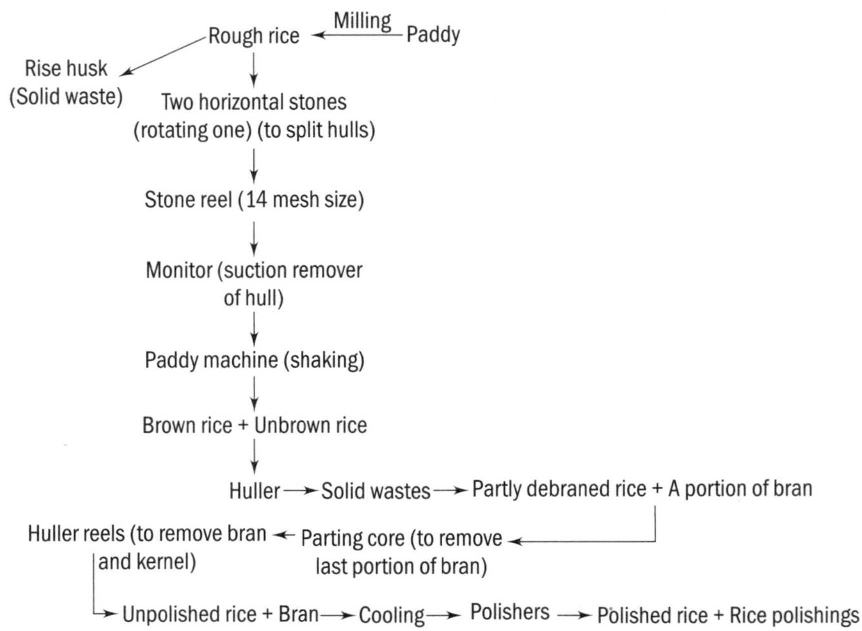

Figure 3.2 Paddy rice milling and its products and co-products as waste

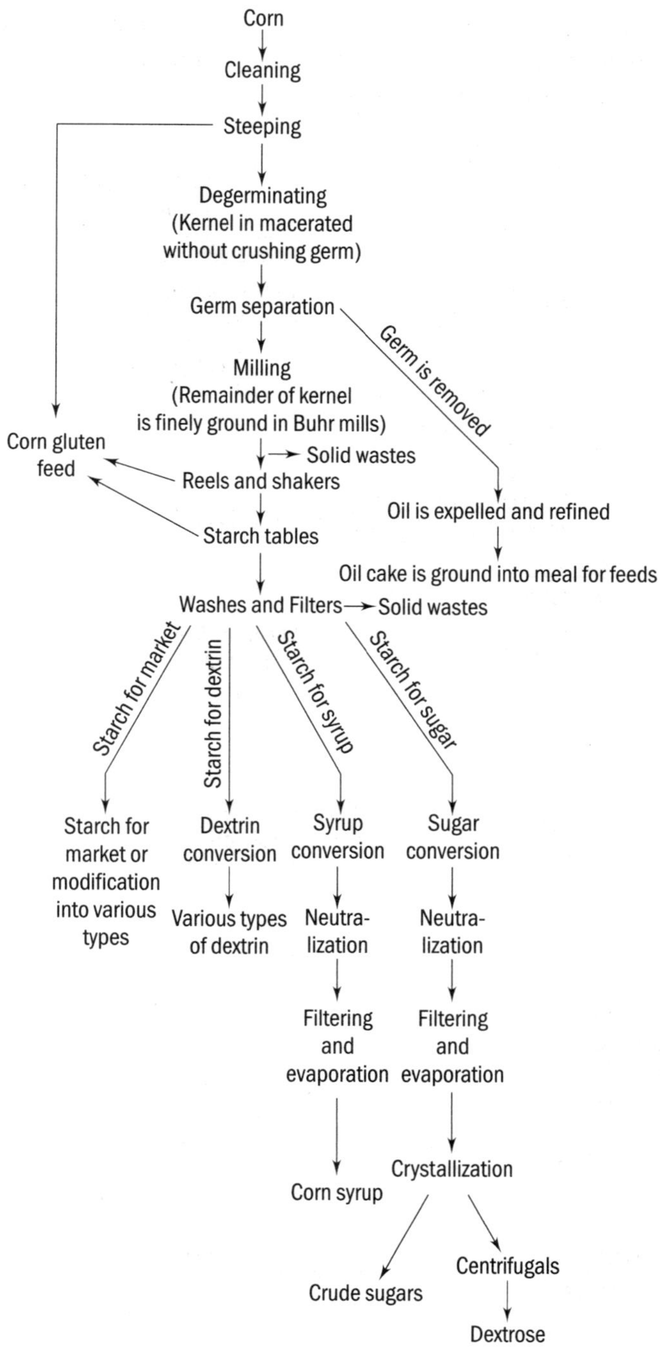

Figure 3.3 Wet milling of corn, its products and co-products including solid wastes

3.4.1 Basis

In fermentative bioconversion, pre-treated and enzymatically (cellulose) saccharified rice straw cellulose is used as feed. The calculation steps are as follows:

Basis: 1 KlEtOH (95% v/v/day) = 741 kg/day

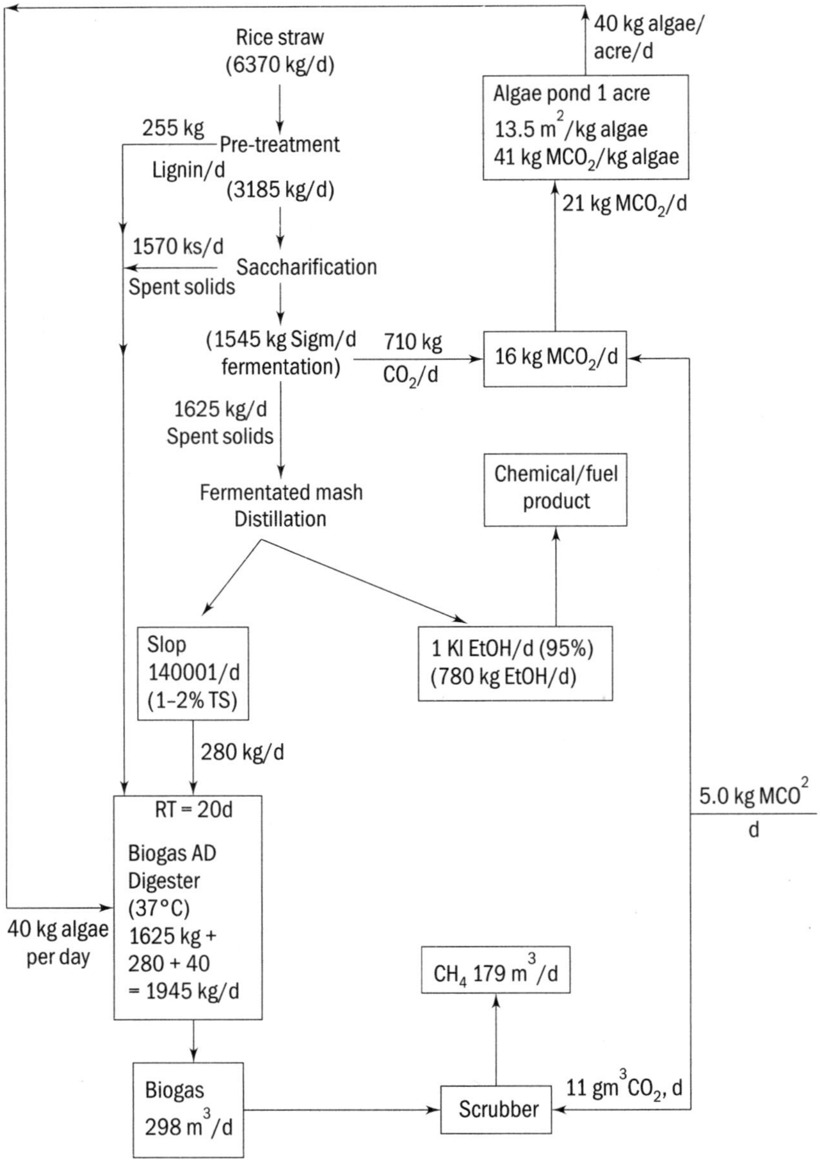

Figure 3.4 KlEtOH production scheme from rice straw integrated to anaerobic digestion process flow sheet of biomethanation

Sugar required in the fermentation $= \dfrac{741 \times 1.01}{0.95 \times 0.51}$ kg/day = 1550 kg/day

Delignified rice straw required $= \dfrac{1550}{0.70 \times 0.70}$ = 3185 kg/day

Whole (cellulose) rice straw required = 3185 × 2 = 6370 kg/day

Spent solids available after pre-treatment and saccharification = 1625 kg/day

Slop available after distillation (containing 1–2%) total solids = 14,000 L/day

Slop TS/day = 280 kg/day = 16 kg/mole/day

3.4.2 Digester Material Balance

Figure 3.4 gives the integrated flow sheet where the digester is filled by algal biomass produced.

In India and many other countries, pulses are consumed in the form of dehusked split grain. Dehusked pulse is produced by milling, either by wet or dry method as shown in Figure 3.5.

3.5 WAYS TO MANAGE SOLID WASTES

With rapid urbanization, India faces massive waste management challenges. Over 377 million urban people, 31% of the population, live in 7935 towns and cities, and generate 62 million tonne of MSW per annum. Only 43 million tonne is collected, 11.9 million tonne is treated, and 31 million tonne is dumped in landfill sites. MSW generation is estimated to reach 165 million tonne/year by 2030. So there is a crying need to manage waste. There

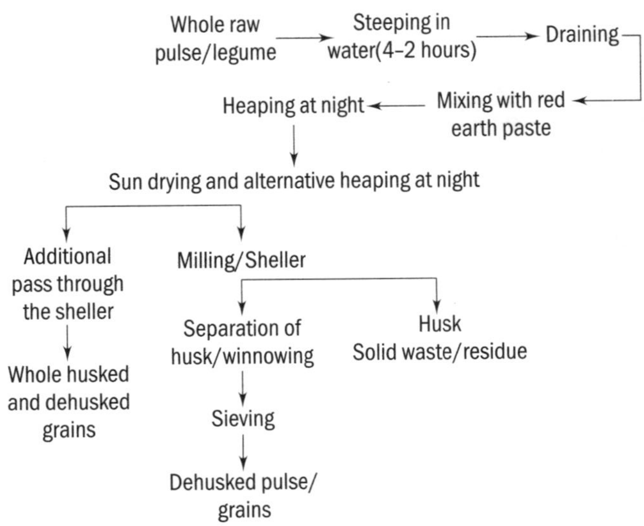

Figure 3.5 Wet milling of pulses/legumes generating solid fine particulates/wastes

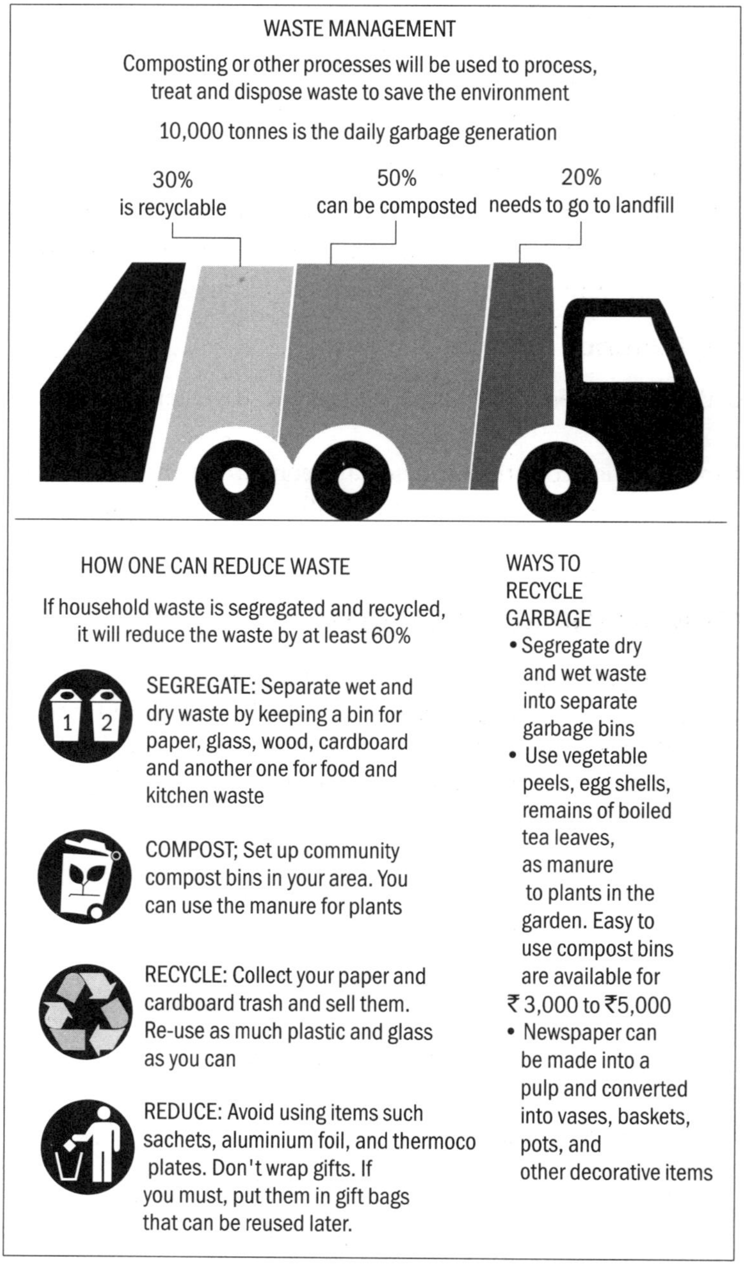

Figure 3.6 Municipal Corporation of Delhi's proposed means of waste management

are many ways by which wastes and garbage can be reduced. Figure 3.6 shows channeling of waste management as proposed by Municipal Corporation of Delhi. Another plan to reduce, reuse, recycle generated wastes for managing solid waste had been proposed by six municipalities in Hoogly district of West Bengal. The uniqueness of this project is that

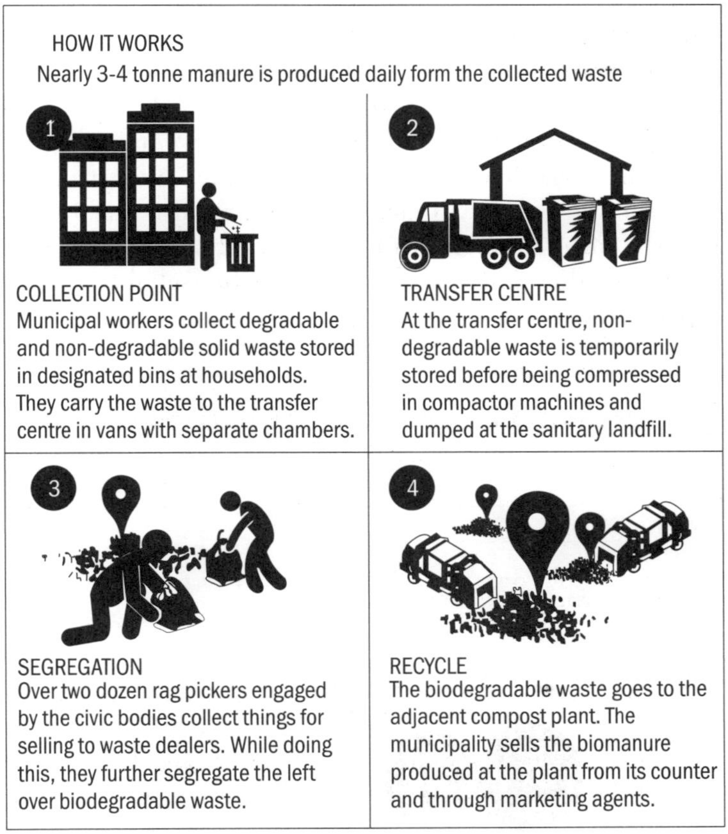

Figure 3.7 Biowaste management project plan of Hoogly district

it had baged the global award for solid waste management, beating Auckland and Milan. This plan is projected in Figure 3.7.

In rural India, more than 67% of households depend on firewood and 10% on dung cake for cooking. Wood, straw or dung are used for burning apart from fossil fuel. Alternate energy resources are needed for the rural environment. Box 3.1 discusses about rural resources being linked to rural inflation while Box 3.2 explains how green wastes can be used.

Ref: J. Gandhi: The TOI, New Delhi, April 14, p7m 2017.

3.6 BIOMASS SOLID WASTES REUSE/RECYCLING ROUTES

3.6.1 US Scientists Results for Waste-to-energy

As mentioned earlier, the impact of solid wastes are enormous. Among solid wastes generated by human activities, cellulosic biomass wastes constitute a substantial part and their reuse by anaerobic digestion using process biotechnology (PB) may deliver value-added products (VAPs).

Box 3.1 Dung cake and firewood—a rural inflation linkage in India

In recent times, increase in the price of dung cake, firewood, and chips have hit the rural population in India. In 2015/16, the price of dung cake rose by 10.8%. In the same period, inflation in firewood and chips went by 7.4%. Overall rural fuel inflation was 6.8% which was two-and-half times more than for urban areas with 2.7%. One of the reasons for higher inflationary pressure in rural areas is the gap prevailing in the public distribution system. About 37% of urban households use petrol and 2% use diesel. In contrast, rural households use 17% petrol and 0.8% diesel, respectively.

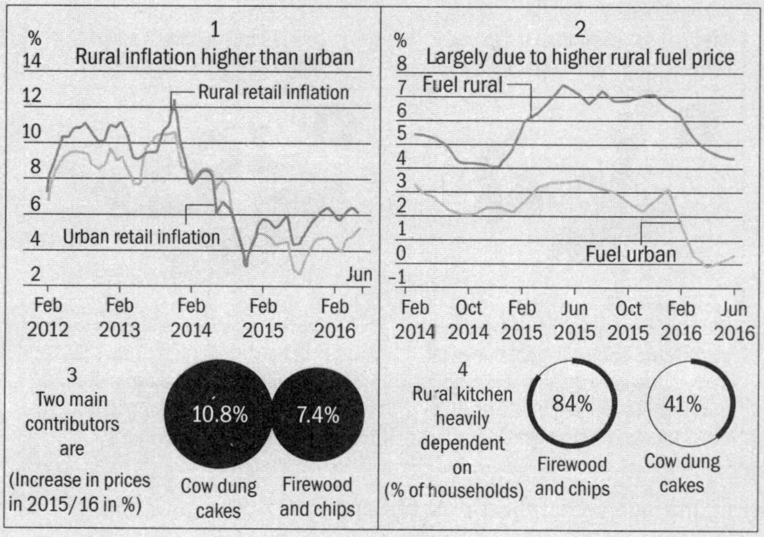

Box 3.2 Making use of green wastes

The National Green Tribunal in 2015 banned burning of dry leaves. Composting was an efficient alternative for getting rid of the foliage. Despite burning of leaves attracting a fine of ₹5000, little has been done in terms of promotion or creation of compost pits in parks. Only South Delhi Municipal Corporation has set up 21 centres to manage green waste. While earlier 20–25 tonne of green waste was sent to landfill area, now it is used to produce natural fertilizer. However, north and east corporations have not have been able to replicate this mechanism of managing organic waste. India can produce around 4.3 million tonne of compost from municipal solid waste. Moreover, if municipal solid waste is not segregated from other types of waste, it can cause methane-fuelled landfill fires. As per new regulations, corporations have to ensure that different kinds of household waste are not mixed.

Source J. Gandhi: The TOI, New Delhi, April 14, 2017.

3.6.2 Cellulosic Biomass Recycling

Biomass and organic wastes (discarded biomass or biomass-derived material) via gasification or other conversion schemes may supply considerable energy needs by the end of the 21st century. A bio-conversion process that is expected to play a role in producing methane from biomass and wastes is anaerobic digestion. Commercial production of substitute natural gas (SNG) and medium Btu fuel gas by anaerobic digestion of waste materials have already started in the US and other countries. Increased usage of anaerobic digestion process for methane production is, however, hindered because of the low reactor rate and conversion efficiency of the conventional digestion process. New digestion techniques are needed to improve the conversion rates and efficiencies so that the full potential of anaerobic digestion as a methane-producing process can be realized. Apart from the crude septic tank, other improved process configurations, such as "standard-rate" digestion, stage digestion, "high-rate" digestion, and anaerobic contact process, have come into existence in the anaerobic digestion process of sewage sludge stabilization. However, the design requirements of these anaerobic sludge stabilization modes, especially high-rate digestion, result in large expensive plants that are difficult to justify for commercial SNG production. Moreover, in conventional digestion process 40% to 70% of the feed organics remain unconverted and must be disposed of at a substantial cost.

Over the years, considerable work had been done to develop advanced digestion methods and process configurations for the conversion of various organic feeds of high Btu fuel gas. The term 'advanced digestion' includes unconventional fermentation modes and process configurations for improved methane production rate and yield.

Researchers have done laboratory evaluation of several advanced digestion methods and a selected biomass–waste blend with the objective to find optimum conversion to the production of SNG. Specifically, digestion of pretreated feed, recycling of digester effluent and product gas, acerbic post-treatment and recycling of digester effluent, and two-phase digestion were evaluated.

3.6.3 The Experiment

A two-phase digestion was conducted. The conventional high-rate digestion was done under baseline operating conditions of 35°C digestion temperature, 0.1 lb VS/t^3-day loading, and a 12-day detention time. Mesophilic (35°C) and thermophilic (66°C) digestion of feed was subjected to mild alkaline (sodium hydroxide) pretreatment with recycling of spent caustic for fresh feed treatment and naturalization of treated feed with digester gas to minimize acid neutralizer requirement. Mesophilic (35°C) digestion recycling was done for product gas and for posttreated digester effluent.

3.6.3.1 Digester feeds

Attempt was made to bio-convert feedstocks comprising water hyacinth (*Eichhornia crassipes*) and Coastal Bermuda grass (*Cynodon dactylon*) indigenous to the Gulf Coast, and sewage sludge and MSW to methane. The biomass and MSW components were finely ground before blending. Coastal Bermuda grass, and mixed activated-primary sludge were blended in the mass ratio of 32.3:32.3:32.3:3:3.1 on a volatile solids (VS) basis. The biomass–sludge blend prepared under mesophilic (35°C) high-rate digestion showed that the mixed feed was superior in terms of nutritional balance, methane yield, and production rate. Biomass–waste blend has many advantages. The mixed feed can be supplied on a year-round basis, and is superior to biomass alone in terms of process economics. In addition, simultaneously energy can be recovered and waste stabilized in an optimized integrated system when a biomass-waste blend is used. The feed used for the work is a mixture of sewage lagoon effluent-grown water hyacinth.

The median particle size of the mixed feed was 0.25 mm. The feed contained moisture, VS, carbon, nitrogen, phosphorus, sulphur, hydrogen, calcium, sodium, potassium, magnesium, cellulose, hemicelluloses, lignin, and crude protein and a high heating value of 88.97, 82.78 (of total solids), 43.14, 1.64, 0.43, 0.31, 5.60, 1.23, 0.78, 1.05, 0.22, 37.5, 31.8, 4.6 and 10.1 wt% and 7.445 Btu/lb (dry), respectively. The empirical formula of the mixed feed was $C_{3.595}\ H_{5.545}\ O_{1.979}\ N_{0.117}\ P_{0.014}\ S_{0.010}\ Ash_{17.22}$ at a molecular weight of 100, and a theoretical methane yield on anaerobic digestion of about 6.4 SCF/lb VS added.[1] The anaerobic biodegradability or volatile solids destruction efficiency of this mixed feed was 66%.

3.6.3.2 Digesters

The reactor vessel are normally flat-bottomed, cylindrical Plexiglass having four vertical baffles mounted 90° apart on the inside walls to prevent vortexing of the culture during mechanical agitation by two stainless steel impellers at 130 rpm. The digesters can be heated by placing them in a chamber that has constant temperature or thermistor-controlled heating tapes. All digesters are geometrically similar in construction. Digesters are generally manually fed once per day after withdrawing an equal volume of digester effluent. Exception is the two-phase system which is equipped with an automated feeding system.

3.6.4 Advanced System Configuration and Operation

3.6.4.1 Digestion of pretreated feed

To improve feed biodegradability and methane yield, the blended mix can be pretreated with caustic soda solution at mild temperatures and pressures. Lower reaction vessel costs, reduced energy and chemical inputs, and low salt concentration in the digester can be achieved by keeping the treatment temperatures at 5°, 25°, 55°, 100°, 121°C and pressures at atmospheric and 30 psig, and selecting dilute caustic. Cost-effective methane production from cellulosic feeds can be increased through alkaline treatment.

Figures 3.8 and 3.9 depict the digestion of the caustic-treated feed at mesophilic (35°C) and thermophilic (55°C) temperatures. In this the undiluted feed is mixed with caustic solution, pretreated at the chosen temperature, pressure, and time. The treated feed is diluted with distilled demineralized water and the digester feed slurry is neutralized with hydrochloric acid or digester gas, Digester gas is used to neutralize the digester feed slurry

[1] Assumes 30% of volatile solids converted to cells on one pass through the digester.

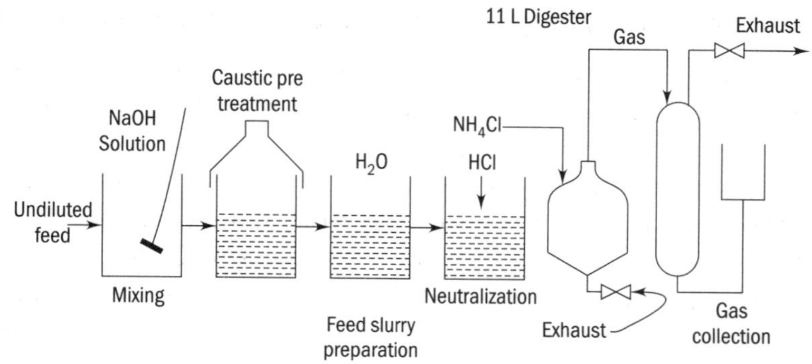

Figure 3.8 Mesophilic digestion of caustic-treated feed

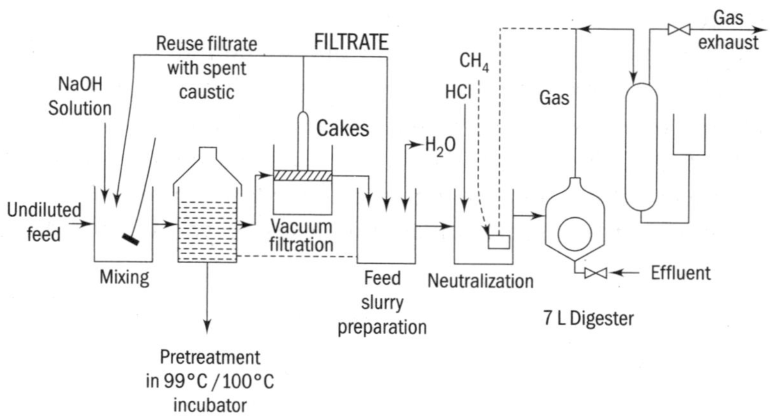

Figure 3.9 Thermophilic digestion of caustic-treated feed

to reduce neutralizer acid requirement and digester salinity, and to increase the bicarbonate alkalinity (buffer and capacity). In some cases, the caustic-treated feed is vacuum filtered, and a portion of the filtrate containing the spent caustic solution is recycled for fresh feed pretreatment. Filtrate recycling reduces the caustic requirement for feed pretreatment, the acid requirement to neutralize the pretreated feed, and the salt concentration in the feed slurry and digester.

3.6.5 Digestion with Product Gas Recycling

Figure 3.10 shows the system used to conduct gas recycling. Digester gas is cleaned and dried by passing it through a glass-wool particulate trap, a water-cooled condensate trap, and a $CaSO_4$-filled gas drying column. The clean, dry gas is recycled at a selected recycle ratio into the digesting culture through a glass diffuser. Recycle ratio is the ratio recycled gas flow rate in dry standard cubic feet at 60°F and 30 inch Hg to the digester gas production rate in dry standard cubic feet.

3.6.6 Digestion with Recycling of Aerobically Post-treated Digester Effluent

The effect of aerobically and biologically post-treated digester effluent on the biodegradability of recalcitrant feed components passing unconverted or partially converted through the digester should be examined. Presuming that aerobic treatment of the digester effluent improved biodegradability, recycling of the treated solids is expected to increase methane production rate and yield, with simultaneous and enhanced reduction of the effluent solids and soluble organics load discharged from the overall anaerobic–aerobic system.

The process configuration in Figure 3.11 shows the digester effluent aerobically treated in a 14 L culture volume semicontinuously or in a 2 L culture volume batch activated sludge unit. Aeration and mixing are done by diffused aeration. Dissolved oxygen is monitored by a lead–sliver galvanic probe inserted in the culture. The 14 L activated sludge unit is a two-compartment tank. Its aeration chamber is separated from the adjacent settling zone by a vertical plate. This unit can be used for semicontinuous operation at aerator detention times greater than 2 days. Settled sludge withdrawn from the bottom of the settler is recycled to the digester. The 14 L unit operated at 35°C with an air flow rate of 1 L/min.

The 2 L batch unit operates under limited aeration (0.12 L/min) conditions at an ambient temperature of about 25°C. Daily, the fill-and-draw activated sludge unit is fed with 1667 mL of fresh digester effluent after withdrawing an equal volume of mixed liquor aerated for 24 hours. Selected volumes of the aerator mixed liquor are recycled of the digester to

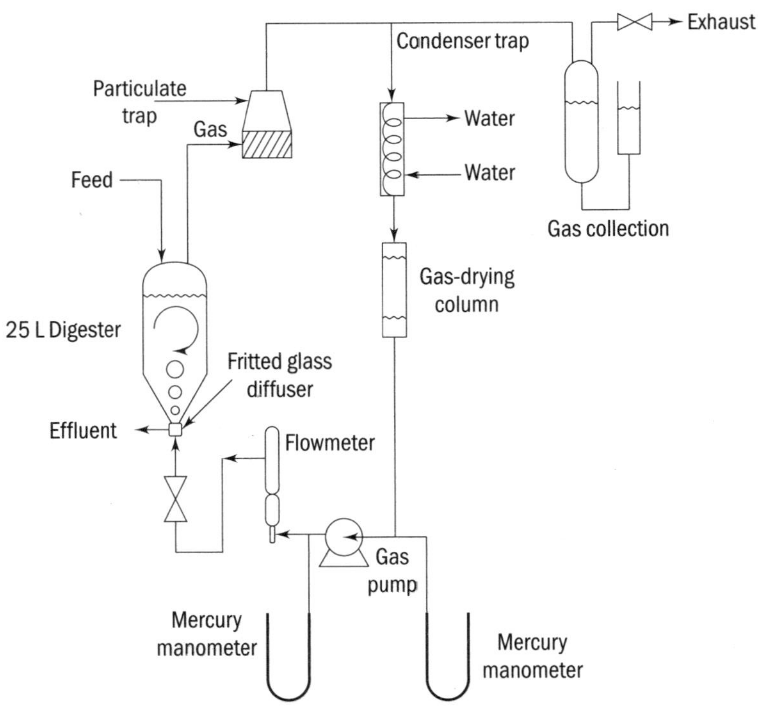

Figure 3.10 Gas recycling system for mesophilic digestion

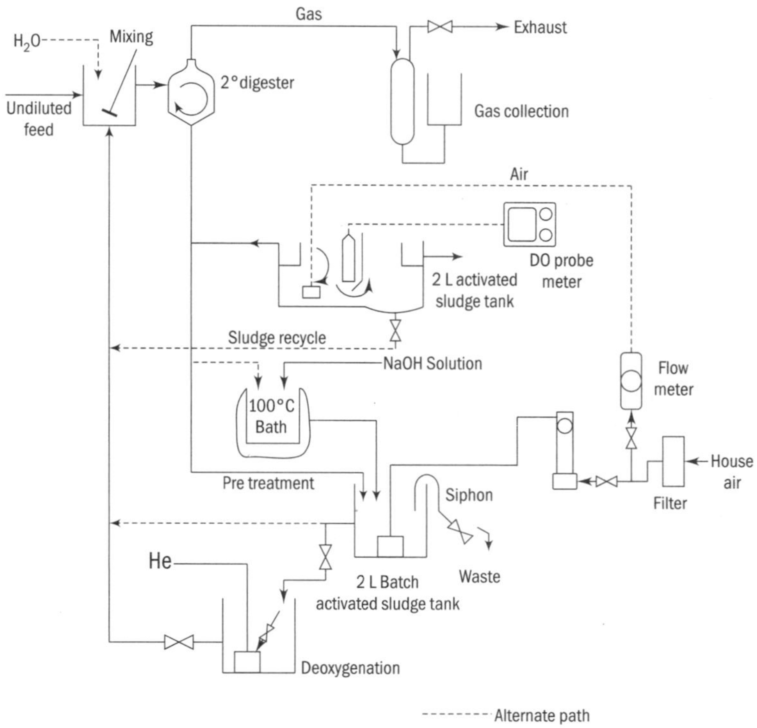

Figure 3.11 Recycling system for aerobically post-treated digester effluent

obtain the desired recycle ratio (defined as the ratio of the volume of mixed liquor to the volume of fresh feed). In some cases, the aerator feed (fresh digester effluent) is pretreated with dilute sodium hydroxide solution for 24 hours at 100°C. The recycle sludge is deoxygenated with a helium purge in some runs before charging it to the digester.

3.6.7 Two-phase Digester

Two-phase digestion has considerable potential for increasing methane production rate and yield. It is a multi-stage, high-rate digestion process in which acidogenic and methanogenic fermentations are optimized in separate digesters. Figure 3.12 shows a two-phase system consisting of a completely mixed acid-phase digester and a completely mixed methane-phase digester or a methane-phase anaerobic filter packed with Raschig rings. The acid digester is gravity fed from a sealed overhead feed reservoir having a helium or argon blanket above the feed slurry. Caustic-treated feed is delivered to this digester in small slugs up to 70 times per day by a timer-operated valve to obtain operating conditions closely approaching those of continuous feeding. Effluent from the acid digester overflowed directly to the completely mixed methane-phase digester. Alternatively, part of this effluent is vacuum filtered, and the filtrate is fed continuously to the anaerobic filter from a constant-head Mariotte bottle supplied with helium to fill the reservoir gas phase.

The gross volume and the void ratio of the packed-bed anaerobic filter are 185 L and 0.63, respectively. Filter effluent is recirculated continuously to the inlet end at a recirculation ratio of 5.1 (defined as the ratio to the recycle flow rate to the daily feed flow rate) to dilute the incoming feed and accelerate the transport of digestion products out of the culture. The hydraulic detention time of the filter is about 2.33 days (defined as the gross volume divided by the daily feed flow rates).

The two-phase system feed can be supplemented with external nitrogen, phosphorus, and magnesium to ensure that feed hydrolysis acidification and gasification are not nutrient limited.

Table 3.2 gives the steady-state performance of conventional high-rate mesophilic digestion of the biomass-waste blend at the baseline loading and detention time. The methane yields from replicate baseline runs ranged between 3.4 and 3.5 SCF/lb VS added. These yields were between 53.1% and 54.7% of the theoretical methane yield. As expected for digesters of similar geometry and mixing configuration, culture volume had no discernible effect on methane production (compare Runs 5M, 6M, and 3MA). Nutritional studies, show that conventional digestion of the biomass–waste feed is not nutrient or growth-factor limited.

[a] Runs 5M and 6M were replicates

[b] Calculated by formula suggested by Klass and Ghosh

Methane yield decreased to 2.4 SCF/lb VS added (43.8% of the theoretical yield) when the loading rate was increased to 0.18 lb VS/ft^3-day and the detention time decreased to 9 days (Table 3.2). These data indicate that conventional digestion process efficiency would

Table 3.2 Conventional mesophilic (35°C) digestion of hyacinth-grass-MSW-sludge blend under baseline and non-baseline conditions

	Baseline runs			
	5M	6M	3MA/2	Run 3MA/7
Operating conditions				
Culture volume (L)	5	5	20	20
Loading (lb VS/ft^3-day)	0.1	0.1	0.1	0.18
Detention time (days)	12	12	12	9
Gas production				
Rate (std vol/vol culture-day)	0.550	0.550	0.570	0.868
Methane content (mol %)	62.0	63.9	60.7	58.7
Methane yield (SCF/lb VS added)	3.40	3.53	3.45	2.88
Effluent quality				
Volatile acids (mol/L as acetic)	42	35	27	6
Efficiency				
VS reduction (%)[b]	33.3	33.6	34.5	29.9
Energy recovery in collected methane (%)	38.3	39.7	38.8	32.4

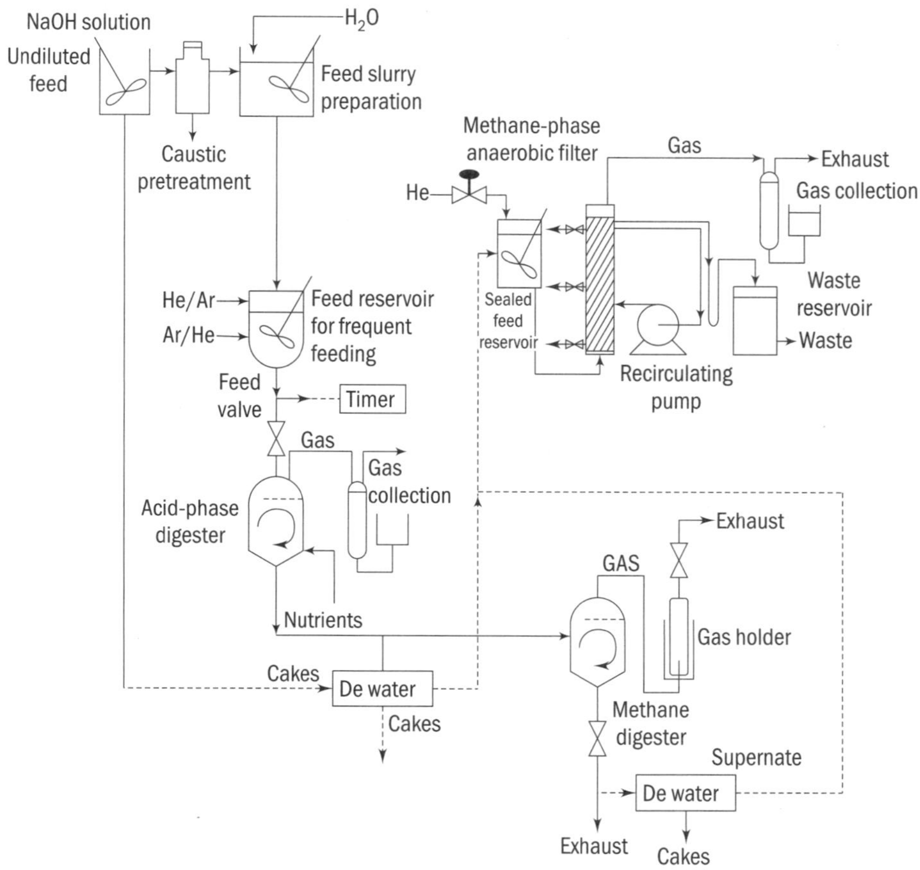

Figure 3.12 Two-phase system for biomass–waste blend

decrease substantially as the loading rate is increased and the detention time is decreased beyond the baseline values of these parameters. Unconventional or advanced digestion methods are, thus, needed to overcome limitations of conventional high-rate process.

3.6.8 Digestion of Pretreated Feed

Under long-term batch digestion conditions, only 66% of blend feed VS is determined to be biodegradable. About 52% of the biodegradable organics and 32% of the cellulose component of the feed are gasified at the baseline operating conditions of 0.1 lb VS/ft^3-day loading and a 12-day detention time. These data suggest that 34% of the feed organics resist anaerobic degradation and that only about one-half or less of the biodegradables can be gasified by conventional high-rate digestion. Hydrolysis and acidification limit the digestion of the highly fibrous lignocellulosic feed which causes low bioconversion efficiency. Improvement can be expected when the feed is pretreated chemically to hydrolyze the complex polymeric substances or to convert them to a form suitable for subsequent enzymatic hydrolysis.

Acid or alkaline treatments of particulate feeds have been shown to improve digester gas yields (17.22–25). Dilute alkaline pretreatment should be evaluated as alkali has shown to be more effective in promoting hydrolysis of cellulosic biomass than acid. It is postulated, for example, that sodium hydroxide breaks down the cross-linked lignin macro-molecules surrounding the cellulose fibers into alkali-soluble lower-molecular-weight units. In this way, the cellulose fibers are exposed for enzymatic hydrolysis during anaerobic digestion. Also, it has been suggested that alkali treatment hydrolyzes ester bonds between the uronic acids of hemicellulose and lignin, thereby enhancing the biodegradability of hemicellulose.

Digestion runs with caustic treated feed have been conducted under a variety of operating conditions. The results of a few selected runs with pretreated feed are presented in Table 3.3. The data in this table show that the highest mesophilic (35°C) methane yield, 4.10 SCF/lb VS added, from digestion of the pretreated feed at a loading of 0.1 lb VS/ft^3-day and a 12-day detention time was obtained with wet feed pretreated with 3 wt% caustic solution. However, the mesophilic methane yield (3.92 SCF/lb VS added) from feed treated with 1 wt% NaOH solution was not significantly different. Also, neutralization of the pretreated feed or addition of external nitrogen to the digester did not affect the mesophilic methane yield. These observations indicate that an increase in mesophilic methane yield up to 20% may be expected at a loading of 0.1 lb VS/ft^3-day and a detention time of 12 days with alkaline pretreatment.

Thermophilic (55°C) digestion of the caustic-treated feed at 0.4 lb VS/ft^3-day loading and a 6-day detention time showed the same methane yield of about 4 SCF/lb VS added as observed during mesophilic digestion of the pretreated feed at a 0.1 lb VS/ft^3-day loading and a 12-day detention time (Run 5T/24, Table 3.4). However, the thermophilic gas production rate was 4–5 times the mesophilic rate. Also, the caustic requirement for feed pretreatment was lower for the thermophilic run owing to the recycling of the spent caustic solution.

A thermophilic methane yield of about 3.7 SCF/lb VS added, which was still larger than the baseline yield, was observed at a loading rate of 0.43 lb VS/ft^3- day and a detention time of 5.5 days when the caustic-treated feed was neutralized to pH 10 with digester gas instead of to pH 9 with hydrochloric acid (Run 5T/26, Table 3.5). This reduced yield resulted from increased loading and decreased detention time. As expected, a higher bicarbonate alkalinity could be maintained when the digester was charged with pretreated feed neutralized with digester gas. In addition, neutralization of the alkaline feed with product gas eliminated the need for neutralizing acid and provided a method of carbon dioxide removal from the digester gas. This technique should provide a reduction in the cost of feed treatment and digester gas cleanup.

Table 3.4 Mesophilic (36°C) and thermophilic (66°C) digestion of hyacinth–grass–MSW–sludge blend pretreated with caustic soda solution

	Mesophilic digester Feed		Thermophilic digester feed	
	Run 6MA/13[a]	Run 6MA/22[b]	Run 5T/24[c]	Run 5T/26[d]
Neutralization method	No neutralization of caustic- treated food	Whole feed to pH 8.84 with acid	Mixed vacuum filter cakes plus filtrate to pH 9 with acid	Mixed vacuum filter cakes plus filtrate to pH 10 with acid
Operating conditions				
Culture volume (L)	10	10	5	5
Loading (lb VS/ft^3-day)	0.1	0.1	0.4	0.43
Detention time (days)	12	12	6.0	55
Gas production				
Rate (std vol/vol culture day)	0.606	0.670	2.851	2.911
Methane content (mol %)	59.5	58.1	56.4	56.0
Methane yield (SCF/lb VS added)	4.10	3.92	4.00	3.68
Effluent quality				
pH	6.84	6.71	7.35	7.43
Volatile acids (mg/L as acetic)	20	21	99	140
Total alkalinity (mg/L as $CaCO_3$)	2618	2433	6075	6298
Bicarbonate alkalinity (mg/L as $CaCO_3$)	2592	2412	5993	6240
Efficiency				
VS reduction (%)	41.9	41.0	43.1	40.0
Energy recovery in collected methane (%)	46.1	44.1	45.0	41.4

[a] The fresh feed was treated with 140 mL of 3 wt% NaOH solution at 55°C for 24 h. The treated feed was not neutralized and NH_4Cl was added.

[b] Feed pretreatment same as Run 6MA/13 except that 140 mL of 1 wt% NaOH solution was used at 25°C for 24 h. The treated feed was neutralized with HCl to the indicated pH and supplemented with 10 mL of 120 g/L NH_4Cl solution.

[c] Fresh feed was treated with 160 mL of 3 wt NaOH solution at 100°C for 24 h. The treated feed was vacuum filtreated and about 158 g of filter cake and 200 mL of filtrate were fed to the digester after dilution to the proper volume and neutralization with HCl to pH 9.

[d] Fresh feed was treated in the same way as in Run 5T/24 except that the alkaline feed slurry was neutralized by bubbling a portion of the digester gas (Ave. 92 L, range 7.19–11.55 L) through it. The average pH of the digester feed after feed gas neutralization was 9.98 (range 9.31–11.55)

Table 3.5 Effect of product gas and post-treated effluent recycling on mesophilic (35°C) digestion of hyacinth–grass–msw–sludge blend

	Run 2MA/11	Run 1MA/13	Run 1MA/18
Recycled material and amount	Digester gas (39 gas recycle ratio)[a]	Aerated effluent (56 vol% recycle)[b]	Aerated digester effluent after deoxygenating (257 vol % recycle)[c]
Operating conditions			
Culture volume (L)	20	20	20
Loading (lb VS/ft^3-day)	0.1	0.1	0.1
Detention time (days)	12	12	12
Gas production			
Rate (std vol/vol culture-day)	0.596	0.630	0.687
Methane content (mol %)	60.5	61.5	56.7
Methane yield (SCF/lb VS added)	3.95	3.83	3.88
Effluent quality			
pH	7.06	6.69	6.80
Volatile acids (mg/L as acetic)	8	9	20
Total alkalinity (mg/L as $CaCO_3$)	3997	2826	4667
Bicarbonate alkalinity (mg/L as $CaCO_3$)	3997	2818	4645
Efficiency			
VS reduction (%)	39.7	37.9	41.6
Overall VS reduction (%)	39.7	–	55.0
Energy recovery in collected methane (%)	44.4	43.1	43.7

[a] Recycle ratio is std *l* recycled gas/std *l* digester gas produced-day

[b] Daily digester effluent was aerated (air flow rate 0.12 L/min) in a 2.0 *l* culture volume batch activated sludge unit operated at a mbient temperature. A volume of 1200 mL of aerated content was recycled with 467 mL of fresh feed daily.

[c] Same operating conditions as in Run 1MA/13, except that the recycle sludge was deoxygenated by bubbling helium through it.

The data suggest the following sequence of operations for improved digestion of blend feed.

1. Caustic treatment of undiluted feed for 24 h at 100°C with 3 wt % NaOH and recycling of spent caustic solution. Fresh caustic is added at the rate of about 3 meq per gram of VS. Spent caustic recycle flow rate is 75 vol % of the feed slurry flow rate.
2. Dewatering of the pretreated feed. Dewatered cakes and the balance of the filtrate after recycling are fed to the digester.
3. Neutralization of alkaline feed slurry with digester gases.
4. Thermophilic digestion of the feed at a loading rate of 0.4 lb VS/ft^3-day and a detention time of 6 days.

3.7 DIGESTION WITH PRODUCT GAS AND POST-TREATED EFFLUENT RECYCLING

The second most advanced system is concerned with the recycling of product gas and liquid effluent to the digester. The effluent should be post-treated before recycling. The product gas is dehydrated before recirculation to the digesting culture. The aqueous effluent is subjected to various forms of post-treatment including sonication, simple heat treatment for 30 min at 270°F and 25 psig, and activated sludge treatment to improve the biodegradability of the undigested solids. Recycling of sonicated and heat-treated sludge to the digester does not increase methane yield above the baseline yield.

3.7.1 Gas Recycling

'Product gases' recycling through digesting culture is expected to increase methane production. There will be additional methane fermentation from increased reduction of carbon dioxide. The sweeping action of the recycled gas might accelerate the removal of gaseous products surrounding the microorganisms, thereby minimizing end-product repression.

Digester gases can be recycled at various recycle ratios from 2 to 125. The best methane yield of 3.95 SCF/lb VS added, which was about 15% higher than the baseline yield, was observed at a gas recycle ratio of 39 (Table 3.5). The gas-phase dilution rate at this recycle ratio is about 0.2 h^{-1}. The existence of an optimum gas recycle ratio is rationalized as follows: as the gas recycle ratio is increased, the opportunity for carbon dioxide reduction and sweeping of the gaseous products out of the digester increases thereby stimulating methane production. However, as the gas recycling ratio is increased further, the culture is increasingly saturated with gaseous end products; this would tend to repress or inhibit additional methane production. The opposing effects of increasing gas recycle on methane production equal each other at the optimum gas recycle ratio which maximizes methane yield.

In gas recycling, the test digester is mechanically mixed so that the effect of the recycled gas on test digester methane production could be evaluated. Mild mechanical mixing is beneficial during gas recycling because scum formation, which arises due to gas flotation of digester solids, is surprisingly not a problem even at very high gas recycle ratios. The reason for this was that mechanical agitation at a mixing Reynolds number of about 9,000 dispersed the surface solids and moved them down into the culture by the folding action of the two propeller mixers placed at heights of one and two impeller diameters above the digester bottom. The results of the gas recycling experiments showed that modest increases in methane yield can be obtained by recycling product gas at an optimum gas recycle ratio. Dual mechanical and gas mixing eliminated the scum problem associated with gas mixing alone.

3.7.2 Aerobic Sludge Post-treatment of Digester Effluent

Aerobic post-treatment treats the "refractory" effluent organics and makes them biodegradable. Recycling the post-treated material for further digestion also increases the methane production. Post-treatment only treats the recalcitrant residue remaining after the biodegradable material is gasified and not the total solids in the feed as in the case of pretreatment. The organic loading rate on the post-treatment process is substantially lower in comparison to a similar pretreatment process. Post-treatment does not affect the digestibility of feed components that are easily gasified in their original forms.

Several species of fungi and aerobic organisms can hydrolyze and degrade complex lignocellulosic substances to simpler substances. *Trichoderma viride, Cellulomonas, Cytophaga hutchinsonii, Cytophaga vulgaris* are examples of aerobic microorganisms that mediate these reactions. Hence, aerobic biochemical and chemical–biochemical post-treatments can be examined. As some aerobes cannot efficiently attack lignocellulosic cellulosic complexes, aerator feed (digester effluent) have been predigested with hot caustic to make the fibers available for aerobic decomposition. Under "limited" aeration, activated sludge post-treatment of the digested residue has been done to arrest cellulose breakdown before monomeric products (hexose, pentose, etc.) are formed, which are readily oxidized aerobically and thus, become unavailable for gasification. The products of partial aerobic digestion can be then gasified anaerobically upon recycling to the digester. Thus, the advantage of limited aeration post-treatment is that it should limit degradation of the residual solids to forms that are lost by oxidation before recycling to the digester.

Studies conducted on post-treatment of digester effluent in the 14-L mesophilic (33°–34°C) aerobic sludge unit at detention times of 2.4 and 5.3 days, and an air flow rate of 1 L/min produced settleable sludge. When this sludge is recycled to the digester at recycle ratios of 5.3 and 33.7%, it affected digester methane yields based on fresh-feed volatile solids ranging between 2.81 and 3.51 SCF/lb VS added. These yields were lower than or equal to the baseline methane yield observed under the same digester operating conditions. This means that aerobic biological post-treatment at long detention times and a high air flow rate (producing residual aerator dissolved oxygen concentration of 0.7 to 2.8 mg/L) do not increase methane production. However, modest increase in the digester methane yield was witnessed in the residue post-treatment in the 2-*l* batch unit operated under limited aeration conditions (Run 1MA/13, Table 3.5). The methane yield did not increase further when the recycle sludge was deoxygenated, the recycle ratio was increased from 56 to 257% (Run 1MA/18), or the feed to the aerator was pretreated with hot caustic. It is interesting to note that the volatile solids reduction increased when the aerobic post-treatment method was used as shown by the overall VS reduction of Run 1MA/18.

3.8 TWO-PHASE DIGESTION OF SLURRY SYSTEMS

Multistage high-rate process superior to conventional digestion can be developed by two-phase digestion. Tables 3.6 and 3.7 present results of two-phase runs.

Table 3.6 Effect of feed pretreatment on two-phase mesophilic (36°C) digestion of hyacinth–grass–MSW–sludge blend

	Untreated Feed	Caustic-treated Feed	
	Run A30/M16	Run A3C/M3C	Run A7C/M7C
Operating conditions[a]			
Acid-phase culture volume (L)	16	16	16
Methane-phase culture volume (L)	45	45	45
Acid-phase loading (lb VS/ft^3-day)	1.0	1.0	1.0
Methane-phase loading (lb VS/ft^3-day)	0.32	0.32	0.32
Acid-phase detention time (days)	2.0	2.0	2.0
Methane-phase detention time (days)	5.6	5.6	5.6
Overall loading (lb VS/ft^3-day)	0.26	0.26	0.26
Overall detention time (days)	7.6	7.6	7.6
External nutrient additions to acid-phase feed	N.P. Mg	N.P. Mg	N
Gas production			
Rate (std vol/vol culture day)			
Acid phase	1.346	–	–
Methane phase	0.924	–	–
Total	1.035	1.484	1.443
Methane content (mol %)			
Acid phase	492	–	–
Methane phase	59.0	–	–
Total	49.7	52.8	55.9
Gas yield (SCF/lb VS added)			
Acid phase	1.34	–	–
Methane phase	2.91	–	–
Total	3.98	5.63	5.50
Total methane yield (SCF/lb VS added)	1.98	2.87	3.07
Effluent quality			
pH			
Acid phase	5.86	6.72	6.82
Methane phase	6.44	7.03	7.15
Volatile acids (mg/L as acetic)			
Acid phase	1047	2440	1860
Methane phase	19	287	148
Efficiency			
VS reduction (%)	24.2	34.2	33.4
Energy recovery in collected methane (%)	22.3	33.4	34.5

[a] No alkali was used for digester pH control. The acid-phase feeds for Runs A3C/M3C and A7C/M7C were pretreated for 24 h with 2.56 L of 1 wt% NaOH solution in a total volume of 48 L under ambient conditions. The product gasses from gasses were collected together for these two runs.

Table 3.7 Thermophilic (66°C) acid-phase and mesophilic (35°C) methane-phase digestion of pretreated hyacinth–grass–MSW–sludge blend

	Run A36/M21	Run A37/M23
Operating conditions[a]		
Acid-phase culture volume (*L*)	5	5
Methane-phase culture volume (*L*)	20	20
Acid-phase loading (lb VS/ft^3-day)	2.0	2.5
Methane-phase loading (lb VS/ft^3-day)	0.44	0.56
Acid-phase detention time (days)	1.0	0.80
Methane-phase detention time (days)	4.0	3.2
Overall loading (lb VS/ft^3-day)	0.40	0.50
Overall detention time (days)	5.0	4.0
External nutrient additions		
Acid phase feed	N	N
Gas production		
Rate (std vol/vol-culture day)		
Acid phase	3.189	3.472
Methane phase	1.553	1.740
Total	1.880	2.085
Methane content (mol %)		
Acid phase	60.3	58.5
Methane phase	59.6	59.1
Total	59.8	57.9
Gas yield (SCF/lb VS added)		
Acid phase	1.53	1.40
Methane phase	3.35	3.11
Total	4.70	4.25
Total methane yield (SCF/lb VS added)	2.81	2.46
Effluent quality		
pH		
Acid phase	6.99	7.05
Methane phase	7.05	7.05
Volatile acids (mg/L as acetic)		
Acid phase	1734	2255
Methane phase	347	591
Efficiency		
VS reduction (%)	28.6	25.8
Energy recovery in collected methane (%)	31.6	27.7

[a] No alkali was used for digester pH control. The acid-phase digester feed was pretreated for 24 h with NaOH solution ambient conditions. The caustic concentration in the slurry was 142 meg/L.

Comparison of acid production yield coefficients for the runs in Tables 3.6 and 3.7 indicate that volatile solids conversion to volatile acids is significantly increased by alkaline feed treatment. Furthermore, the VS-to-acid conversion efficiency of the pretreated feed remains the same as the system detention time is decreased from 7.6 to 4 days, the gas production rate increased by 50% compared to a 17% decrease in methane yield. These observations indicate that it is desirable to operate the two-phase system at a detention time of 4 days or less, and to couple this system to a cell mass recycling device (for example, anaerobic settler) to prevent methane yield reductions and volatile acids accumulation associated with short detention times.

3.9 PACKED-BED ANAEROBIC DIGESTER

The packed-bed anaerobic digester operates with filtrates from the vacuum filtration of the acid- digester effluent. Filtrate is used because unfiltered acid-digester effluents tend to clog the packed bed. The feed to the packed-bed digester has volatile acid concentrations between 1500 and 2000 mg/L as acetic and a solids content of about 0.5 wt%. The digester is operated at an HRT of about 2.3 days and a loading rate of about 0.15 lb VS/ft^3-day based on the gross

Table 3.8 Steady-state performance of continuously fed mesophilic packed-bed methane digester

	Run 8MA/4	Run 8MA/5	Run 8MA/6
Operating conditions[a]			
Temperature (°C)	35	33	38
Loading (lb VS/ft3-day)	0.13	0.15	0.16
Detention time (days)	2.30	2.33	2.34
Effluent recycle ratio (%)	0	510	510
Gas production			
Rate (std vol/vol culture-day)	0.495	0.872	0.698
Methane content (mol %)	73.5	72.3	68.7
Methane yield (SCF/lb VS added)	2.82	4.08	3.05
Effluent quality			
pH	6.73	6.86	6.86
Volatile acids (mg/L as acetic)	186	131	68
Efficiency			
VS reduction (%)	23.3	34.4	27.0
Energy recovery in collected methane (%)	31.7	45.9	34.3

[a] The packed-bed digester had a gross volume of 18.5 L. The bed had a void ratio of 0.63. Filtrate from a slurry-phase acid digester effluent having a volatile acid content of 1550–2000 mg/L as acetic acid and a solid content of about 0.5 wt % was used as feed. No pH control was used.

filter volume (1.5 days and 0.24 lb VS/ft^3-day when based on the void or culture volume). Effluent recirculation and culture temperature had significant effects on gas production (Table 3.8). Recirculation of the effluent from the top to the bottom of the bed increased gas production rate by 76% while also increasing the methane yield by 45%. Methane content was above 70 mol % with or without recirculation. Increase in mean digester temperature from 33°C to 38°C decreased gas production rate by about 20% and methane yield by about 25%. Depending on the operating condition, volatile acids concentrations in the filter effluent varied between 90 and 190 mg/L (Table 3.8), which were still lower than those in the slurry methane digester operated at a higher HRT of 5.6 days (Table 3.7).

It should be pointed out that the performance of the packed-bed methane digester is more dependent perhaps on the influent volatile acids concentration than the overall VS loading rate. Thus, filter performance with acid-digester filtrate might be expected to be better than that obtained with methane-digester effluents having comparable volatile acids content.

Box 3.3 discusses Georgia Institute of Technology's attempt to make cellulosic ethanol over corn-based ethanol.

Box 3.3 Georgia Institute of Technology's attempt to make cellulosic ethanol

Georgia Institute of Technology had been, for a long time, keen to extract nature's energy using PB. Most of the USA's ethanol can one day be supplied from the ethanol made from trees, grass and other common plants. If the USA combines cellulosic ethanol with that made from crops such as corn, the country could cut gasoline slash yearly fossil-fuel emissions. Georgia aspires to be the centre of this green-power movement. The state has been focusing on biotechnology. Much of the state's bioenergy effort is focused on cellulosic ethanol as it has potential advantages over corn-based ethanol. To make fuel from corn, a great deal of power is required which offsets some of its green-energy benefits. Moreover, world food supplies can also get affected as a result of food crops being redirected for fuel. For cellulosic ethanol to be a success, scientists need to find ways to extract plants fuel bearing sugars in an efficient and environmentally sensitive manner.

A thermo-chemical process converts wood biomass into biofuels. The fermented biomass broth that eventually becomes ethanol contains about 90% water and only about 10% alcohol. Earlier, the liquid is strained through a series of sieves and heated to cook out the water. This affects its energy efficiency. Efforts are underway to develop membranes that would filter water from ethanol using little or no heat.

The ethanol broth will be flowed through a water-loathing membrane containing tiny particles that allow the alcohol to pass through while blocking the water. Then, a second membrane—hydrophilic or water loving—will filter out any trace amounts of water left in the alcohol. The result will be pure ethanol.

3.10 TECHNOLOGY OF CELLULOSE PROCESSING

Cellulose is one of nature's most abundant macromolecules and generally exists as lignocelluloses. Nature's photo-bioreactors produce enormous amounts of lignocellulosic biomass and materials like agro-residues as primary products of photosynthesis. Scientific investigations on the renewable ligncellulosic material and its bioprocessing have been carried out by investigators all over the world for development through technology. A potential substrate for bioconversion, using biochemical engineering and process biotechnology principles and strategy, useful base chemicals and co-products may be produced from lignocellulose. In countries such as India, where this raw material resource is abundantly available, resource-based technology development is of great relevance. The development of multipurpose flexible facility can demonstrate its economic importance. In recent years, the possibility of microbial cellulase biosynthesis and its scale-up is also an important development in cellulose bioprocessing. Cellulase biosynthesis can be eukaryotic and prokaryotic. Industrial alcohol production by bioprocessing of agroresidues instead of from molasses has drawn the attention of biotechnologists.

3.10.1 Approaches of Cellulose Bioprocessing

Lignocellulose pretreatment is followed by saccharification and subsequent fermentation of fermentable sugars. The involved steps in this bioprocessing include the following:

(a) Lignocellulosic pretreatment
(b) Microbial production of cellulases for biocatalytic saccharification
(c) Cellulose saccharification for fermentable sugar
(d) Bioconversion of fermentable sugars for products

Cellulose is directly bioconverted by pretreatment by anaerobic bacteria such as *Clostridium thermocellum*. Bioconversion of cellulosics can be done through bimolecular redesigning of cells.

3.10.1.1 Lignocellulose pretreatment

Physicochemical and biological delignification methods are possible routes for pretreatment of lignocellulosics. However, pretreatment methods vary to a great extent depending on the nature of the raw lignocellulosics, ease of appliance, and cost effectiveness. For example, chemical processing with 1–15% NaOH with simultaneous mild thermal treatment is the most commonly used method. Steam treatment of materials like straw and especially hard wood has also been used. The major objective is to produce pure cellulose by easier processing at a reasonably low cost.

3.10.1.2 Progress in microbial production of cellulases

A large number of micro-organisms can utilize/degrade either one or more of the major components of lignocellulosic materials. These organisms are known to be cellulolytic. Many cellulolytic organisms, during utilization of lignocellulosic materials or in some cases sugar, can produce intracellular enzymes called cellulases. It has been possible to increase microbial

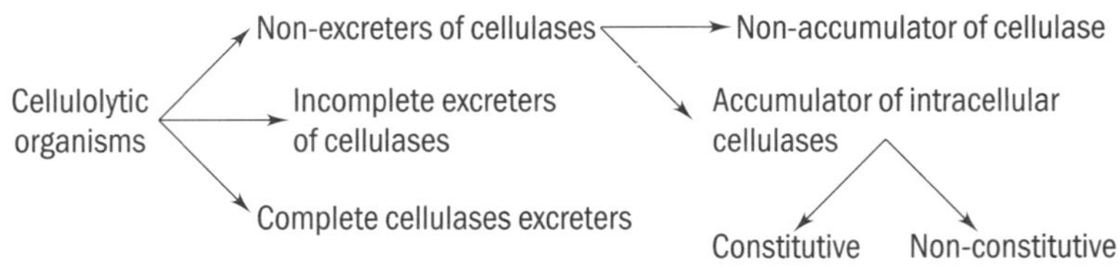

Figure 3.13 Classes of cellulase producers

cellulase yield by induced mutation, selecting suitable cultivation process conditions and using appropriate process strategy.

3.10.1.3 Cellulolytic organisms and cellulase producers

Figure 3.13 gives the classes of cellulose producers.

3.10.1.4 Synthetic pathway and mode of action of cellulases

The degradation of crystalline cellulose is a complex process requiring participation of many enzyme components. The following concepts have been used to elucidate the mode of action of cellulases:

(a) Two-component enzyme action
(b) Concept of multi-component enzyme action
(c) Concept of synergistic action of enzymes

For enhancement of cellulase enzyme action, the availability of a large quantity of active cellulase is very important. *Trichoderma reesei* is an important producer of cellulases; through its mutational/genetic upgrading, its strains can give as high as 14.2 IU/mL activity. Biochemical mechanisms controlling cellulase synthesis and activity should be understood as also the regulatory mechanism in cellulase production. For high yield of cellulase by microbial strains, minimization of catabolite repression and end-product inhibition are required.

3.10.1.5 Empirical strain redesign engineering and scale-up

Induced mutation and genetic manipulation techniques have been used to develop hyper-producing strains by overcoming many of these regulatory inhibitions. The genetic lineage of current hyper cellulase yielding mutants showed that about 10–15 fold increase in cellulase yield could be obtained by genetic upgrading of the original strain. The use of proper production methods, suitable cultural conditions and an appropriate process strategy may result in further improvement in the biosynthetic yield of cellulase. Therefore, the aspect of microbial engineering and technology of cellulase production should be integrated with microbial and biophysico-chemical aspects to obtain a highly active cellulase system. In this regard attempts have been made to produce cellulase using the batch, fed batch, and continuous cultivation techniques and different strategies.

3.10.2 Biotechnological Aspects of Cell Cultivation and Cellulase Production

T. reesei has been used extensively in large-scale production of cellulases.

3.10.2.1 *Batch cultivation—microbial parameters*

(i) *Morphological*: Cellulase production by *T. reesei* QM 9414 has been carried out in SF, STR, and SSF in relation to morphological variations of the fungus. In a low-level glucose-containing cellulose medium, pellet form produced cellulase with an activity of 2.0 FPA and 65 IU/mL Cx activity. Filamentous form on the other hand produced only 1.1 FPA and 45.50 IU/mL Cx activity. It indicated the desirability of maintaining the pellet form in cellulase production.

(ii) *Inoculum age and volume*: Inoculum age and volume greatly influence cellulase productivity. It was shown (Mukhopadhyay, *et al.* 1979) that use of 36 h old inoculum (10% v/v) with 0.01–0.06 IU/mL FPA on 1% beech wood cellulose medium containing 0.05% yeast extract and no peptone yielded maximum cellulase. Older inoculum with higher FPA resulted in decreased cellulase productivity. Spore inoculum in shake flask showed excellent growth, but on the third day enzyme production declined due to pH inactivation. In the laboratory fermenters, very young mycelia showed the same behaviour like that of spores, but third-day-old mycelia gave very good yields. In the pilot plant, 8–10% v/v inoculum gave good yield (Nystorm and Diluca, 1977). It was also shown that one-day old inoculum required 12.6 mL/L of 1.0 (N) NaOH to adjust pH on 1% cellulose and gave reproducible and consistent results. One-day old inoculum released 1.24 mg/mL soluble protein as compared to 1.04 mg/mL with three-day old inoculum.

Longer inoculum age and higher FPA presumably provided high initial sugar level in the broth. Productivity is reduced by repressing enzyme-forming system.

3.10.2.2 *Batch cultivation—physical and chemical environments*

(i) Temperature: Results of batch cultivation experiments at different controlled temperatures showed highest growth and cellulase production at 29°C on 1% beech wood medium. At this temperature μm and cellulose uptake rate constant (K_T) were 0.08 h^{-1} and 0.008 h^{-1}, respectively. Time lag to start growth (8 h) and appearance of extracellular enzyme (4 days) were also minimum at 29°C in the range (27–32°C) studied (Mukhopadhyay, Ghose, Fiechter, *et al.* 1979).

(ii) pH: The influence of pH on cellulase production by *T. reesei* has been reported by many workers (Mukhopadhyay and Ghose, 1977). At cellulose concentration of 1% or greater if pH is uncontrolled, the enzyme is inactivated. It is necessary to control the pH between 3.0 and 3.5 to prevent enzyme inactivation.

(iii) Aeration: An adequate oxygen transfer rate (OTR) in the culture liquid must be maintained during growth and enzyme production. During the enzyme production phase, however, there is a requirement for at least the maintenance level of oxygen uptake rate for energy metabolism, and the corresponding amount of oxygen must be supplied to the system. If the OTR is low or insufficient, that is less than the maintenance ratio of oxygen, cellulase biosynthesis ceases. Using aeration rate of 3.21 min^{-1} (fixed) and agitation 400 rpm 12.0 and 21.5 IU $L^{-1}h^{-2}$ FPA were obtained in non-enriched (no yeast extract) and enriched media respectively (Mukhopadhyay, Ghose, Fiechter, *et al.* 1979). Maximum oxygen uptake rate at this aeration agitation (K_La 120 mM $O_2L^{-1}h^{-1}$) was 8.4 $mMO_2^{-1}h^{-1}$.

(iv) **Medium components:** Replacement of peptone with cheaper materials like CSL (corn steep liquor) and using freeze dried cells of Rut C 30 have reduced the cost. There have been attempts to increase cellulose concentration but because of its low bulk density it results into thick slurries, causes serve foaming problem, and creates difficulties in adequate mixing and aeration. The optimum cellulose concentration for small fermenters upto 15 L is about 2% although in the pilot plant as high as 8% (w/v) has been used (Nystrom and Diluca 1977). However, a C/N ratio of 8.0 is to be maintained. Above 9.0 C/N the batches become severely nitrogen limited. Maintenance requirement of C/N 5.5 in cellulase production phase has been reported (Mukhopadhyay and Malik, 1980).

3.10.2.3 Continuous cultivation

It has been reported that productivity in continuous cultivation of *T. reesei* QM 9414 is higher than in batch cultivation (Peterson, 1977; Sahai and Ghose, 1977).

Single-stage cultivation: Maintenance of high biomass is equally important in single-stage continuous culture. If pH is maintained low, growth and enzyme productivity are poor. In single-stage continuous culture, long residence time of cells in the reactor is required to obtain reasonable cellulase yield. A productivity of 21.2 IU/L/h at a dilution rate 0.2–0.4 h^{-1} maintaining enzyme level 1.7 IU/mL at pack with cell mass 2 gL^{-1} at a steady state in 51 fermenter with KLa 250 h–1 were obtained (Sahai & Ghose 1977). To overcome the problem in the single-stage continuous culture, two-stage cultivation is attempted.

Two-stage cultivation: In this approach, the growth phase is the first stage and cellulase production is the second stage (Mitra and Wilke 1975). With mutant Rut NG14 from QM 69 in two-stage cultivation, productivity as high as 95 IU/L/h in the second stage at D = 0.028 h–1 has been claimed. Also it was predicted that, if cell mass could be maintained at 2% level one may approach the productivity as high as 300 IU/L/h on fermentable capacity basis. With the same culture on 6% cellulose in batch process only maximum productivity of 45 IU/L/h at 0.055 h^{-1} and FPA, 151 IU/L could be obtained.

Solid-state cultivation: In Japan most of the commercially available cellulase was produced by solid-state cultivation (Toyama and Ogawa 1972). Solid cultures were prepared by growing the fungus *T. reesei* QM 9414 on rice straw, wheat bran, newspaper, sawdust, bagasse, and so on. Cellulosics were mixed with an organic nitrogen source at a ratio of 8.2 and cultivated on trays for one day at 25°C. Stronger enzyme activity was claimed to be produced than in submerged culture.

The advantage of this process over submerged fermentation is that highly pure and concentrated cellulase can be obtained.

3.10.3 Developments in Process Engineering Strategies

Cell recycle: It is a means of avoiding glucose repression while enhancing biomass in continuous cellulose fermentation by recycling the mycelia back into the fermenter as the broth is harvested (Wilke, *et al.* 1976). The use of cellulose grown cells resulted in dramatic improvement of enzyme productivity and of enzyme concentration in the harvest.

pH cycling and temperature profiling: It has been reported that cultivating in Mandel's medium containing 1% beech wood cellulose and 0.05% yeast extract showed fall of pH from 5.7 to 3.0 in 18 h. The pH was maintained between 3.0 and 5.0 with NaOH. In a 141 Ghemap fermenter (working volume 81) maximum FPA productivity 21.5 IU/L/h was obtained with 120 mMO_2/L/h K_La, 4 g/L cell mass and 36 h old inoculum (10% v/v). The pH cycling continued for three days and from the fourth day pH started rising and was controlled at 5.7 to avoid inactivation of enzyme at higher pH, due to heavy sporulation (Mukhopadhyay, *et al.*, 1978 1979). Cell mass, cellulase yield, productivity were further increased to 10 g/L^{-1}, 7.2 IU/mL and 44.0 IU/L/h respectively, by simultaneous pH cycling and temperature profiling strategy (Mukhopadhyay and Malik 1980). Increase of cellulase yield could be seen in both stirred tank and horizontal rotary using this strategy (Mukhopadhyay 1981). Influence of control of pH cycling limit of cellulase production and growth of *T. reesei* have been reported (Mukhopadhyay 1981). It is shown that maximum cell mass of *T. reesei* will not necessarily produce maximum cellulase yield.

Immobilized whole-cell reactor: Cellulase production by *T. reesei* mutant RUT C30 immobilized on 4% Carrageenan beads was carried out for 13 days. Average cellulase productivity was 26.0 FPU/L/h. Carbon and nitrogen requirements per FPU produced were reduced to 1/4 to 1/2 those of conventional cultures (Eveleigh, *et al.* 1982).

3.10.4 Cellulose Saccharification Concerns and Progress

In India, the renewable lignocellulose biomass that deserve significant considerations are rice straw and cane sugar bagasse. Rice straw is not currently sufficiently utilized to obtain value-added products. It is available in all parts of India. The annual production of rice straw is close to 90 million tonne. Attempts need to be made to find technological feasibility of its utilization through bioconversion. Bioconversion of rice straw to ethanol and co-products may serve the following purposes:

1. Better scope of utilizatingrenewable lignocellulosic and other material.
2. Training of technology students to understand upstream and downstream processing and their integration of a bioprocess technology.
3. Examining the possibilities of utilizing of the facility for multipurpose bioprocessing at a later stage.

3.10.5 A Glance at Process Biotechnology

The biotechnology process of bioconversion of lignocellulosic rice straw to ethanol and co-products has five technological process components.

1. Physico-chemical biomass refining and separation of lignin.
2. Cell culture engineering and technology for cellulase enzyme production and separation to use it in saccharification and its excess as co-product.
3. Simultaneous saccharification and fermentation of refined biomass using vacuum cycling strategy for ethanol production.
4. Low-energy separation of ethanol from aqueous fermentation mash medium and recycling residual biomass/hemicellulose.
5. Bioenrichment of residual biomass/hemicellulose for single-cell protein (SCP) production as fodder.

The major co-production of the integrated technology components are lignin, cellulases, residual biomass, CO_2, and possible slop sol biosurfactant.

Besides minimizing loss of the components, the co-products of this integrated process biotechnology may have substantial value addition opportunities. The noted feature of the process integration has, therefore, been the total recycle of substrate at each step, in order to minimize the loss and reduce environmental pollution.

3.10.6 Commercial Value Addition to Cellulases and Celluloses

Cellulase enzyme has a great potential for commercial value addition for the following sectors:

1. **In food-processing industry:** Cellulase and macerozyme (a cell separating enzyme) as a complement to cellulase been used in the processing of vegetables. The edible tissue of vegetables consists of parenchymatous cells. These cells are bound to each other by a cementing substance, particularly protopectin. When parenchyma was treated with macerozyme, the tissue was degraded into individual cells. This is due to decomposition of insoluble protopectin to soluble pectinic acid. Unicellular vegetables have been obtained from carrots, spinach, cabbage, potatoes, garlic, and several others by using this enzyme. Unicellular fruits and vegetables have the advantage of digestibility, easy preservation, easy transportation, compatible combination, compatible mixing with seasoning, good flavours, various vitamins, and easy cooking.
2. **For laboratory biomaterial extraction:** Agar is a widely used biomaterial in the microbiological laboratory. Cellulases have been used in extracting agar from ogonori (*Gracilaria converoides*) in Japan.
3. **In daily life pharmaceutical and clear agent:** A few daily life application of celluloses are given in Table 3.9.

Table 3.9 A few daily life application of celluloses

Eukaryotic cellulose	Prokaryotic cellulose
• Fibers for fabrics • Chemical derivatives • Membrane fabrication • Microcarrier • Material in R & D • Disposable utensils	Sensitive diaphragms for stereo headphones. Food product called 'Nata' in the Philippines. 'Cellulon' has many commercial uses like: • binders for ceramic powders • thickener for paints, ink, adhesives and even foods • paper coating • to protect burned skin

3.11 BIOMASS CONVERSION AS AN ALTERNATIVE TECHNOLOGY FOR FUEL PRODUCTION

Many other countries in the world have started taking preventive measures for the possible future oil emergency. In order to achieve such a target, other energy sources were developed. Governments have started investing in the development of new petroleum alternative energy sources.

In order to develop new technologies for obtaining such petroleum alternative liquid fuels, efforts are being made to (1) develop technology for new liquid fuel production from syngas, (2) upgrade technology of tar sand bitumen and shale oil, and (3) develop technology for biomass conversion and utilization.

3.11.1 General Scheme of Biomass Conversion Projects

3.11.1.1 Selection of raw materials

Resources of cellulosic raw materials are quite different from one country to another. Chemical compositions of bagasse, rice straw, and wood saw-dust are shown in Table 3.10. As shown in Table 3.10, bagasse and rice straw contain less amount of lignin than wood saw-dust.

3.11.2 General Scheme of Ethanol Production from Cellulosic Biomass

Figure 3.14 shows the enzymatic saccharification of cellulose.

First, cellulosic biomass must be pretreated by some physico–chemical measures. As cellulosic biomass is usually cemented by lignin component and also natural cellulose has more or less crystalline structures in its molecule, so these characteristics often prevent access of cellulolytic enzyme, cellulase, to degrade cellulose molecule to sugar. So, the lignin component should be removed and the crystalline structures of cellulose loosened in order to make the enzymatic saccharification proceed more efficiently. Cellulase production step is also very important in order to supply cellulase with much more reasonable prices. In the saccharification step, the pretreated cellulose and hemicellulose are converted to mono-saccharides, that is glucose and xylose, by cellulolytic enzymes. Sugar solution thus obtained is fermented by yeast to produce ethanol. The ethanol produced in the fermented broth is recovered by either distillation or other new techniques, and finally converted to fuel alcohol after dehydration step.

In this general scheme, the most efficient and energy saving unit processes are required. And, in the total system, waste treatment process and effective recovery of heat produced in the total system are also required.

3.11.3 Production of Cellulase

In order to succeed in the enzymatic conversion of cellulosic biomass to fuel alcohol, improvement of cellulase or cellulolytic enzyme production is a very important problem. A list of cellulose-producing microorganisms is given here.

Trichoderma reesei

Trichoderma viride

Trichoderama harzianum

Aspergillus fumigatus

Aspergillus terreus

Penicillium funiculosum

Sporotrichum thermophile

Thielavia terrestris

Irpex lacteus

Cellulomonas fimi

Cytophaga dissolvens

Among them, *Trichoderma reesei* shows best productivity of cellulase. Therefore, artificial mutation works using *Trichoderma reesei* QM 9414 strain as the parent strain. Genealogy of artificial mutants of *Trichoderma reesei* is shown in Figure 3.15. Cellulase productivities of some mutants were examined using 5 L-jar fermentors, and the results are shown in Table 3.11. As shown in Table 3.11, productivity of a mutant, KDR-27 strain, showed about 3 times higher titer of cellulase activity than the parent strain, QM 9414 strain. Further mutants, KGD-3 and PC-3-7 strains, showed much better titers of both cellulase and β-glucosidase activities. Crude enzymes were prepared from fermented broths of different mutants, and the various enzyme activities are shown in Table. 3.12. Among them, the crude enzyme from PC-3-7 strain showed about two times higher filter paper unit (FPU) than other enzyme preparations.

3.11.4 Pretreatment of Cellulosic Biomass

Pretreatment of cellulosic biomass is a very important step in the enzymatic conversion of biomass. However, the effectiveness of pretreatment processes is quite different depending upon the kinds of biomass resources. The steam-explosion process is quite effective in treating wood chips. Moreover, treatment of used paper by two-roll mills was quite effective for the subsequent enzyme saccharification. Various pretreatments such as mechanical and chemical treatments, electron beam irradiation, ozonization and so forth have also been tried. However, in the case of bagasse and rice straw, treatment with hot diluted alkali solution was found to be the most effective and the simplest. When bagasse chips are treated with 0.3N sodium-hydroxide solution at 120°C for 15 minutes, the remaining solid material saccharified with high rate by the subsequent enzymatic treatment.

Figure 3.16 illustrates alkali treatment of bagasse and rice straw. From the figure, it seems treatment with 0.3N NaOH is best for the maximum recovery of sugar from raw materials. And, it seems that bagasse is better raw material than rice straw from the standpoint of weight recovery of sugar from the unit amount of raw material.

3.11.5 Enzymatic Saccharification

Enzymatic saccharification might be affected by the difference of pretreatment conditions and the amounts of celluloytic enzyme used in the saccharification step. Figure 3.17 shows saccharification of pretreated bagasse in a 1-kL size reactor.

In this case, pressed pretreated bagasse (70% water content) was fed in reactor in which celllulase concentration was kept at 150 U/mL as CMCase activity. The saccharification reaction was carried out at pH 5.0 and at 45°C for 24 h. Then, 15.7% of the total sugar was accumulated after 24 h. The saccharification rate against initial cellulose and hemicellulose was about 80%.

The recovery of cellulolysic enzyme by the ultra filtraction membrane is also investigated in order to decrease the cost of enzyme.

3.11.6 Continuous Alcohol Fermentation with Immobilized Yeast Cells

Alcohol fermentation itself is an already established technology. However, in order to minimize the initial investment cost and increase the productivity of alcohol fermentation, various trials, for example, cell circulation system, continuous fermentation with flocculated yeast cells, Zymomonas fermentation and so forth, have been carried out. There is also a process by which continuous alcohol fermentation is carried out with immobilized yeast cells. In this process, living cells of *Saccharomyces cerevisiae* are entrapped in calcium–alginate beads.

As shown in Figure 3.18, preparation of immobilized cell beads can be easily attained by showering drops of sodium-alginate solution containing intact yeast cells into calcium chloride solution from the nozzle attached to the top of the reactor. In the case of 1 kL-size reactor, the preparation can becompleted within several hours. Therefore, no special equipment is needed for the preparation of beads.

As shown in Figure 3.19, three types of prototype reactors can be constructed in order to obtain basic data for designing a pilot plant. The total volume of each reactor was 1 kL. The first reactor is a tall tower type with adjustable L/D ratio from 7 to 10. The second reactor is a short column type with adjustable L/D ratio from 1.5 to 2. The third reactor is a rectangular type whose interior is divided by a number of vertical plates between which the substrate solution flows laterally in a zig-zag course from one end of the reactor to the other.

It was recognized that at least two columns must be connected in a series to obtain higher conversion yield owing to the strong turbulent effect caused by carbon dioxide gas evolved during fermentation especially in the first column. From these results, a designed process is shown in Figure 3.20.

In the early stage of bench-scale experiments, the acid-producing bacteria often got contaminated and yeast cell activity deteriorated. The contamination problem can be eliminated by adjusting the initial pH of the inlet substrate solution at 4.0 with sulphuric acid. Maintenance of yeast cell viability also much improved by supplying a small amount of air continuously and by adding small amount of sterol and an unsaturated fatty acid in the initial cell beads.

Figure 3.21 shows a simplified version of the process. In the new process, sterilizer for inlet medium and also seed culture fermentor were eliminated in order to minimize the initial investment cost.

In this operation, about 9 (V/V) % of ethanol can be continuously produced form diluted cane molasses for more than 4,000 hours, that is more than half a year. The conversion yield of ethanol against available sugar can be kept as high as 95%. This process is controlled automatically by a computer system. The biggest merit of this process is a very high productivity of ethanol per unit volume of reactor, that is about 20 times higher productivity than a conventional batch-wise fermentor. Other big merits of this process are significant decrease of labour power and energy caused by the simpleness of operation and non-sterilization system.

The process can be applied for any type of ethanol production from conventional raw materials such as cane molasses, beet molasses, and starch hydrolysate. This process is being used for the fuel alcohol production from cellulosic biomass.

3.11.7 Alcohol Fermentation from Xylose

Another important problem in the alcohol production from cellulosic biomass is the conversion of xylose of ethanol. Regular yeasts such as *Saccharomyces cerevisiae* cannot convert xylose to ethanol because of lack of xylose–isomerase in the cells. Therefore, an effective microorganisms needs to be found which can convert xylose to ethanol. Trials have been done using *Pachysolen tannophilus*, *Candida tropicalis*, and so on. However, due to lower alcohol tolerance of such microorganisms, it seems still impractical to make use of such microorganisms for the conversion of xylose to ethanol.

The yeast strain, *Kluyveromyces cellobiovorus* KY 5199 has been found to be effective. As shown in Figure 3.22, this strain can convert xylose and also cellobiose to ethanol as well as glucose. Usually, enzymatic hydrolysates of cellulosic biomass contain some amounts of cellobiose besides xylose. So, much assimirability of this strain seems favourable.

Alcohol fermentation by *Kluyveromyces cellobiovorus* KY 5199 was tried by using bagasse hydrolysate containing 9.1% of total sugar. The result is shown in Figure 3.23. *Kluyveromyces cellobiovorus* can produce about 30 g of ethanol from 91% of sugar mixture in semi-anaerobic conditions. However, *Kluyveromyces cellobiovorus* still has some defects in the conversion yield and alcohol tolerance. Therefore, in order to make use of this strain in practical fermentation, special measures such as combination use with *S. cerevisiae*, combined technology with vacuum fermentation, and immobilized cell technology must be applied. D-xylose isomerase gene is also being added into *S. cerevisiae* cell by recombinant DNA technology (Figure 3.24).

3.12 FUTURE ASPECTS OF BIOMASS CONVERSION PROJECT

In order to make the total system more economical, there are several problems that have to be solved. These are as follows:

(1) How to minimize the total construction cost.

(2) How to supply raw materials with reasonable prices.

(3) How to optimize the materials balance from the stand point of production cost. As a part of raw materials and wastes from the process are used as heat-energy source, so the highest yield of ethanol does not mean the lowest production cost.

(4) How to simplify the total process. For example, simultaneous saccharification and alcohol fermentation may be possible.

(5) How to make use of xylose fraction. At present, the aim is to increase the alcohol yield from xylose. However, if it will not be economical, then alternate way of using xylose has to be discovered.

Table 3.10 Chemical compositions of cellulosic biomass

Biomass	Cellulose	Hemi-Cellulose	Lignin	Ash	Others
Bagasse	41	24	18	2	8
Rice Straw	35	35	6	8	16
Wood Saw Dust	40–55	20–35	25–30	0.2–2.0	

Table 3.11 Cellualse productivities by *T. reesei* mutants

Mutants	CMCase U/mL	FPU U/mL	β-glucosidase U/mL	Extracellular protein mg/mL
QM 9414	60	6.1	3.2	10
KY 746	88	7.2	3.5	11
K – 14	106	12	1.1	17
KDR-27	150	18	2.9	20

(5L-jar Fermentor, 6% Avicel, pH 3.0, 28°C, 7 days)

Table 3.12 Enzyme activities of crude enzyme preparations produced by different mutants of *T. reesei*

Mutant	CmCase (U/mg)	β-glucosidase (U/mg)	F P U U/mg)
KY 746	15.0	0.10	0.39
KDR-27	14.5	0.23	0.45
KDG-3	15.5	0.18	0.45
PC-3-7	11.4	0.17	1.05

Note Enzymes were precipitated by $(NH_4)2SO_4$, and then desalted by Sephadex G-25 treatment, and freeze-dried.

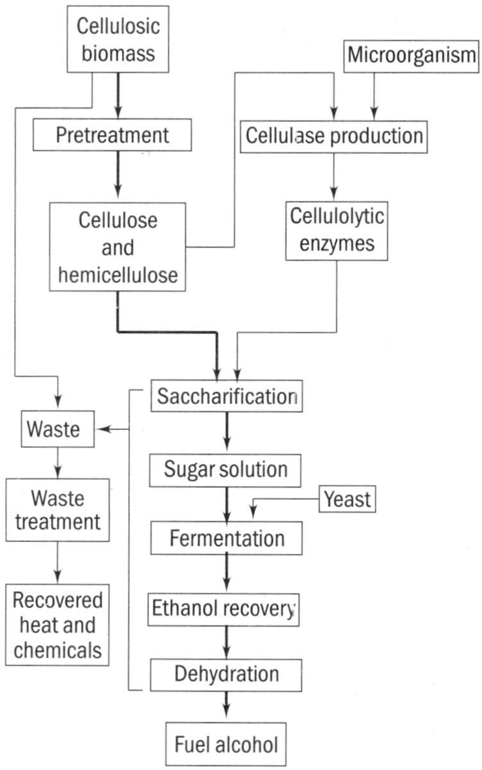

Figure 3.14 Flow diagram of ethanol production from cellulosic biomass

Saccharification of alkali-treated biomass was carried out with CmCase as 25 U/mL at 45°C for 48 h (pH 5.0).

- Bagasse
- Rice straw

Saccharification conditions:

45°C, pH 5.0; CM Case 150 U/mL

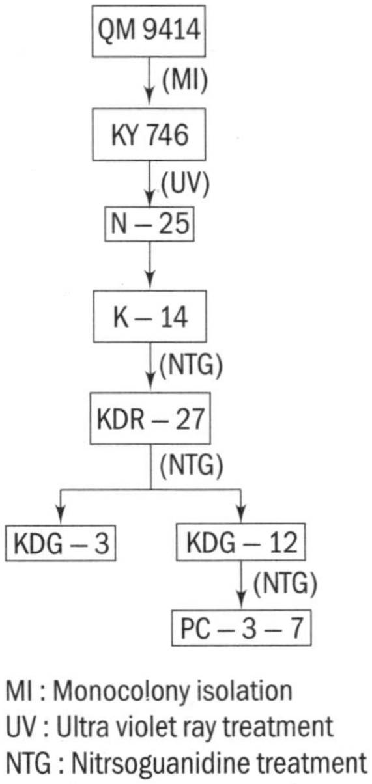

Figure 3.15 Genealogy of artificial mutants of *Trichoderma reesei*

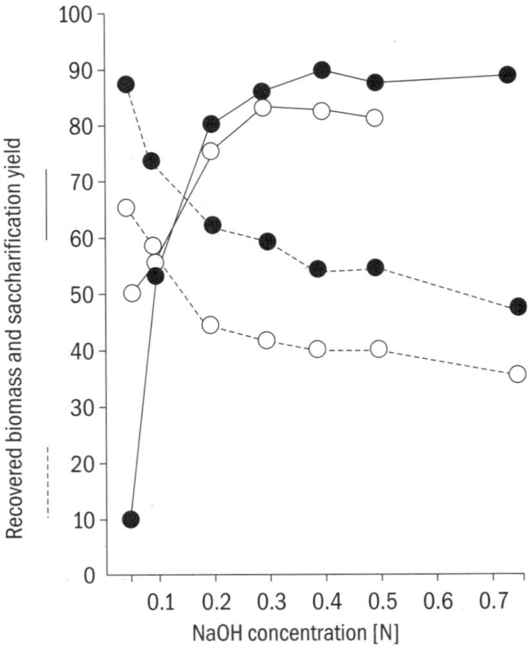

Figure 3.16 Weight recovery and saccharification rate in alkali treatment of bagasse and rice straw

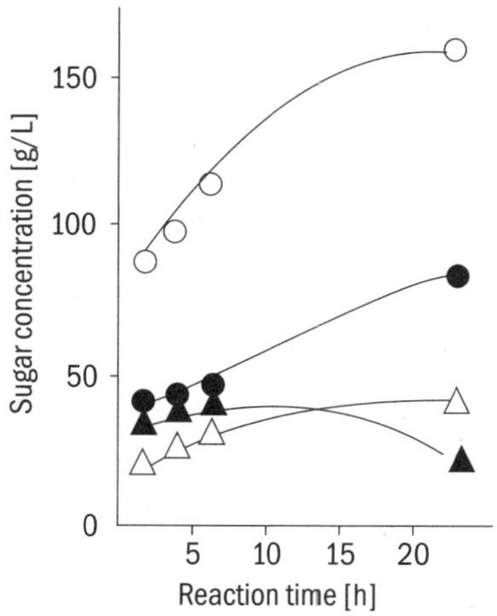

Figure 3.17 Saccharification of pretreated bagasse in a 1 kL-size reactor

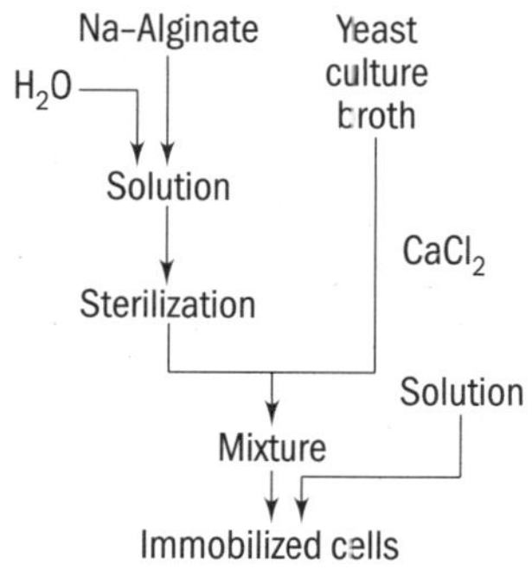

Figure 3.18 Preparation of immobilized yeast cells

• Total sugar • Glucose

Δ Xylose ♦ Cellobiose

CL-1 column reactor L/D = 7–10, H = 6 m, capacity 1 kL

CL-2 column reactor L/D = 1.5 – 2, H = 2.5 m, capacity 1 kL

CL-3 rectangular reactor H = 1 m, capacity 1 kL

(gas–liquid separation type)

The medium used for cell growth and fermentation had the following composition per litre: yeast extract, 4.5 g; peptone, 7.5 g; and carbon sources, 80 g. Ethanol fermentations were carried out at 28°C under semi-anaerobic conditions (pH 5.0).

Sugar composition of bagasse hydrolysate:

Glucose 55 mg/mL, Cellobiose 4 mg/mL

Xylose 18 mg/mL, Xylobiose 14 mg/mL

(total sugar 91 mg/mL)

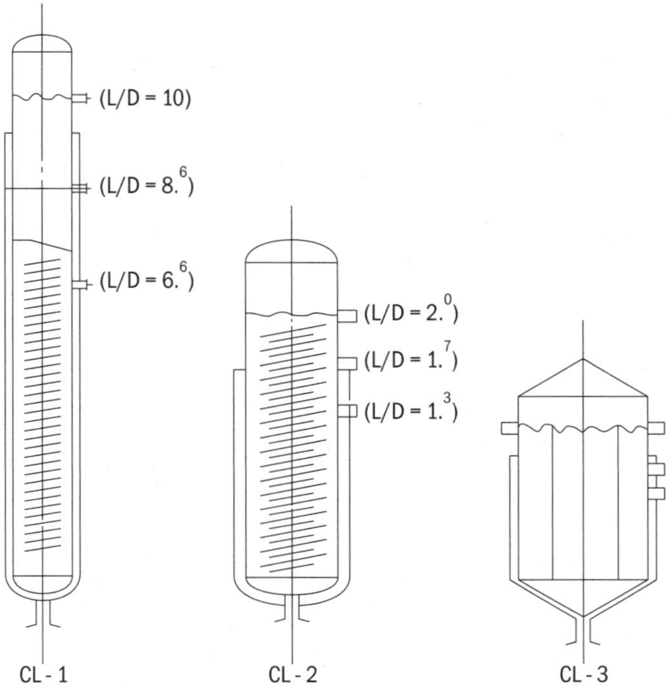

Figure 3.19 Designation of prototype reactor

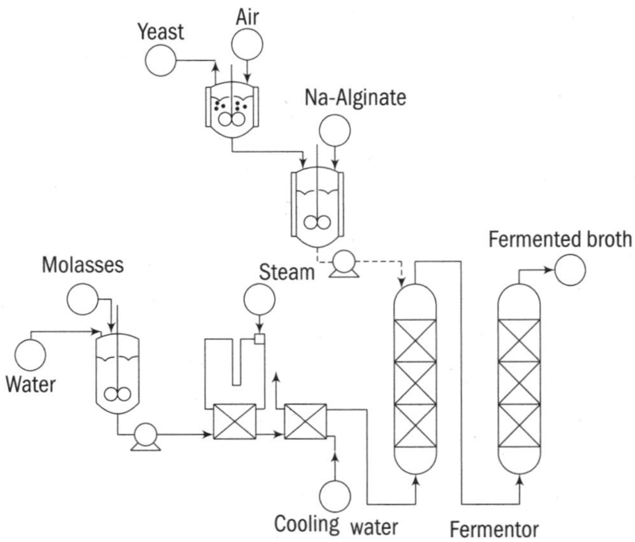

Figure 3.20 A tentative process flow sheet

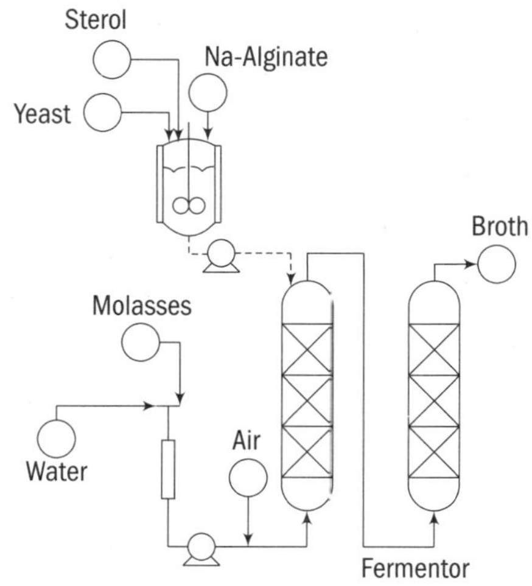

Figure 3.21 Process flow diagram of the pilot operation

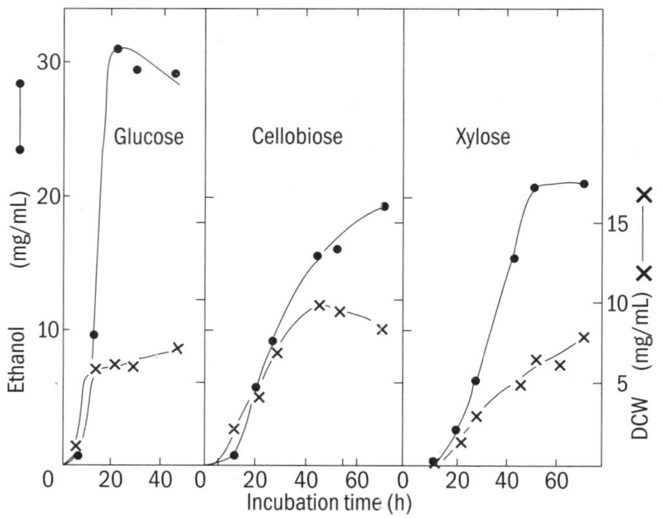

Figure 3.22 Ethanol fermentation from various sugars by *Kluyveromyces cellobiovorus* KY 5199

3.13 BIOTECHNOLIGICAL ROUTES FOR THE PROVISION OF BASE CHEMICALS

The demand for chemical feedstocks such as methanol, ethanol, ethylene, propylene, butadiene, and aromatics is in tens of millions of tonne per annum for Europe alone. These materials are currently derived from three major natural resources: natural gas, crude petroleum, and coal; none of which are renewable resources within any useful time-scale. All three of these resources are also used as major sources of energy, either directly as such or by the provision of electricity. Some husbanding of them becomes possible as alternative

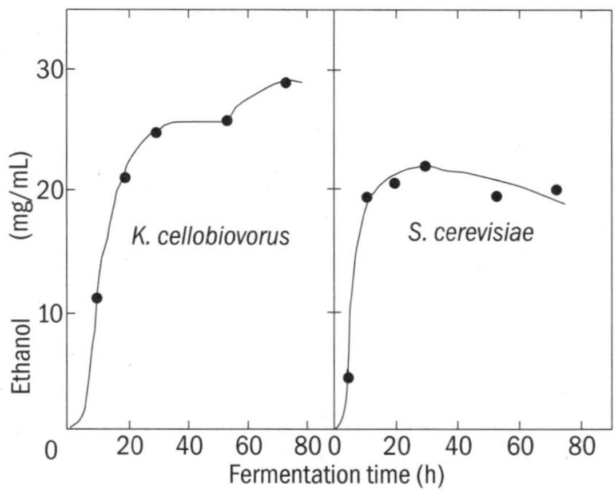

Figure 3.23 Ethanol fermentation from alkali-treated bagasse hydrolysate by *K. cellobiovorus* and *S. cerevisiae*

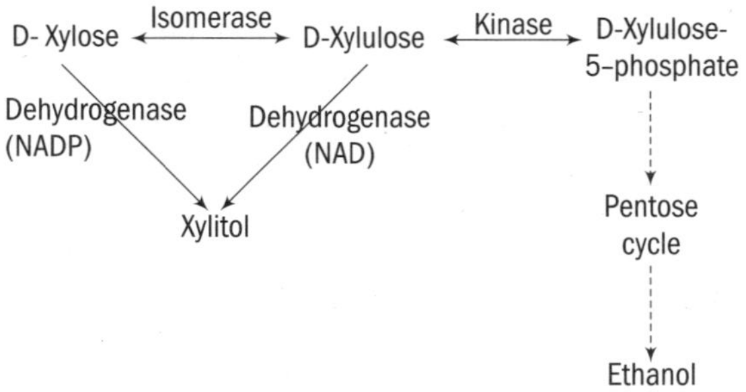

Figure 3.24 Metabolic pathway of xylose in yeasts

source of energy become available: nuclear (fission and eventually fusion), hydro-electric, geothermal, tidal, etc. Alternative sources of energy will therefore prolong the arrival of the era when coal, oil, and gas have largely disappeared. Natural gas will probably be the first of the three resources to become exhausted though processes for the gasification of coal should be highly developed at that stage and will, therefore, provide us with adequate supplies of methane. The chemical industry in its perpetual search for the cheapest starting materials is a remarkably resilient one as it is entirely capable of effecting major changes in its supply of feedstocks. As oil resources begin to dwindle or become too expensive, chemical industries will switch back to their previous dependence upon coal as their natural starting point. As has been said elsewhere of the chemical industry "oil (is) an interlude in a century of coal".

Thus it is against a relatively confident background, and against an industry which already operates on very large scales of manufacture, that biotechnology, if it is to compete or contribute, must be measured as a potential provider of base chemicals.

3.13.1 Potential Substrates

The main opportunities for biotechnology to provide large-scale, low added-value products would appear to reside in:

(a) Improved use of ligno-cellulosic materials: These include straw, wood and wood waste, vegetable and plant material produced as necessary but waste adjuncts to existing food crops, domestic refuse including paper, and peat which is available in some countries in the world and is a major resource in a few of them.

These materials, apart from peat, can be considered of zero value and their cost to a biotechnology industry would therefore reside only in their collection. It is unlikely that starch or sucrose could ever be used to provide low cost chemicals as these commodities already have an intrinsic value and their proper use would be as food materials. To divert them on any large scale into biotechnology would put pressure on existing food prices.

It is now thought unlikely that direct utilization of woody materials, with its combination of cellulose, hemicellulose and lignin, will be achieved by fermentation processes alone; partial, if not complete, hydrolysis using chemical/mechanical means would seem to be the most likely choice. Combined chemical-enzymatic processes have though been suggested. The products of degradation include both hexoses and pentoses and neither sugar can be ignored though each could be utilized by different organisms if necessary. Lignin, which can account for up to 30% of woody material, must also be used in the most beneficial manner; at the present time little production beyond that of vanillin has been achieved. In theory, a whole range of chemically useful aromatics could be derived from lignin if the appropriate organisms could either be found or genetically engineered.

(b) Development of autotrophic microorganisms: Figure below summarizs the basic strategy of the preceding section. However if appropriate microorganisms could be developed, alternative scenarios become possible.

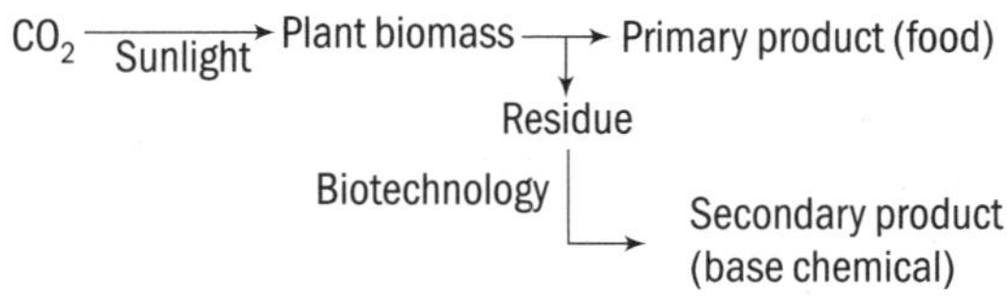

Extensive scenarios have been suggested using algae grown photosynthetically either in sea, brackish, or freshwater locations to provide supplies of chemicals. The number of useful chemicals so derived is currently limited to glycerol and some hydrocarbons plus associated lipid perhaps being suitable for use as a diesel fuel.

It may be expected that algae could play an important role in the provision of some chemicals in those countries of the world with available land and a warm ambient temperature coupled with abundant sunshine. It has been estimated that an oil-bearing algae could yield approximately 25 tonne/hectare/year assuming only a very modest 3.3% photosynthetic conversion efficiency.

The principal, but long-term hope for countries in more temperate climates would however appear to reside in the development of chemolitho-trophic bacteria which could grow non-photosynthetically on H_2 and CO_2. These substrates could be produced simultaneously from coal, in which some CO would also be present though bacteria have already been described which will grow efficiently on such a substrate. Alternatively, H_2 could be produced by a variety of means. CO_2 would then be either atmospheric or derived from mineral carbonates.

An exciting possible scenario arising from these considerations has been development of using a mutant of *Alcaligenes eutrophus* which would grow on CO_2 plus H_2 and would produce high yields of a metabolic intermediate such as pyruvate. This pyruvate could then be used by a second microorganism as a precursor for higher value added products. Alternatively, other types of mutants could be produced which would accumulate other chemically useful intermediates.

A tremendous amount of work has been done and written on ethanol. It can be produced quasi-economically in most countries: all that is required is appropriate government legislation to be enacted to erect the necessary trade barriers and remove the others which are restrictive. Biotechnological production of ethanol is therefore mainly a socio-politico-economic problem, not a biotechnological one.

Production of acetone and butanol by *Clostridium* spp. and butandiol by *Klebsiella* spp. are technically feasible and were, of course, carried out before the advent of the petrochemical industry. Its large scale re-adoption depends on either producing (probably by genetic means) microorganisms with a higher tolerance to butanol, or a cheaper means of recovering butanol (which is the desired product) from the fermentation broth.

Poly-β-hydroxybutyrate (PHB) is a biotechnological product of some uncertainty. Its applications are both unique and limited: it has many similarities with chemical polymers such as polyethylene but is biodegradable. It was first produced by ICI process in a pilot-scale process. In 1996 Monsanto (who sold PHB as a copolymer with PHV under the trade name Biopol) bought all patents for making the polymer from ICI/Zeneca. Strictly, PHB is not a base chemical unless it proved possible to modify it in some simple manner, probably chemically, which would then permit it to be used alongside, or instead of, some of the more expensive chemical plastics. PHB is currently produced from carbohydrate feedstocks but could be produced from CO_2 plus H_2.

Microbial oils and fats are probably feasible biotechnological products though whether they could be produced economically enough to compete against existing plant or animal commodities is doubtful unless zero-cost substrates can be found in bulk. Fatty acids are used extensively in the chemical industry to produce various technological products though their main use is as a food substance. However, it is important to point out, as with all biotechnological products, the subsequent use of a produce is governed by the choice of substrate used to produce it. Thus, for products intended for animal or human consumption, it is imperative that there is no dubiety over the acceptability of the substrate used as feedstock. Should the feedstock be such that it itself could not be introduced into the food

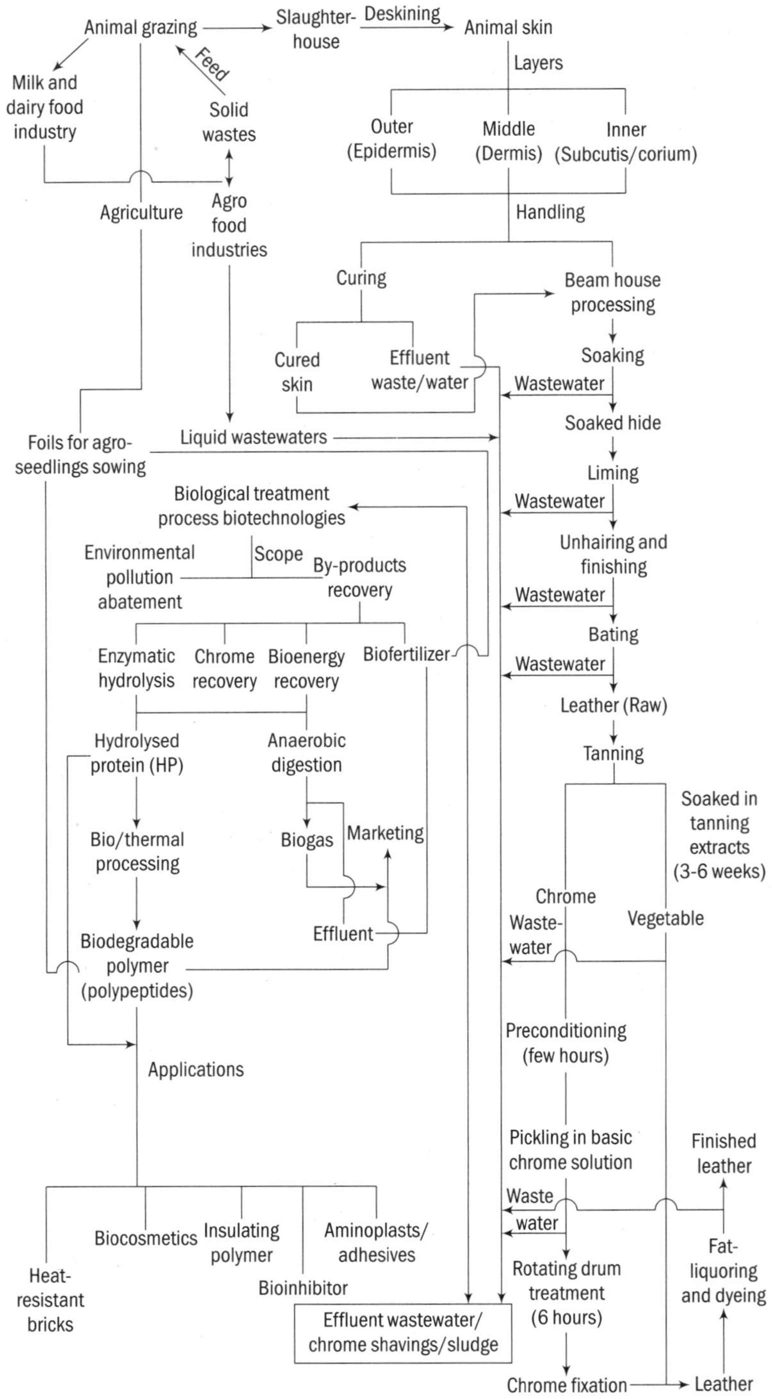

Figure 3.25 Total recycle integration approach in solid food waste related industries

chain (for example domestic refuse, industrial wastes), then it is most unlikely that the product could not be so introduced. With these substrates, therefore, products destined for technological applications must be produced and these must obviously be chosen with care to affect the best return on capital invested. Such products not only include fatty acids but organic acids such as itaconate as well as butanol and butanediol which have already been mentioned.

Possibilities of producingpropylene oxide, have been considered. In a single step oxidation of propylene, alkene epoxides is produced

$$\begin{array}{ll} CH_3\text{–}CH{:}CH_2 \text{ --------> } & CH_3 - C{=}CH \\ & \qquad \backslash\,/ \\ & \qquad\ O \end{array}$$

in which immobilized enzyme/cell catalysis is used.

3.14 PROTEINOUS BIOWASTES RECYCLING

3.14.1 Meat Processing Industry Wastes

A total recycle approach of meat processing solid-liquid wastes treatment is shown in Figure 3.25. It reduces pollution to a large extent.

3.14.2 Fish Processing Waste

Processing fish waste for extraction of value added products such as protein, lipid, gelatin, amino acids, collagen, and oil has become one of the most intriguing researches due to its valuable properties. Recently, extraction of lipid from sardine fish waste was reported using microwave-assisted extraction (MAE) and compared with Soxhlets and Hara and Radin methods. A mixture of two organic solvents in proper proportion can be used.

Bibliography

Apaire, V., J. P. Guiraud, and Galzy, P. 1983. Selection of yeasts for single cell protein production on media based on Jerusalem artichoke extracts. *Z. Allg. Mikrobiol.* 23: 211–218

Banerjee, U. C. and S. N. Mukhopadhyay. 1990. *Indian Chemical Engineer* 29(4): 43–46

Bellamy, W. D. 1974. *Biotechnol. Bioeng.* 16: 869–80

Bisaria, V. S. and T. K. Ghose. 1981. *Enz. Microbial. Technol.* 3(2): 89

Boulton, C. A. and C. Ratledge. 1978. *Topics in Enzyme and Fermentation Biotechnology*, A. Wiseman (ed.), Vol. 9, pp. 11–77. Chichester: Ellis-Horwood

Brown, O. 1981. *Chemy. Ind.* pp. 95–97

Bu'Lock, J. D. 1979. In Symposium of the Society for General Microbiology A. T. Bull, D. C. Ellwood and C. Ratledge (eds), Vol. 29, pp. 309–325. Cambridge: Cambridge University Press

Bu'Lock, J. D. and A. J. Bu'Lock (eds). 1980-1983. Acetone-butanol fermentation and related topics. *Science & Technology Letters*, Northwood

Cannon, R. E. and S. N. Anderson. 1991. Biogenesis of bacterial cellulose. *Critical Review of Microbiology* 17: 435

Carter, G. B. 1981. *New Scientist* 86: 216–218

Cysweski, G. R. and C. R. Wilke. 1977. *Biotechnol. Bioeng.* 19: 1125

Dubinsky, Z., T. Berner, and S. Aaronson. 1978. *Biotech. Bioeng. Symp.* 8: 51–68

Duncan, W. 1981. *Chemy. Ind.* pp. 311–316

Engelabart, W. and H. Dellweg. 1976. Abstract. Fifth International Fern. Symp. p. 376

Eveligh, D. E., *et. al.* 1982. *Biotechnol. Letters* 4 (5)

Frost, G. M. and D. A. Moss. 1987. Production of enzymes by fermentation. In *Biotechnology*, Vol. 7 a, J. F. Kennedy (ed.), pp. 65–211. Weinheim: VCH

Ghosh, S. and D. L. Klass. 1976. Clean fuels from biomass, sewage urban refuse, and agricultural wastes. Symposium sponsored by Institute of Gas Technology, Orlando, Florida, 27–30, January

Ghosh, S. and D. L. Klass. 1981. Advanced process development for methane production from biomass-waste blends. *Am. Chem. Soc.* pp. 251–278

Ghosh, S, J. R. Conrad, and D. L. Klass. 1974. Research Symposium of the 4 Annual Conference, Water Pollution Control Federation, Denver, Colorado, 6–11 October

Ghose, T. K. 1981. Cellulose conversation, In *Advances in Food Producing System for Arid and Semi-arid Lands*, pp. 225–66. New York: Academic Press

Ghose, T. K., *et al.* 1979. *Biotech. Bioengy.* 21: 283

Ghose, T. K. and V. S. Bisaria. 1979. *Biotechnol. Bioeng.* 21: 131–46

Ghosh, S. and D. L. Klass. 1978. *Process Biochem.* 13: 15–24

Ghosh, S., M. P. Henry,and D. L. Klass. 1979. Second Symposium on Biotechnology in Energy Production and Conservation, Gatlinburg, 3–5 October

Ghose, T. K., P. V. Pannirselvam, and P. Ghosh. 1983. *Biotechnol. Bioeng.* 25: 1577

Ghose, V. K., P. V. Pannirselvam, and P. Ghosh. 1984. *Biotechnol. Bioeng.* 24: 241

Ghosh, V. K., T. K. Ghose and K. S. Gopalkrishanan. 1982. *Biotechnol. Bioeng.* 24: 241

Gibbs, D. F. and M. E. Greenlalgh. 1982. Biotechnology in the production of chemical feedstocks and derived products; strategic issues and options for the European Community, Vol. 2, pp. 180. Brussels: FAST bio-Society Network of the European Economic Community

Haug, R. T. 1977. *J Water Pollut. Control Fed.* 49 (7): 1713

Heertjes, P. M. and R. R. Van der meer. 1979. Industrial Waste Conference, West Lafayette, 8-10 May

Howells, E. R. 1982. *Chemy Ind.* 8: 508–511

Humphrey, A. E.1979. The hydrolysis of cellulosic materials to useful products. *Adv. Chem. Ser.* pp. 25–53

Jeffries, T. W. 1981. *Biotechnol. Letters* 3: 213

Klass, D. L. 1979. Abstracts, Biomass for Energy sponsored by the *U.K. Sec. lnt. Solar Energy Sec,* London, 3, July

Klass, D. L. 1980. Proceedings in Bio-Energy World Congress and Exposition. Atlanta, 21-24, April, Bio-Energy Council, Washington. D.C.

Klass, D. L. and S. Ghosh. 1977. Symposium Papers in *Clean Fuels from Biomass and Waste,* Institute of Gas Technology Orlando, 25–28 January

Klass, D. L. and S. Ghosh. 1979. Biomass as a nonfossil fuel source. *American Chemical Society and the Chemical Society of Japan Joint Chemical Congress,* 24 (2): 414–28

Klass, D. L. and S. Ghosh.1980. Fuels from biomass.*American Chemical Society National Meeting,* 27–29 August, San Francisco

Klass, D. L., S. Ghosh, and J. R. Conrad. Symposium papers in *Clean Fuels From Biomass. Sewage. Urban Refuse and Agricultural Wastes,* Symposium sponsored by the Institute of Gas Technology, Orlando, 27–30 January

Kierstan, M. 1980. *Proc. Biochem.* 15: 2, 4 & 32

Kirk, K. 1971. *Anny. Rev. Phytopathol.* 9: 185

Le Duy, A. 1979. *Proc. Biochem.* 14: 5–7

Mandels, M. 1981. *Annual Report on Fermentation Processes* 5: 35

Marchessault, R. H. and J. St-Pierre. 1978. *Proceedings of Chemrawn Conference I,* p. 613, Toronto, Canada, 10–13 July

Mitra, G. and C. R. Wilke. 1975. *Bioengy.* 17: 1

Montenecourt, B. S. and D. E. Eveligh. 1979. *Adv. Chem. Ser.* 181: 279

Mukhopadhyay, S. N., P. Ghose, and V. H. Potty. 1973. *J. Food Sci. Technol.* 12(3): 120

Mukhopadhyay, S. N., T. K. Ghose, and S. Kundy. 1985. Substrate and product inhibition problems in cellulose fermentation for production of ethanol: approach to its solution, pp. 450–50. *Proc. Biotech.*

Mukhopadhyay, S. N. and T. K. Ghose. 1997. *Proc. Biocon, Symp.,* IIT Delhi, pp. 97–109

Mukhopadhyay, S. N., T. K. Ghose and A. Fiechter. 1979. *Biotechnol. Letters* 7: 203

Mukhopadhyay, S. N. and R. K. Malik. 1980. *Biotechnol. Bioeng.* 22: 2237

Mukhopadhyay, S. N. and R. K. Malik. 1981. *Proc. Biocon. Biochem. Eng. Symp.,* IIT Delhi, p. 295

Mukhopadhyay, S. N. 1981. *Adv. in Biotechnol.* 3

Mukhopadhyay, S. N. 1981. *J. Frenent. Technol.* 59: 305

Nystrom, J.M. and A.L. Allen. 1976. Pilot-scale investigations and economics of cellulase production in enzymatic conversion of cellulosic materials: Technology and Applications. *Biotechnol. Bioeng. Symp.* (No. 6), pp. 55–74, Wiley, New York

Nystrom, J. M. and N. P. Diluca. 1977. *Proc. Biocon. Symp. III*, Delhi

Pohland, F. G. and S. Ghosh. 1971. *Biotechnol. Bioeng. Symp. No. 2*, 85–106

Poulsen, O. M. and L. W. Petersen. 1992. Degradation of microcrystalline cellulose: Synergism between different endoglucanases of cellulomonas sp. ATCC 21399. *Biotechnol. Bioeng.* 39: 121–23

Ratledge, C. 1982. Microbial oils and fats: an assessment of their commercial potential. *Prog. Ind. Microbiol.* 16: 119–206

Ratledge, C. 1984. Microbial conversion of alkanes and fatty acids. *J. Am. Chem. Society* 61: 447–453

Ratledge, C. and J. D. Bu'Lock. 1982. Biotechnology in Europe, pp. 101–109. Frankfurt: Dechema

Reese, E. T. 1976. *Biotechnol. Bioeng. Symp.* 6: 9

Rich, L. G. 1973. Solids waste management in *Environmental System Engineering*, pp. 265–271. Tokyo: McGraw-Hill

Riege l, E. R. 1949. Industrial Chemistry: Solids Disposal, 5th Edn, pp. 268–271. Mumbai: Reinhold Publishing House

Rogers, P. L., D. Phil, K. J. Lee, and D. E. Tribe. 1980. *Process Biochem.* 15: 7

Sahai, V. and T. K. Ghose. 1977. *Proc. Biocon. Symp.* pp. 260–90

Samejima, H. The use of biomass conversion as an alternative technology for the production of fuels such as ethanol, CHEMRAWN III Proceed., Hague, Netherlands

Saxena I. M. and R. M. Brown, Jr. 1989. Cellulose biosynthesis in *Acetobacterxylinum*: a Genetic Approach in Cellulose and wood Chemistry and Technology, C. Schuerch(ed.). New York: Wiley

Sexena, I. M., F. C. Lin, and R. M. Brown, Jr. 1990. Cloning and sequencing of cellulose synthase catalytic submit gene of Acelobacterxylinum (Abstract), *Meeting. Am. Soc. Microbiol*, p. 277

Schlegel, H. G. 1969. *Fermentation Advances*, D. Perlaman(ed.), pp. 807–832. New York and London: Academic Press

Schneider, H., P. Y. Wang, Y. K. Chan and R. Maleszka. 1981. *Biotechnol. Letters* 3: 89

Tassinari, T. and C. Macy. 1977. *Biotechnol. Bioeng.* 19: 1321

Toyama, N and T. K. Ogawa. 1972. *Proc. IV IFS Ferment. Tech.* p. 743

Wilke, C. R., *et al.* 1976. *Biotehnol. Bieneg.* 18: 1315

4 Environmental Variance and Effects of Pollution on Living Human Genomic Body/Bioreactors

4.1 INTRODUCTION

With continuous expansion of economic, cultural/social, and industrial/commercial activities, the hazard to different forms of life has increased. Resultantly, survival of living resources on the earth, including humans has been threatened. Humans pollute environment in multiple ways. All over the world, celebrations such as cultural events, national days, marriages, and festivals are accompanied with large-scale burning of firecrackers, this causes high-level air and noise pollution. Religion and nature are closely linked. Various rites and rituals performed under different religions of the world show how nature/environment and various life forms are intimately connected. Such linkages provide public's non-genetically acquired information on environmental variance, leading to biological evolution, as depicted in Figure 4.1.

4.2 QUANTIFICATION APPROACH OF ENVIRONMENTAL VARIANCE

By analysing Figure 4.1, it may be possible to express environmental variance in terms of the following equation:

$$V_E = V_{DE} + V_{PE} + V_{AC} \qquad \ldots(4.1)$$

Where V_E = Environmental variance

V_{DE} = Direct effect of environment

V_{PE} = Non-genetic parental effect

V_{AC} = Additive cultural inheritance

4.3 ENVIRONMENTAL VARIANCE IN THE FUNCTIONS OF LIVING HUMAN GENOMIC BODY/BIOREACTORS

If environmental variance, by cultural/social events, can be associated with atmospheric temperature variation, by seasonal change, may lead to severe air pollution, especially, in local region. This is true, particularly, in the case of many festivals and celebrations when

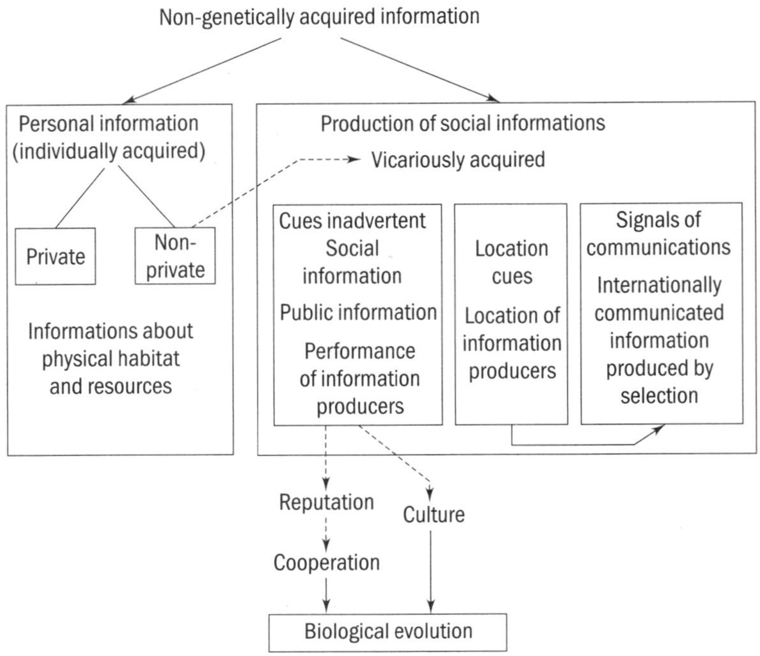

Figure 4.1 Various forms of non-genetically acquired information in variation of biological evolution

firecrackers are burned on a mass level. Such a large-scale burning of firecrackers releases considerably amounts of air pollutants such as carbon monoxide, nitrogen dioxide, sulphur dioxide, and particulate matters containing lead, cadmium, copper, potassium, aluminium, barium, iron, strontium, and mercury. These air pollutants are hazardous to almost every life form on earth. Not only this, such events have severe effects on humans. A human body can be divided into three principal layers, with mind being the inner layer, senses constitute the middle layer, and the flesh can be considered as the outermost layer. Exposure to these pollutants may result in spread of a number of diseases. In addition, pollution has a strong influence on thinking pattern of living human genomic body/bioreactors (LHGBRs). Our thoughts and behaviour are directly linked to our outside environment. We cannot think constructively when constantly surrounded with multiple pollution sources. For instance, our health deteriorates when exposed to continuous air pollution. In addition, chronic noise levels have a negative effect on the brain and behaviour of LHGBRs.

4.4 EFFECTS OF ENVIRONMENTAL VARIANCE ON LHGBRS

4.4.1 Good Effects

For LHGBRs, good is restraint in their deeds. Good is restraint in speech, in mind. So, good is restraint in almost everything. Therefore, environmental variance which has an

Table 4.1 Stress management and healing

S.No.	Action
1	Practice Yoga and silence every day. This is preferably followed at the same time and place.
2	For linking body with mind, breathing should be practised.
3	Associate Yoga with meditation
4	Change in approach: cultivate spirit of rendering service

influence on mind and soul can guide an individual towards goodness. Within LHGBRs, there lies a centre of tranquillity, serenity, self-knowledge industry (SKI), and true awareness. This is regulated by genes and genomic library of individual LHGBR which may get good effects of environmental variance by practising soul management (SM) in a pollution-free environment. The SM helps LHGBRs to experience true freedom of mind, in respect of fears, desires, stresses and tensions, insecurities, and material complexities. Every LHGBR is in search for peace and happiness. Good effects of environmental variance on LHGBRs provide relaxation which helps in dealing with stresses and strains of modern life. Other measures that can considerably help in stress management include breathing exercises, Yoga, and meditation (Table 4.1).

4.4.2 Ill Effects

It can be observed from Equation 4.1, how environmental variance is effected by factors, such as V_{DE}, V_{PE}, and V_{AC}. For example, when a person accuses others or imposes blames on them, by getting mentally agitated from their actions, V_{DE} is said to have played its role. If this persists for a long duration, V_{PE} may ultimately change the person's perception about the society and he may become more inclined towards wrongdoings, for example, involvement in crimes. Such perceptible ill effects of environmental variance have become pronounced in today's modern world, of which stress is the root cause. It is important to mention here that V_{DE} may not be of much concern if its ill effects are temporary and non-repetitive, but repetition of ill effects may lead to permanent sickness of the mind of the individual. In order to avoid ill effects of environmental variance, stress management is essential (Table 4.1). Young generations who are exposed to various environmental variances are more vulnerable to psychological disorders.

4.5 ENVIRONMENTAL VARIANCE IN RELATION TO CONSTANT EVOLUTION FOR BIOLOGICAL EVOLUTION

Everything in the universe is the result of a cosmic phenomenon called evolution. Nothing in the universe skips evolution. Continents and species evolve because of the concept of evolution. All life forms are constantly involved with an evolutional course. Even LHGBRs' mind and knowledge are influenced by evolution. Charles Darwin gave ample evidence about 'origin and evolution' in his Theory of Evolution, based on biological species. Evolution of LHGBRs and other species, the earth, and the entire cosmos is a continuous and inevitable process. Existence of each life form is an outcome of evolution. Environmental variance enables all living species, sub-species, ecosystems, all biomass, and the entire biosphere evolve. Interconnection of human-plant-microbes (HPM), under environmental variance, is also an influential factor in the process of constant evolution. In this connectionist system, knowledge is expressed in terms of the values of numerical coefficients, assisted with links between pairs of simple agents called 'neural units'. Application of such knowledge is based on operations that closely resemble with linear threshold functions, which in turn, are equivalent to Boolean functions. It involves many patterns and needs expert systems to accomplish the evaluation process. Fuzzy logic is principally utilized in development of control systems. Artificial neural networks are being developed to recognize multi-patterns. Models of human mind machines (HMM) utilize this kind of intelligence for solving problems related to health.

4.6 ENVIRONMENTAL VARIANCE-LINKED CHEMIOSMOTIC DISTURBANCE

4.6.1 Proton Motive Force Mediation

In LHGBRs, oxidative phosphorylation may get affected by environmental variance. This takes place primarily due to disturbance in respiratory system. Respiratory complexes, constituting respiratory chain, translocate protons on oxidation, and H^+-ATPase/ATP synthase (F_0F_1) dehydrate ADP and P_i to form ATP using back flow in the presence of sufficient proton motive force (PMF). In 1966, Nobel Laureate, Peter Mitchell propounded chemoosmotic (coupling) hypothesis to provide new thinking in this direction which states that oxygen-free radicals may participate in the form of reactive oxygen species (ROS). Molecular oxygen is an attractant for which chemosensing is linked to metabolism. The attractant's response to oxygen is processed through cytochrome oxidase, this reduces oxygen supply to stimulate electron transfer in order to boost PMF. This may change the molecular operating environment (MOE), affecting metabolism and causing ailments/diseases. Also, ROS including O_2^-, OH^-, and H_2O_2 in MOE in LHGBRs may lead to oxidative stress. These ROS are also responsible for causing several degenerative diseases in LHGBRs.

4.6.2 Translocation Mediation

In LHGBRs, multi-cellular system of transportation of nutrients in cellular metabolism is very important. Nutrient transfer mainly occurs through an active transport mechanism in cellular membranes. The phosphotransferase system is involved in the group translocation mechanism. Sugars are often transported by this mechanism, which involves phosphorylation, thereby, trapping those nutrients which membrane carriers can send to the appropriate phosphorylase. For efficiency in such a translocation mediation mechanism, appropriate MOE is needed. In case, oxidative stress alters appropriate MOE in group translocation, it results into occurrence of ailments in LHGBRs.

4.7 POVERTY-DRIVEN ENVIRONMENTAL VARIANCE EFFECTS

4.7.1 Genome Engineering Changes

In LHGBRs, changes in functions and regulations are guided by many factors (Table 4.2). They undergo changes or adaptation because their genomes are linked to their activities.

Table 4.2 Facets of LHGBRs between birth and death and imaginary beyond

S.No.	Functions performed			Humanism
1	Male/boy (Donor pole)	Like/love (Biochemotaxis)	Female (acceptor pole)	Social benefits
2	Conceiving	Cells/fusion/biofertility	Birth	LHGBRs' functions
3	Birth (B)	On	Earth	Creation/population explosion
4	Pro	To	Grow	Biomass
5	Crawler	To	Stroller	Biomechanics
6	Grow	To	Show	Bioefficiency
7	Learn	To	Earn	Bioefficiency
8	Name	For	Fame	Bioefficiency
9	Vision	For	Mission	Bioefficiency
10	Karma (Deeds)	With	Dharma (Religion)	Bioefficiency
11	Finish	And .... (Death/D)	Vanish	Cremation (departed soul)
12	Soul	(Play)	Role	In fragmented form
13	Role	(To find)	Pole	For rebirth
14	Subject	(Fragmented)	Objective	In hypercycles
15	*Nirvana*	For	*Paritrana*	To get into ASN
16	*Paritrana*	For	*Mahanirvana*	To get into ASN
17	*Mahanirvana*	For	*Mahaparitrana (Moksha)*	For no rebirth and beyond ASN

ASN - almighty supra nature; LHGBRs - living human genomic body/bioreactors

Scientists and bioengineering GOEs working on genes and genome regulation give due importance to this as daily activities of LHGBRs is directly linked to their stress level. Some genes within genome are known to determine behavioural patterns of LHGBRs, for example, mood swings, faithfulness to companions, and many other phenomena. As stress is directly related to activities of LHGBRs, toxic materials such as free radicals of ROS are formed in the fuel cells, for example, mitochondria. Formation of ROS and possible damages that can be caused by these radicals in DNA of LHGBRs are shown in Figures 4.2 and 4.3, respectively.

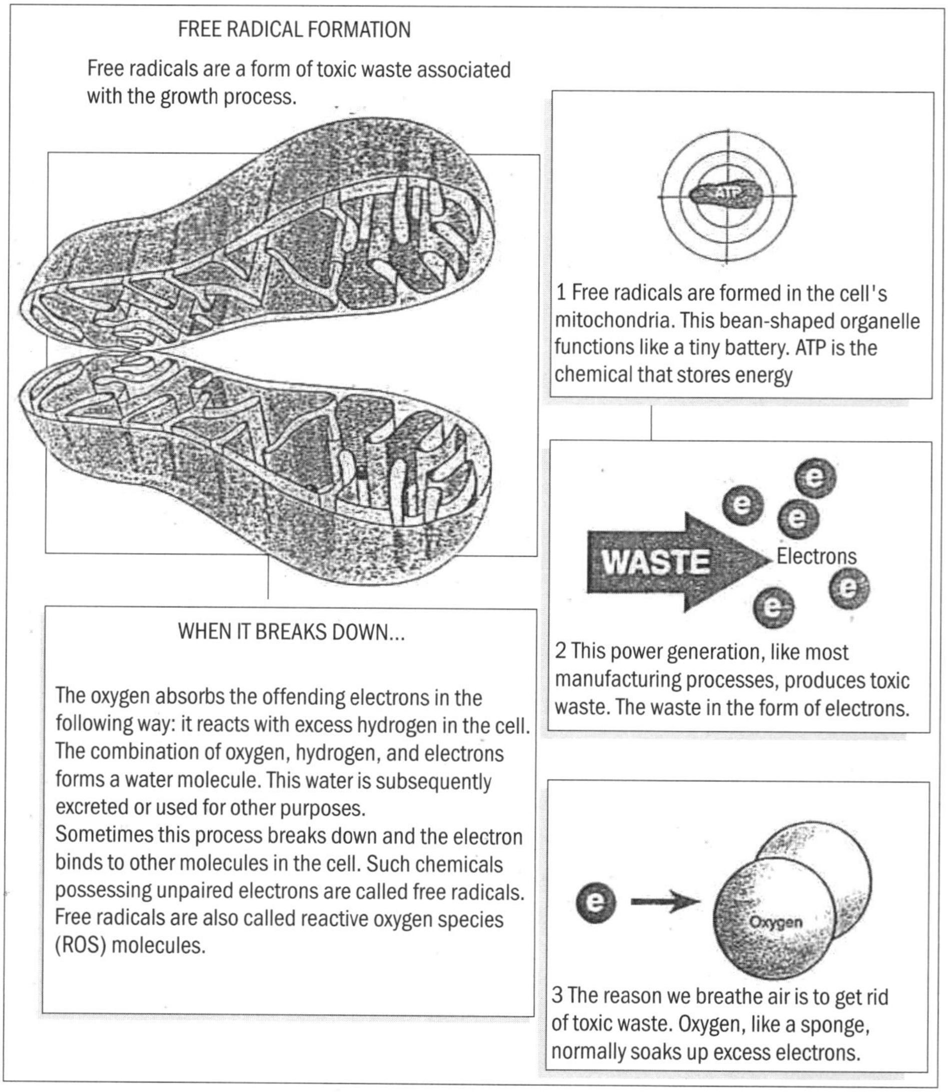

Figure 4.2 Schematic of free radicals of ROS formation in LHGBRs' fuel cell/mitochondria

DEMONETIZATION IN INDIA

Demonetization of ₹500 and ₹1000 currency notes on 8 November 2016 by Government of India was done with the aim of stabilizing the country's economy. However, demonetization with it brought a lot of inconvenience to the general public, senior citizens were most affected. Long queue of people for cash outside ATMs was a common scene during that period. With limits imposed by Reserve Bank of India on cash withdrawal, the transactions were hit further. Demonetization had a profound effect on country's gross domestic product (GDP). However, demonetization was not all about cash crunch, it has a number of positives sides too. It came out as an effective weapon against black market, counterfeit notes lost their value, and digitization was encouraged for secured transactions (cashless).

Another factor such as poverty-linked environmental variance may cause changes in LHGBRs' DNA. Living in poverty is a challenge—from both economic and biological points of view. Economically, LHGBRs are deprived of very basic facilities, necessary for sustenance of life. Biologically, this can cause changes in people's DNA that can develop a depressing attitude towards life. To overcome this depression, they may get diverted towards addictions such as alcohol and drugs.

Functions and regulations in LHGBRs for anatomical scale-up dynamics in the facets of life performances are listed in Table 4.2.

4.7.2 Pollution Effects on Living Human Genomic Body/Bioreactors

In LHGBR's, anatomic/structure-related health hazards/disorders/ailments are caused mainly due to occurrence of genomic aberrations/mutations/infections and damages. The aberrations result in prevention of normal functions of LHGBRs. This is mainly caused by introduction of some kind of pollutant to the body. A pollutant may enter an LHGBR through different sources, such as water, air, soil, and radioactive material.

4.8 RADICAL DAMAGE

Molecules of ROS can do a lot of damage to interior of cells. The cells, however, have evolved mechanisms to fight them. This is illustrated in Figure 4.3.

4.9 ENVIRONMENTAL VARIANCE/CLIMATE CHANGE AND GLOBAL WARMING

Environmental variance with the climate change has become a global concern. Figure 4.4 suggests how on the world level the underlying global warming trend is becoming stronger each year.

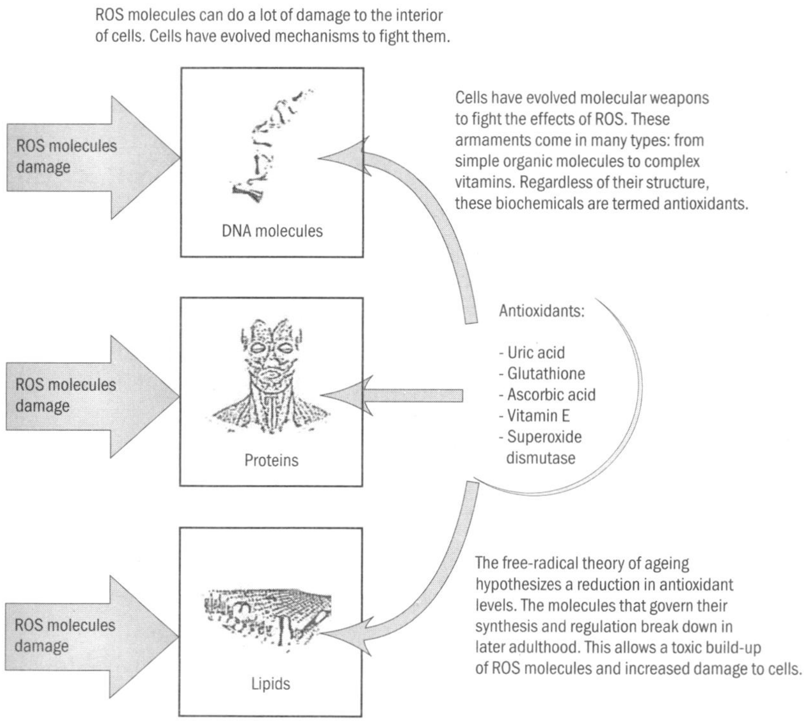

Figure 4.3 ROS radical damages in LHGBRs

4.10 INDIA'S CONCERNS ON ENVIRONMENTAL VARIANCE AND GLOBAL CLIMATE

Realizing the seriousness of ill effects of environmental variance and global climate change, India has taken steps to mitigate pollution caused by hydrofluorocarbons (HFCs).

India's decision to ratify the Paris Agreement has come after ensuring compliance of domestic legal requirements, internal discussions and after obtaining clarity from Nations Framework Convention on Climate Change (UNFCCC) with regard to transparency and participation of countries in future processes till freezing year (2031): Here 'freezing year' has been meant for the year when use of HFCs will peak before being rapidly scaled down and phased out while the baseline is instrumental in fixing levels. Though rich nations want an early freezing year (2021), India has been insisting to use HFCs till 2031 before phasing out. This is to make sure that development is not compromised and must get enough time before the phasing out for HFCCs starts. Recently, in the UK, the women protested against climate change by following 'birth strike'. Idea is not to have kids at all due to existence of unfavourable conditions that are likely to arise as a result of climate change.

Figure 4.4 Effect of EV with climate change and global warming

Bibliography

Danchin, E., L. A. Giraldeau, T. J. Valone, and R. H. Wagner. 2004. Public Information: from noisy neighbors to cultural evolution. *Sci.* 305: 487

Mitchell, P. 1966. *Biological Rev.* 21: 445–502

Mitchell, P. 1978. In Les Prix Nobel en 1978, Nobel Foundation, Stockholm

Mukhopadhayay, S. N. 2017. Products, facets, and curses of LHGBRs. *INAE monthly e-newsletter* 8(8)

Riegal, E. R. 1959. *Industrial Chemistry*, 5th edn. New York: Reinhold Publishing Corporation

Sone, N. 1992. Chemiosmotic Coupling in Bioenergetics. In *Int. Sympo. Adv. Computing for Life-Science*, July, pp. 19–24, Iizuka, Fukuoka, Japan

The Times of India. 2015. Let Our Gods Be In Constant Evolution by V. Singh, New Delhi, 14 April, p. 18

The Times of India. 2019. There is birth strike over climate change, New Delhi, 17 March, p. 18

The Times of India. 2015. Needed Air quality curbs on fire crackers, New Delhi, 29 October, p. 4

The Times of India. 2016. 2 Oct. New Delhi, p. 21

The Times of India. 2016. Poverty changes DNA of people may make them drug addicts by Ian Johnson. 27 May, p. 25. New delhi

5 Pollution: Parameters, Abatements, and Treatments

5.1 INTRODUCTION

In recent years, dependency of wastewater treatments on biotechnological processes has increased considerably. In biological treatment procedures, microorganisms break down organic pollutant matter, resulting in the stabilization of putrifiable wastes (or refuses). These microorganisms are broadly grouped under aerobic (or oxic, facultative) and anaerobic (or anoxic) microorganisms. Some of the characteristics of these biological decomposers are listed below:

(i) Oxic microorganisms need molecular oxygen for carrying out metabolic activities.

(ii) Facultative microorganisms are capable of working in both oxic and anoxic conditions.

(iii) Anoxic microorganisms derive energy from organic molecules and function in the absence of oxygen.

Heterotrophic microorganisms are usually prominent species in mixed treatment process biotechnology. These microorganisms need an organic carbon source for both energy and cell mass synthesis. Unlike heterotrophs, autotrophic microorganisms use an inorganic carbon source such as carbon dioxide or carbonate for carrying out energy and cell mass synthesis. These types of microorganisms derive energy either from the oxidation of inorganic compounds such as nitrogen and sulphur (chemosynthetic) or from solar energy (photosynthetic).

5.2 GENERAL CONCERNS AND PRINCIPLES

Bioprocessing treatment of wastewaters basically involves either of two mechanisms to accumulate and store biomass. The first is a flocculated suspension of biological growth called activated sludge which is mixed with wastewaters. In the second mechanism, a biological film is fixed to an inert surface support medium. The wastewater is made to pass through this biological film. The applied process biotechnology methods, developed earlier, for treating wastewaters, were composed of maximum of three systems: (i) activated sludge

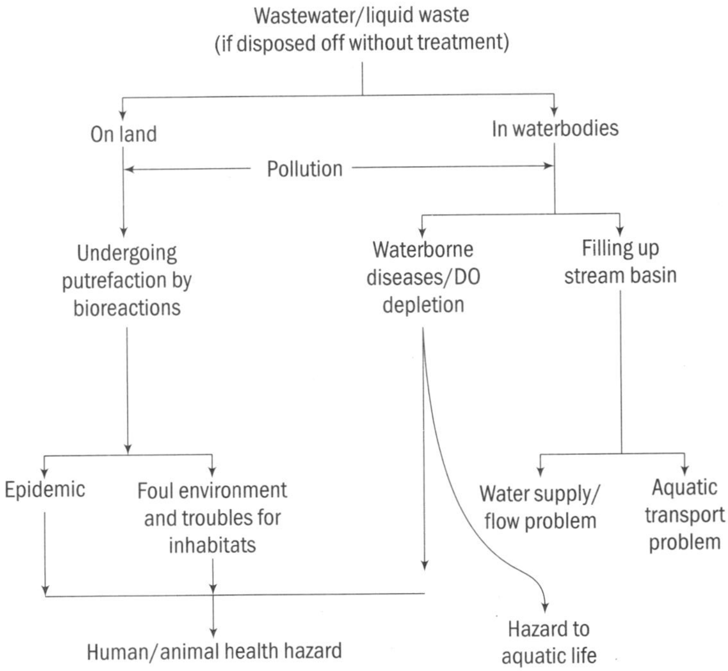

Figure 5.1 Possible fates of disposal of untreated wastewater or liquid waste
DO - dissolved oxygen

process (ASP), (ii) trickling filter (fixed film process or FFP), and (iii) anaerobic digestion (AD) of waste organic materials.

The principal aim of treating wastewaters or liquid wastes is to avoid their improper disposal. Possible results of untreated wastewater, as shown in Figure 5.1, indicate the possible hazards that can arise due to inappropriate disposal. The extent to which the untreated wastewater can be hazardous to the environment highly depends on its source of origin (Figure 5.2). Therefore, the treatment of wastewaters for removal of undesirable materials is a multi-purpose activity (Boxes 5.1 and 5.2).

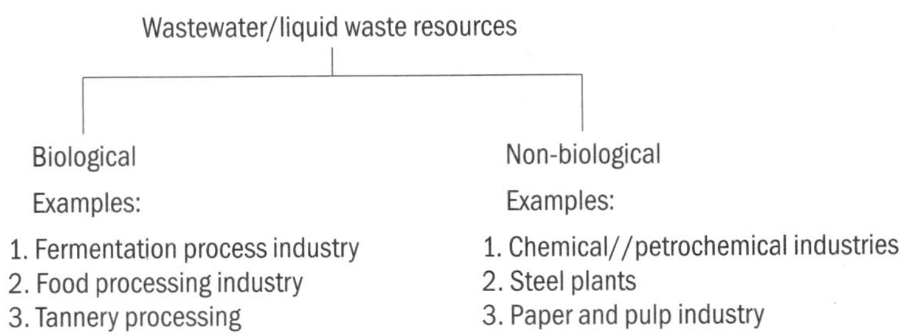

Figure 5.2 Types of wastewater resources

Box 5.1 Biological treatment of wastewaters is mainly carried out for:

- Reducing contained pollutants
- Increase water reuse by preventing pollution of receiving water
- By-product recovery
- Minimizing increase in pollutants' discharge as a result of plant expansion and production increase
- Taking care of society (considering plant location)

Box 5.2 Types of undesirable materials that can be removed from wastewater through biological treatments:

- Soluble organics responsible for depleting dissolved oxygen (DO) contents
- Suspended solids
- Trace organics such as phenols, responsible for imparting bad taste to drinking water
- Heavy metals, cyanides, and toxic organics
- Colour and turbidity
- Nitrogenous and phosphorous compounds
- Refractory materials
- Oil and volatile materials
- Radioactive components

Recent advances in biotechnology treatment processes are primarily focused at improving basic biotechnology, innovations, broadening of the application of the process, and improvements in the design, analysis, and operational procedures.

In bioprocessing treatment, microbial cell populations may be specifically adapted to certain compounds in order to successfully achieve bio-oxidation. Examples of such compounds include phenol, cyanide, polyphenols, and thiocyanates. Most organic materials present in industrial wastes are amenable to bio-oxidation. Presence of toxic or inhibitory waste in the total stream hampers the process of biological treatment.

In such cases, design of biological wastewater treatment procedure, especially, for treating complex industrial wastes plays an essential part. It should include characterization of various waste streams, contributing to overall flow, for example, river or other flowing systems. The two most important characterization parameters kept in the mind while designing treatment procedures include biochemical oxygen demand (BOD) and chemical oxygen demand (COD). BOD and COD are in correlation, thus, influence each other.

5.2.1 Parameters/Indices

5.2.1.1 Biochemical oxygen demand test

Biochemical oxygen demand is usually defined as the amount of oxygen required by bacteria while stabilizing decomposable organic matter under aerobic conditions. The BOD test is essentially a bioassay procedure, involving the measurement of oxygen consumed by living organisms (mainly bacteria) while utilizing the organic matter present in waste. This test is widely employed for determining pollution strength of sewage and industrial wastes, in terms of oxygen required by marine microorganisms to decompose waste material discharged into natural water courses in which aerobic conditions exist.

Biological degradation of organic matter under natural conditions is brought about by a diverse group of organisms that carry the oxidation process essentially to completion, that is, carbon dioxide and water are formed as the end matter. Therefore, it is important that mixed group of organisms, commonly called 'seed' be present in the test. The rate at which the reactions proceed is governed to a major extent by population numbers and temperature. Temperature defects are held constant by performing the test at 20°C. Theoretically, an infinite time is required for complete biological oxidation of organic matter but for all practical purposes, the reaction may be considered complete in 20 days. However, a 20-day period is too long to wait for results in most instances. It has been found that a reasonably large percentage of the total BOD is exerted in 5 days; therefore, BOD tests are developed on the basis of a 5-day incubation period.

5.2.1.1.1 The nature of biochemical oxygen demand reaction

Studies of the kinetics of BOD reactions have established that they resemble unimolecular reactions (or first order) in character. In reality, the rate of the reaction is proportional to the amount of oxidizable organic matter remaining at any time (Eqs 5.1 and 5.2).

$$dL/dt = -KL \qquad \text{...(5.1)}$$

where, L = organic matter remaining, that is, organic matter to be oxidized.

$$\int_{L_a}^{L_t} dL/L = -K\int_0^{t} dt \quad \text{or,} \quad \log_e L_t/L_a = -kt \quad \text{or,} \quad L_t/L_a = e^{-kt}$$

or, $\log_e L_t/L_a = -kt$ or, $L_t/L_a = e^{-Kt}$

or $L_t/L_a = 10^{-kt}$, where $k = 0.4343\,K$

so, $$y = L_a - L_t = L_a\,(1 - L_f/L_a) = L_a\,(1 - e^{-Kt}) \text{ or } L_a\,(1 - 10^{-kt}) \qquad \text{...(5.2)}$$

where, y = BOD at any time t; L_a = ultimate BOD

5.2.1.1.2 Phase of biochemical oxygen demand curve

A typical BOD curve is shown in Figure 5.3. The curve shows two distinct phases. The 'seed' which is derived from soil or domestic sewage, contains large number of saprophytic bacteria that utilize carbonaceous matter present in the samples. Seed is also composed of autotrophic

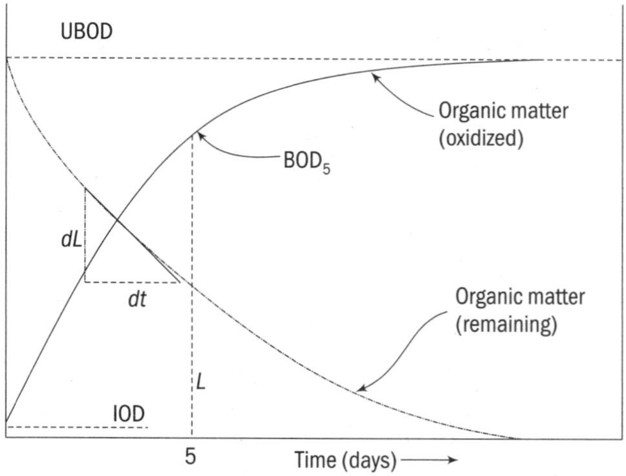

Figure 5.3 BOD curve
BOD - biological oxygen demand; IOD - immediate oxygen demand; UBOD - ultimate biological oxygen demand

bacteria, particularly, nitrifying bacteria, which are capable of oxidizing non-carbonaceous matter. The nitrifying bacteria are usually present in relatively small numbers in unaerated domestic sewage, fortunately their reproductive rate at 20°C is such that their populations do not become sufficiently large to exert an appreciable demand for oxygen until about 8 or 10 days have elapsed in the regular BOD test. Once the organisms become established, they oxidize nitrogen in the form of ammonia to nitrite and nitrate by using oxygen.

$$2NH_3 + 3O_2 \xrightarrow[\text{(Nitrite-forming bacteria)}]{} 2NO_2^- + 2H^+ + 2H_2O$$

$$2NO_2^- + O_2 + 2H^+ \xrightarrow[\text{(Nitrite-forming bacteria)}]{} 2NO_3^- + 2H^+$$

The interference caused by nitrifying organisms is so high that it makes almost impossible to take measurements of the total carbonaceous contents, unless provision is made to eliminate them. It is a major reason for selection of a 5-day incubation period for a regular BOD test. In case, effluents are from biological treatment units, such as trickling filters and activated sludge, populations of nitrifying organisms are fed to the effluents to utilize significant amounts of oxygen during the regular 5-day incubation period. Amount of residual carbonaceous BOD in such cases plays an important role in assessing plant's efficiency. The action of nitrifying bacteria can be arrested by the use of specific inhibiting agents such as methylene blue. Alternatively, the microbial population can be reduced from the samples by sterilization.

5.2.1.1.3 Evaluation of the values of K and L_a

Formulation of the first-stage BOD of wastewater as a first-order reaction yields Eqs 5.4 and 5.5. Graphical evaluation of L_a and k is shown in Figure 5.4.

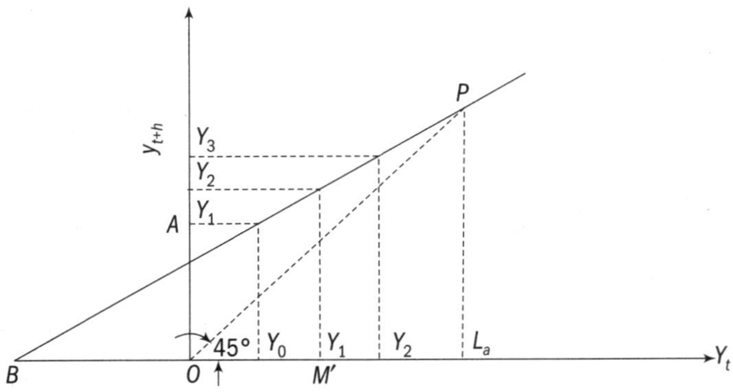

Figure 5.4 Graphical evaluation of L_a and k

$$\frac{dt}{yt} = K(L_a - y) \text{ or } \int_{y_0}^{L_a} dy/(L_a - y) = K\int_0^1 \quad ...(5.3)$$

or $$\log_e \frac{L_a - Y_t}{L_a - Y_0} = -K_t \text{ or } \frac{L_a - Y_t}{L_a - Y_0} = e^{-kt} \text{ or } 10^{-kt} \quad ...(5.4)$$

Rearranging, $$1 - \frac{L_a - Y_t}{L_a - Y_0} = 1 - 10^k \text{ or, } y_t - y_0 = (L_a - y_0)(1 - 10^{kt})$$

or, $$y_t = (L_a - y_0)(1 - 10^{-kt}) + y_0 \quad ...(5.5)$$

Similarly, $$\begin{aligned} y_{t+h} &= (L_a - y_0)(1 - 10^{-k(t+h)}) + y_0 \\ &= L_a - (L_a - y_0)\, 10^{-k(t+h)} \\ &= L_a - (L_a - y_0)\, 10^{-kh} + (L_a - y_0)\, 10^{-kh} \\ &\quad -(L_a - y_0)\, 10^{-k(t+h)} \\ &= L_a - L_a\, 10^{-kh} + \{(L_a - y_0)(1 - 10^{-kt}) + y_0\}\, 10^{-kh} \\ &= L_a - L_a\, 10^{-kh} + y_t\, 10^{-kh} \end{aligned}$$

$$y_{t+h} = L_a (1 - 10^{-kh}) + 10^{-kh}\, y_t \quad ...(5.6)$$

If y_{t+h} is plotted as an ordinate against y_t, as the abscissa, the value of the intercept on the ordinate at $y_t = 0$ of a straight line fitted either by eye estimation or by the method of least squares is

$$L_a (1 - 0^{-kh})$$

and the intercept on X axis, that is, when $y_{t+h} = 0$, is

$$L_a (1 - 10^{-kh})/10^{-kh}$$

$$OA = L_a (1 - 10^{-kh});\ OB = L_a (1 - 10^{-kh})/10^{-kh}$$

Slope of the line = $OA/OB = 10^{-kh}$ and if $h = 1$,

$$k = \log_{10} \frac{OB}{OA}$$

Graphical evaluation of L_a: At P, $y_t = y_{t+h}$ and thus in Eq. 5.4

$$y_t - 10^{-k}.y_t = L_a (1 - 10^{-kh}) \quad \text{or,} \quad y_t = L_a$$

Hence, the particular value of y_t at the point P is L_a.

Evaluation of y_0: Plot y_1 on Y axis and draw a horizontal line to cut BOD at M. Drop a perpendicular from M on Y axis that meets at M′.

That is, OM′ = y_0

Alternatively, Eq. 5.2 may be used after evaluating L_a and k

$$y_1 = (L_a - y_0)(1 - 10^{-k}) + y_0 \quad \text{...(5.7)}$$

or

$$10^{-k} y_0 = y_1 - L_a (1 - 10^{-k})$$

or

$$y_0 = (y_1 - OA).\ 10^{k} \quad \text{...(5.8)}$$

5.2.1.2 Chemical oxygen demand

The amount of oxygen required to oxidize wastewater components through purely chemical reaction under specified condition is defined as COD.

5.2.1.3 Biological oxygen demand–chemical oxygen demand correlation

The BOD measures strength of wastes which are biologically oxidizable and COD measures wastes' strength which are chemically oxidizable. Usually COD is greater than BOD. All wastes—whether organic or inorganic—essentially contain the following components:

(a) (i) Biochemically oxidizable (BO) part
  (ii) Non-biochemically oxidizable (NBO) part

(b) (i) Chemically oxidizable (CO) part
  (ii) Non-chemically oxidizable (NCO) part

This can be represented mathematically as,

$$I = \text{Inorganic} = I_{CO} + I_{NCO} = I_{BO} + I_{NBO}$$

$$O = \text{Organic} = O_{CO} + O_{NCO} = O_{BO} + O_{NBO}$$

Now,

$$COD = I_{CO} + O_{CO} = (I_{CO} + O_{BO}) + (O_{NBO} - O_{NCO})$$

$$BOD = O_{BO} + I_{BO} = (O_{BO} + I_{CO}) + (I_{NCO} - I_{NBO})$$

$$COD - BOD = (O_{NBO} - O_{NCO}) - (I_{NCO} - I_{NBO})$$

It shows that

$$COD > BOD \text{ if } (O_{NBO} - O_{NCO}) > (I_{NCO} - I_{NBO})$$

When waste components resistant to biochemical oxidation are more than the waste components resistant to chemical oxidation:

$$COD > BOD \text{ if } (O_{NBO} - O_{NCO}) > (I_{NCO} + O_{NCO})$$

5.2.1.4 Other indices

Besides BOD and COD, various other indices can be used for measuring strength of wastes and wastewaters. Some of these indices are briefly described here:

(i) **Total oxygen demand (TOD):** It is the stoichiometric amount of oxygen required to completely oxidize a substance into products such as CO_2, H_2OHNO_3, H_3PO_4, and H_2SO_4.

(ii) **Ultimate biological oxygen demand (UBOD):** It is the total BOD measured over an incubation period long enough for the reaction to reach virtual completion.

(iii) **Immediate oxygen demand (IOD):** The amount of oxygen utilized by a waste or wastewater sample immediately (within 15 min.) upon being mixed with air saturated distilled water.

5.2.1.5 Correlation of biological oxygen demand with other indices

From the time profile of BOD, as shown in Figure 5.5, the distinctions between the indices are clear.

It has been shown that

$$y_5 = BOD_5 = (UBOD - IOD)\,(1 - K^t) \quad ...(5.9)$$

where K is rate constant = 0.794 and t is time in days. So,

$$y_5 = BOD_5 = (UBOD - IOD)\,(1 - 0.794^5)$$

or $$y_5 = BOD_5 = 0.68\,(UBOD - IOD) \quad ...(5.10)$$

5.2.1.6 Determination of indices (American Petroleum Institute's Method)

BOD_5: Sample is diluted with distilled water (standard), seeded with bacteria and then incubated at 68°F (20°C) for 5 days. The decrease in oxygen content between original sample and incubated sample gives BOD_5 (20°C).

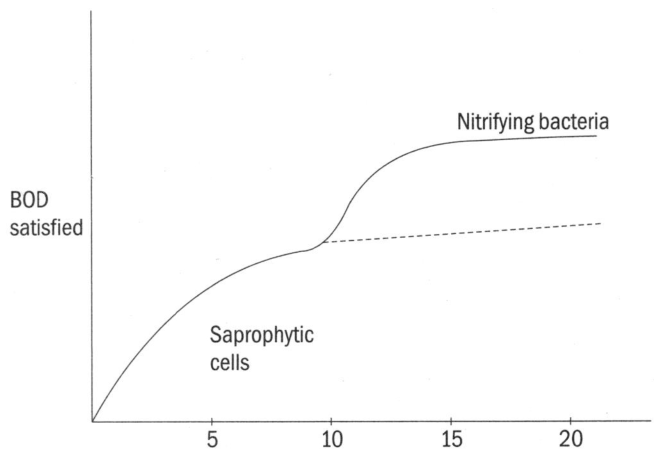

Figure 5.5 BOD reaction profile curve
BOD - biological oxygen demand

COD: Sample is digested for 2 h at 290°F with H_2SO_4 and excess $K_2Cr_2O_7$. The consumption of $K_2Cr_2O_7$ gives the 2-h COD. It is also expressed as 4-h $KMnO_4$ of COD.

IOD: The sample wastewater is diluted with aerated distilled water and allowed to stand for 15 min. The dissolved oxygen (DO) content is determined for the diluted mixture. The oxygen consumed is obtained from the difference between the oxygen added with the distilled water and that remaining in the diluted mixture at the end of 15-minute. reaction period.

5.2.2 Processing Effluent Water

5.2.2.1 Coal processing

The majority of energy requirement, in rural and most industrial areas, in India, is met by coal. About 7308 million tonnes of coal reserves were confirmed in 1986. The smoke of coal, especially, in rural areas, immensely contributes to indoor air pollution. Handling of coal is a challenge, in addition, SO_2 produced by its combustion contributes to acid rain.

Many coal-based industries such as steel, fertilizer, carbon black are in operation in Eastern India, particularly, in Durgapur industrial belt of West Bengal. All of these industries generate large volumes of wastes/residues and wastewaters. These wastes and wastewaters are responsible for causing land as well as water pollution (local potable/usable water). However, proper processing of wastes and wastewaters prior to their disposal can minimize pollution to a very large extent. If they are disposed of without giving adequate treatment/care they are likely to contribute to environmental pollution, either by putrefaction, or by biodegradation. Liquid waste effluents discharged from these industries range between 0.03 and 0.4 M^3/tonne of coal processed. These wastewaters/liquid effluents are toxic, in terms of pH, colour, dissolved solids, phenol, free ammonia, thiocyanate, cyanides, sulphide, BOD, COD.

5.2.2.2 Hazardous components and toxicity

The major toxic components contained in the liquid effluents of coal-based industries are phenols, cyanides, and thiocyanates. Toxicity features of some of these components are given in Table 5.1. When humans unknowingly consume water contaminated with industrial effluents, they become prone to diseases such as diarrhoea. Other symptoms include mouth sore, dark urine, and burning mouth. The toxic effects of phenol may not be so acute with regard to human beings but its presence (in traces) in potable water is objectionable. This is because on chlorination phenol forms chlorophenol which imparts unpleasant taste to water.

The basic phenolic compounds are reported to be carcinogenic in nature. Phenol toxicity to fish life, for the first time, was identified in 1928 when carps of size 8 inches were exposed to phenol-contaminated water (phenol concentration: 400 mg L^{-1}). Carps could not withstand the toxicity levels and were dead within 1 h of the exposure.

Another component of the industrial effluents is the cyanide, an extremely toxic substance. In mammalian system, exposure to hydrocyanic acid disrupts the process of oxidation in

Table 5.1 Toxicity of coal-processing industrial effluents

Component	Nature/symptom of toxicity	Toxicity level
Phenols	In humans: • Diarrhoea • Mouth sores • Dark urine • Burning mouth In potable water: • Unpleasant taste • Harmful to aquatic life (0.03 mg L^{-1})	In traces (0.001 mg L^{-1})
CN^-	• Inhibits cytochrome oxidase action, resultantly, tissue cells are not able to make efficient use of O_2 • Paralyses the respiratory centre or brain cells	0.009 mg kg^{-1} body weight
SCN	Dizziness, skin eruption, running nose, vomiting, nausea, thyroid problem, nervous system disorder	0.015 mg L^{-1}

the tissue cells. It paralyses the respiratory centre of the brain cells and retards ability of tissue cells to utilize oxygen. Cyanide is known to be an inhibitor at the terminal oxidation of the cytochrome (cytochrome oxidase). Cyanide forms a stable complex with cytochrome oxidase. This results in loss of the enzyme activity, often resulting in death. Hydrocyanic acid is toxic, if present, in concentrations of 0.3 mg L^{-1} in the air inhaled. The lethal dose of the toxicant is 10 mg kg^{-1} of the body weight. The fatal dose for animals is 9 mg kg^{-1} of the body weight. Fish has a high sensitivity to cyanide. Survival of minnows, catfish, and carps is threatened at a concentration of 0.3 mg L^{-1}.

In humans, presence of thiocyanates, at concentrations above 15 mg/100 mL in blood, is critically dangerous. Chronic absorption of thiocyanate may cause dizziness, skin eruptions, running nose, vomiting, nausea, and mild to severe disturbances of the nervous system. Thiocyanate-contaminated water on chlorination or irradiation releases cyanide by splitting the sulphur molecule. It inhibits the adsorption of iodine by thyroid gland. Ammonium thiocyanate at 280–300 mg L^{-1} concentration is reported to have lethal effect on spotted shellfish while goldfish exposed to 1600 mg L^{-1} concentration is expected to die within 24 h of exposure.

Free ammonia has also been found to be toxic to fish life, even at concentration level of up to 2.8 mg L^{-1}. High concentration of unionized ammonium ions in the solution is mainly responsible for the toxicity. In addition, as the pH of the solution increases, concentration of unionized ammonium ions also increases.

Wastewater from the coal carbonization industries initially is of straw yellow colour. The colour is intensified upon storage. Several explanations can be given for the development of this colour. The catechol present in the wastewater under alkaline conditions becomes oxidized. The wastewater also contains iron which can form colouring complexes with thiocyanates and cyanides to impart yellow colour. However, exact chemistry behind this colour development is yet unknown. Also, the colour development during exposure to air and light has been shown to exhibit toxicity of algal population (above 600 units).

Very little information is available with reference to the sulphide concentration in the coal carbonization effluents except that in the case of low-temperature carbonization (LTC) process, it lies in 67–74 mg L^{-1} range.

Soluble sulphide in water reacts with hydrogen ions to form HS or H_2S. The toxicity of sulphide is derived from the H_2S rather than the sulphide ions. The lethal concentration of sulphide for fish life is between 0.0032 and 0.355 mg L^{-1}. The value of BOD to COD ratio of different types of wastes from different coal carbonization processes lies between 0.5 and 0.9, indicating, that it is amenable for biological treatment. Several attempts have been made to treat the waste after dilution with domestic sewage; however, no major success has been achieved in this direction.

5.2.2.3 *Major liquid effluents' origin in coal-processing routes*

Coal can be processed through both conventional and non-conventional routes (Figures 5.6 and 5.7). Conventionally, coal is processed for energy generation and to produce coke for industries.

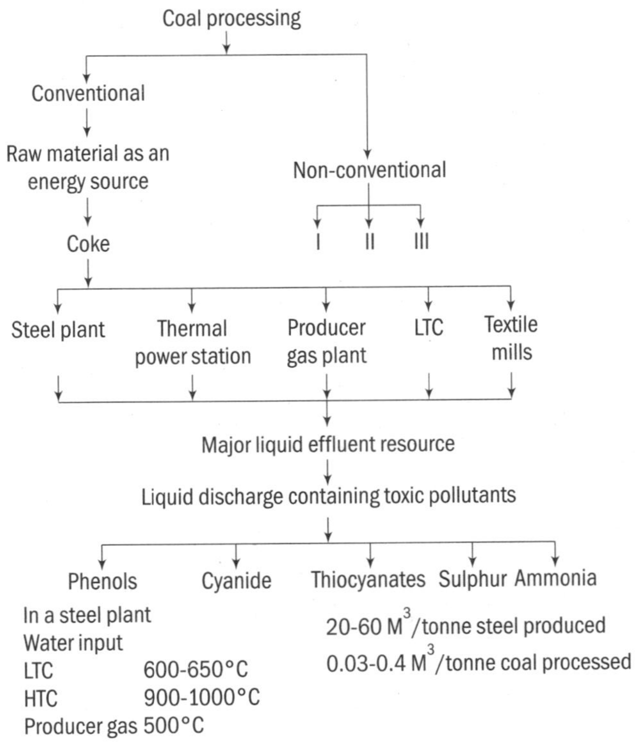

Figure 5.6 Conventional coal-processing
LTC – low-temperature carbonization; HTC – high-temperature carbonization

Note Details of non-conventional bio routes (I, II, and III) are shown in Figure 5.7.

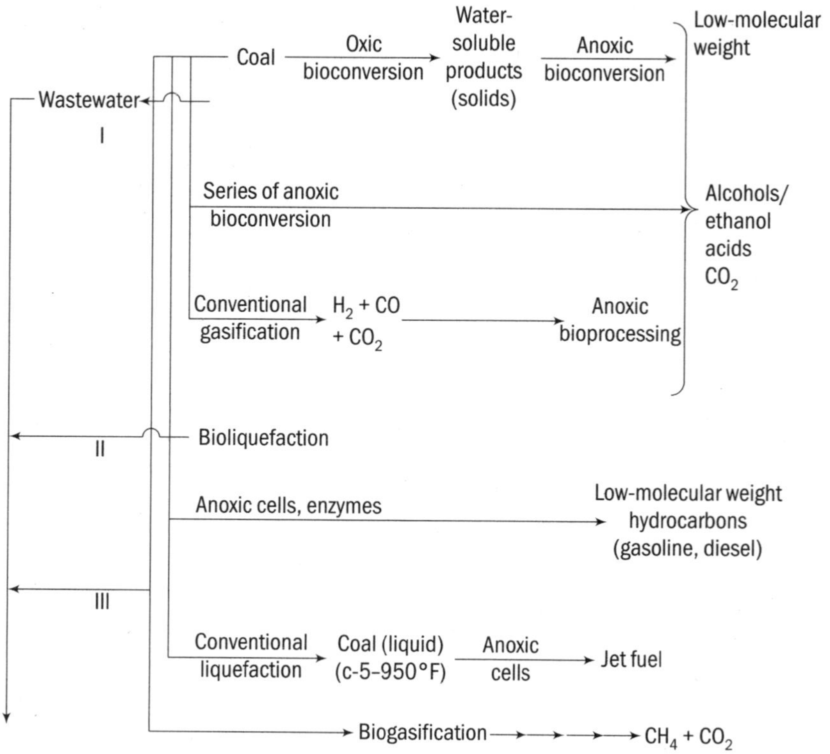

Figure 5.7 Non-conventional routes of bioprocessing of coal to liquid and gaseous fuels and chemicals

Owing to technological advances in coal-processing field, non-conventional bio-processing routes have also been developed (Figure 5.7). However, in both the routes, the principal aim of coal processing is to serve the following purposes:

(a) Produce low-sulphur coal
(b) Reduce smoke generation
(c) Reduce air pollution
(d) Reduce health hazard
(e) Reduce COD:BOD
(f) Higher heat generation
(g) Obtaining possible useful metabolites/chemicals

5.2.2.3.1 Conventional route of coal processing

Conventional route of coal processing is shown in Figure 5.6. It is worth mentioning that conventional processing of coal is expensive in comparison to non-conventional coal processing. For reduction of the expenses, unconventional microbial biodegradation of liquid effluents can be done. Biodepiritization of coal by microbial processing is one such approach. Bioliquefaction and biodesulphurization of coal are other approaches of reducing expenses.

It is necessary to identify potential microbial strains for such purposes. However, microbial processing of coal has its own merits and demerits (Table 5.2). Comparison of conventional and non-conventional processing of coal is summarized in Table 5.3.

Table 5.2 Merits and demerits of microbial bioprocessing of coal

Merits	Demerits
• Less energy intensive • Less capital intensive	• Slow process • In most systems cell population is undefined • Control and regulation are difficult

Table 5.3 Conventional *vis-a-vis* non-conventional processing of coal

Conversion parameters	Thermal/chemical processing			Bioprocessing		
	Gasification	Liquefaction	Desulphurization	Depyritization	Bio-lique faction	De-sulphurization
Temp(°C)	700	400	375	30–35 lab	30–35 lab	30–35 lab
Pressure (PSIG)	1000	2000	–			
Catalyst/ treatment	–	+	Molten caustic	Microbes	Microbes	Microbes
Reaction/ product	–	–	–	*	**	***

PSIG - pounds per square inch gauge

* $4FeS_2 + 15O_2 + 2H_2O \xrightarrow[\textit{ferroxidans}]{\textit{Thiobacillus}} 2\ (Fe)_2(SO_4)_3 + 2H_2SO_4$

** Acids, low-molecular weight alcohols, and CO_2

*** Organic sulphur (C–S) removal (conventional pyrite sulphur)

5.2.2.3.2 Non-conventional bioremediation

(i) Non-conventional bioremediation route makes use of potential activity of microbial strains in degradation of pollution-generating components in coal effluents (Figure 5.7).

(ii) Some eubacteria and archaebacteria have high potential for carrying out bioprocessing of coal for bioremediation. These microorganisms accelerate the oxidation of pyrite, producing ferric sulphate and sulphuric acid. Examples include *Putida fluorescence, Sulfolobus acidocaldarius, Candida tropicalis,* and *Debromyces globosus* (Table 5.4).

5.3 BIODEPYRITIZATION STRATEGIES

Biodepyritization strategy may vary, depending on existing conditions (Table 5.5). In heap or percolation bioleaching, the actual composition of microbial population is still not clearly known. The disadvantage of this method is the long time required for desulphurization. However, there is one advantage associated with the use of biodepyritization; rate of thermophilic operation accelerates with the increasing temperatures. Treatment of the domestic waste without dilution is possible but only when the waste is supplemented with phosphorus. The influent phenolic concentration has an influence on the aeration period. Pilot plant studies have shown that the wastewater from steel mills (coke ovens) could be treated in aerobic activated sludge plants with an organic loading of 0.716 lb/lb of sludge/per day. Ammonia in concentration of above 4000 mg L^{-1} is reported to inhibit severely the biological activity of the sludge. Apart from ammonia, thiocyanate, cyanide, sulphide oxidized polyhydric phenols, calcium and chlorides also affect the biological process. Biological treatment of coal carbonization waste is widely practised in Britain.

Table 5.4 Potential microbial strains for biodepyritization of coal by microbial leaching

1. Eubacteria
   - *Thiobacillus thiooxidans* (oxidizes only sulphur)
   - *Thiobacillus ferroxidans*
   - *Leptospirilllus ferroxidans*
   - *Putida fluorescence*
   - *Acidianus brierleyi*
   - *Pseudomonas alcaligenes*
2. Archaebacteria
   - *Sulfolobus acidocaldarius*
3. Yeast
   - *Candida tropicalis*
   - *Debaryomyces globosus*
   - *Pichia* sp.
   - *Trichosporon cutaneum*
4. Fungi
   - *Aspergillus* sp.
   - *Penicillium* sp.
   - *Neurospora* sp.

Table 5.5 Features of biodepyritization design strategies

Design strategies	Mode/parameters	Features
Heap/percolation leaching	• Recycle: 1 h duration • Run off-mode, lump coal	• Simple and inexpensive • Long-residence times • Indigenous • Cell population use
Tank bioreactor operation	• Length: 8" Depth: 20" • 1/2 angle of base: 30° • Slurry volume : 120 L baffles	• More complicated operation • Short-residence times
Other possible designs	• Lagoon • Slurry pipeline bioreactor	• Inclination to horizontal

Wastewater containing phenols, cyanides, thiocyanates and ammonia is amenable for biological degradation; however, there are many inherent factors that have a profound effect on biodegradation of organic pollutants contained in coal. These factors and their consequences are listed in Tables 5.6 and 5.7.

Table 5.6 Chemical and environmental factors influencing biodegradability of organic pollutants for bioremediation

Structural factors	Consequences
1. Molecular weight or size	Limited active transplant
2. Polymeric nature	Extracellular metabolism required
3. Aromaticity	O_2-requiring enzymes
4. Halogen substitution	Lack of dehalogenating enzymes
5. Solubility	Competitive partitioning
6. Toxicity	Enzyme inhibition cell damage
7. Xenobiotic origin	Evolution of new degradative pathways
Environmental factors	**Consequences**
1. Dissolved oxygen requiring	O_2 sensitive and oxy enzymes
2. Temperature	Mesophillic temperature optimum
3. pH	Narrow pH optimum
4. Dissolved carbon	Organic/pollutant complexes concentration dependent growth
5. Particulates, surfaces	Competition for substrate
6. Light	Photochemical enhancement
7. Nutrients and trace elements	Limitation on growth and enzyme synthesis

Table 5.7 Biological factors influencing the biodegradation of organic pollutants for bioremediation

Biological factors	Consequences
1. Enzyme ubiquity	Low frequency of degradation species
2. Enzyme specificity	Analogous substrates not metabolized
3. Plasmid encoded enzymes	Reacclimation
4. Enzymes regulation	Low frequency of degradation species
5. Competition for habitat selection	Instability
6. Population regulation	Repression of catabolic enzyme synthesis
	Required acclimation or induction
	Extinction of low-density populations
	Lack of establishment of degradative populations
	Low-population density of degradative organisms

5.3.2 Fundamentals of Bioremediation

Phenolytic microorganisms may belong to bacteria, archaebacteria, yeast, or fungi. The metabolism of the aromatic ring can take place in both anoxic and aerobic conditions.

Table 5.8 Factors influencing the analysis of the kinetics and extent of biodegradation for bioremediation

Experimental factors	Problems encountered
1. Analytical method substrate disappearance	Completing abiotic processes
2. Biotransformation	Complex analyses
3. Mineralization	Incomplete biochemical pathways
4. Scale up/down	Comparability among reactor designs
5. Feedstock complexity chemically/biologically undefined complex wastewaters	Poor simulation and predictability difficult in interpretation

Considerable amount of information is available on the oxic metabolism of aromatic compounds. Biologically phenol forms catechol. The enzymes involved in biological pathways are hydroxylases, oxygenases, dehydrogenases, and adulates.

In bioremediation, biological oxidation may occur through different routes such as:

- Removal of electrons by transformation of metallic form of ferrous iron to ferric
- Removal of hydrogen as in the case of alcohol to acetaldehyde
- Addition of oxygen as in the case of oxidation of carbon to carbon dioxide

The role of oxygen in the reactions is to release an electron from the terminal oxidation stage. Oxygen is also required in certain other enzymatic reactions. The biochemical transformation of the aromatic involves the introduction of molecular oxygen to the substrate. One atom of oxygen is inserted into the substrate and the other atom of the molecule is reduced to water:

$$RH + O_2 + 2(H) \longrightarrow ROH + H_2O$$

Here RH represents an aromatic molecule. The aromatic ring is attacked by the enzyme 'oxyganase'. For ring fission, both atoms of oxygen are incorporated into the hydroxylated ring. In the oxidation of a simple aromatic molecule, in addition to the electron transport chain reaction, molecular oxygen is also required for the opening of the aromatic ring.

A variety of microorganisms have been found to metabolize cyanide and phenolic compounds. Most of them have metabolic pathways, leading to the formation of amino acids.

5.3.3 Bioprocess Engineering Analysis

The scheme of detoxification of phenolic wastewater at pilot level bioremediation is shown in Figure 5.8 in which *Candida tropicalis* was used. Inoculum need in the process was developed as described in Figure 5.9. Considering the schematic diagram of Figure 5.9 and operation of completely ASP system, the material balance of the substrate detoxification can be given by:

$$V \cdot dX/dt = QiXi - \{Q_w X_w + (Q_i - Q_w)\, X_e\} + V \cdot r' \qquad ...(5.11)$$

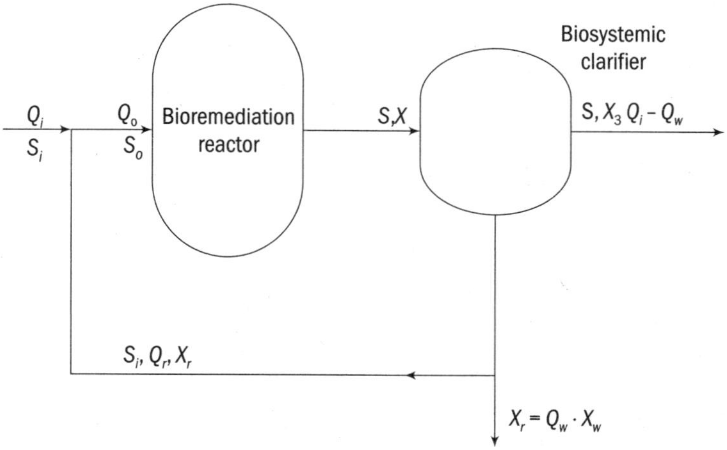

Figure 5.8 Schematic of bioremediation

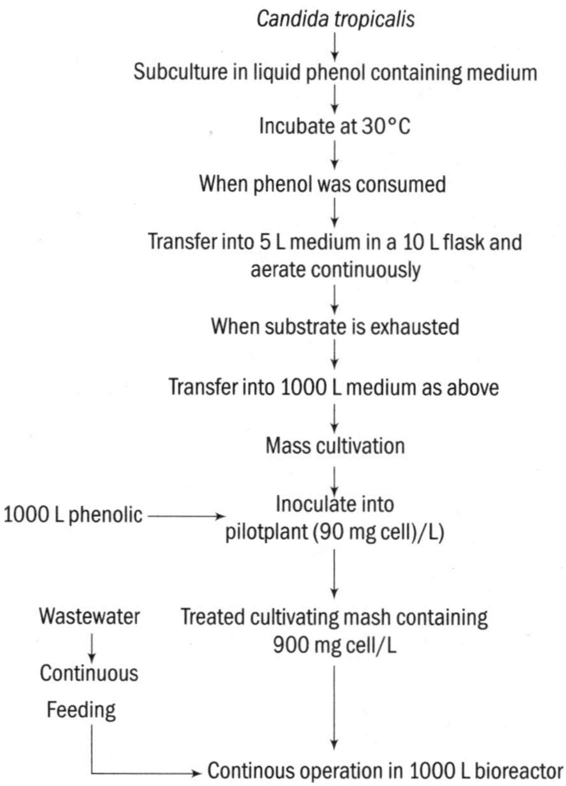

Figure 5.9 Development of inoculum for continuous pilot plant operation of detoxification of phenolic wastewater of coal processing

Symbols used in Figure 5.8 and many instances in the chapter stand for the following meanings:

X_e, X, X_w, X_r = MLSS (mg L^{-1}) in bioreactor, effluent suspended solid, recycle suspended solids

Q_i, Q_r, Q_w = Flow rates (1/h), in influent, recycle, wastage

S_e, S = Substrate concentration (mg L^{-1}) influent and in the bioreactor

t = hydraulic detention time (h)

θ_c = Mean cell residence time (h)

K = System rate constant

K_S = Substrate concentration at half of μ_{max}

V = ASP reactor volume 1

Y = Yield coefficient

μ_m = Maximum sp. growth rate (h^{-1})

k_d = Biological decay rate constant (h^{-1})

k = maximum substrate utilization rate (d^{-1})

From Eq. 5.11

$$V.r^1 = (Q_w X_w) + (Q_1 - Q_w) X_e \quad ...(5.12)$$

where

$$r^1 = \frac{\mu_m X_S}{K_S + S - K_d X} \quad ...(5.13)$$

On dividing by VX, equation obtained is

$$Vr^1/VX = (Q_w X_w) + (Q_i - Q_w) X_e/VX = 1/\theta_c \quad ...(5.14)$$

On simplifying it reduces to $1/\theta_c = Y.U - K_d$ ...(5.15)

By plotting $1/\theta_c$ against U (phenol utilization rate) values for Y and K_d can be obtained. The growth of microorganisms occurs by the utilization of the substrate, according to the following equation:

$$[(\mu_m - X_s (K_S + S)] V = YQ_O (S_O - S) \quad ...(5.16)$$

or

$$k(S/K_S + S) = (S_0 - S)/\theta X \quad ...(5.17)$$

where

$$\mu_m = kY$$

By rearranging Eqs 5.16 and 5.17

$$\theta X/(S_0 - S) = [K_S/k] [1/S] + 1/k \quad ...(5.18)$$

And by plotting $1/S$ against $\theta X/(S_0 - S)$ values for k and K_S can be obtained. Also as oxygen plays an important role in biosystems, the oxygen utilization in the system can be obtained by mass balance in the following way:

Net oxygen = Oxygen utilized for + Oxygen utilized for ...(5.19) consumed substrate oxidation respiration

So, KgO_2 utilize/day = a' kgO_2 utilized + b' kgO_2 utilized for ...(5.20)

for substrate oxidized respiration

The plot Kg substrate removed/day against KgO_2 utilized/day can give values of a′ and b'.

5.4 BIOREMEDIATION REACTOR STUDIES

In recent years, the hazard due to release of waste/liquors from coal-processing units has increased substantially. This is mainly attributed to: (i) increased used of coal and (ii) coal gasification and liquefaction processes which have not advanced much with time and thus are environmental polluting. Some of the most biological hazardous compounds present in coal-derived waste liquors include phenols, thiocyanates, ammonia, and cyanides. Simple elimination of low-level phenols from wastewaters through phenol fermentation is the most frequently adopted method of bioremediation.

The general strategy towards bioremediation of waste liquors is the application of highly evolved symbiotic populations of microbial flora (aerobic or anaerobic) in a classic reactor design so that all hazardous, polluting compounds can be degraded in the process.

For a long time, the basic reactor design continued to be the simple activated sludge aeration vessel, as used in ASP, followed by a clarifier. Many investigators have made systematic studies to ascertain the optimum operation parameters for the activated sludge in a continuous stirred tank bioreactor (CSTBR). Development of a three-phase fluidized bed bioreactor (FBBR) proved very beneficial in bio-processing of wastes. It could provide valuable information regarding techniques of determining limiting reactants, minimum retention time for biodegradation of limiting reactants, and understanding of the relative inhibiting effects of various compounds upon other compounds' degradation rates for different feed rates. It could also elucidate relative merits of each bioreactor for accomplishing bioremediation in terms of degradation of a given daily discharge of phenol-bearing wastewater, having common synthetic feed composition. For comparison of degradation rates (DR) of a bioreactor, following equation is utilized:

$$DR = \frac{Q_L(C_i - C)}{V_R} \quad \text{...(5.21)}$$

Where Q_L is the volumetric flow rate, V_R is the volume of the reactor, and C_i and C are concentrations of phenol in inlet (at any time).

Typical experimental results using active symbiotic bacterial population indicate that if the volumes of the bioreactors were about equal, the CSTBR could treat $2\,V_R$ (volume/day), PBBR $5\,V_R$, and FBBR $8\,V_R$ (undiluted volumes). This information provides preliminary indication on the suitability of FBBR in the effective bioremediation of phenolic hazards from coal-processing liquid effluents.

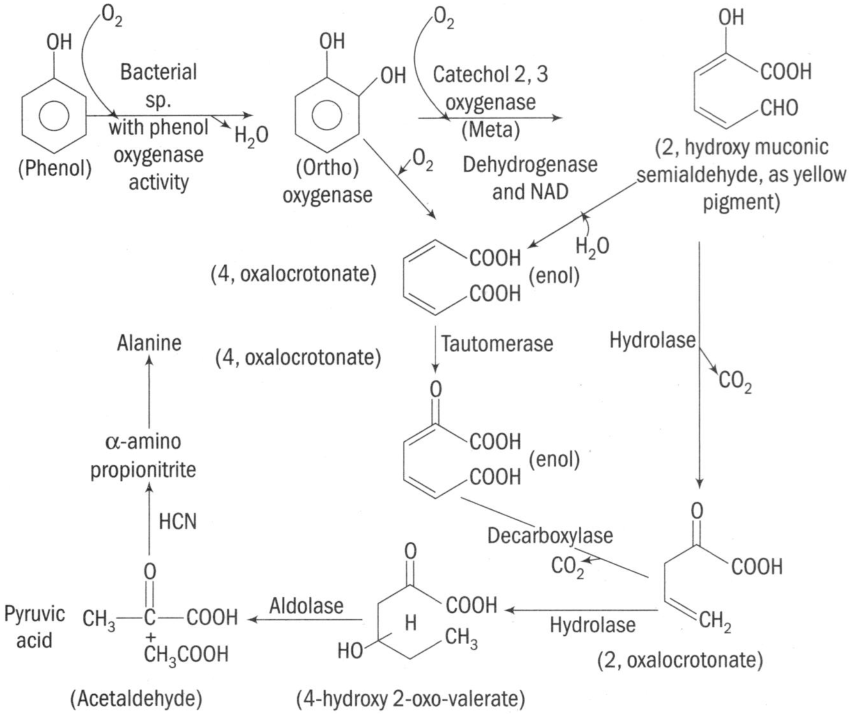

Figure 5.10 Bioremediation of phenolics in coal-processing effluents

More recent studies on phenol fermentation in a small, continuous STR bioreactor using pure culture of *Pseudomonas putida* strain (ATCC 17484) could develop rate equations of phenol biodegradation, the rate of intermediate inhibitor accumulation or degradation, and the rate of cell biomass production. The phenol biodegradation rate constant (k_{phe}) for pure culture system lies in 0.4–1.2 range while in mixed sludge culture system, k_{phe} is about $1.9 \times 10^{-2} \pm 8.8 \times 10^{-3}$. In sludge, it is determined using acridine orange direct counting (AODC) procedure.

5.4.1 Pathways to Useful Chemicals

(1) Meta-ortho pathway: Coal bioprocessing and bioremediation of coal-processing effluents can produce several important metabolites. One example is the formation of useful amino acids through meta-or the bioremediation pathway (Figure 5.10). In this pathway, hydroxymuconic semialdehyde is an important intermediate. However, in terms of its usefulness, it has not materialized fully.

(2) Organic sulphur removal pathway: Bioremediation for microbial desulphurization of coal principally involves removal of sulphur-cleaving C–S bonds in the model compound. Dibenzothiopene (DBT) has been used most extensively as a model compound for removing sulphur (Figure 5.11). This process causes liberation of sulphate without oxidation or

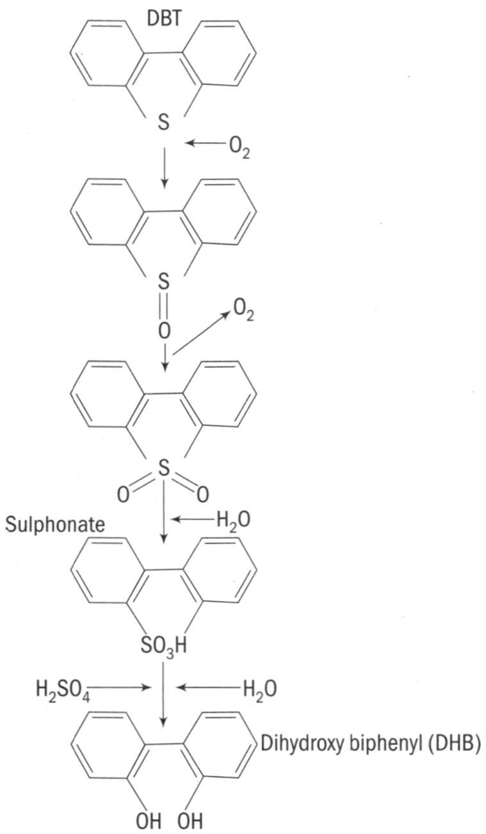

Figure 5.11 Possibility of dihydroxy biphenyl production through organic sulphur removal pathway of bioremediation

degradation of organic carbon matrix. At international level, lot of research is in progress for cloning of responsible genes for carrying out more efficient desulphurization and ability to transfer such genes to other organisms in order to develop potential cells.

These pathways are useful in reducing pollution hazards caused by coal effluents to a great extent. Metabolic pathway of phenol degradation by *Pseudomonas putida* and its entry into tri-carboxylic acid (TCA) cycle in cell biomass formation occurs as per the following metabolic scheme.

$$\text{Phenol} \longrightarrow \text{Intermediate} \longrightarrow \text{Cell biomass}$$

Bibliography

Dandekar, A. 2019. Life, death, rebirth. The Times of India. In Speaking Tree. 17 March, p. 3. New Delhi

Kolchanov, N. and R. Hofestacedt (eds). 2008. *Bioinformatics of Genome Regulation and Structure.* New Delhi: Springer Publishing Company

Li, J. and A. E. Humphrey. 1989. In *Bioprocess Engineering: the first generation,* T. K. Ghose (ed.), pp. 190–206. Chichester: Ellis Horwood Ltd

Mukhopadhyay, S. N. 2018. *Process Biotechnology Fundamentals,* 4th edn. New Delhi: Viva Books Pvt. Ltd

Mukhopadhyay, S. N. 2017. *Food Engineering: process and technology.* New Delhi: Viva Books Pvt. Ltd

Prescott, S. C. and C. G. Dunn. 1959. *Industrial Microbiology,* 3rd edn, p. 788. New York: McGraw Hill Book Co. Inc.

6 Air Pollution and Its Abatement

6.1 INTRODUCTION

According to World Health Organization (WHO), air pollution is among the world's top environmental health risks for living beings including living human genomic body/ bioreactors (LHGBRs). Worldwide, air pollution accounts for 1 out of 8 deaths. This makes air/atmospheric pollution as one of the most alarming global concerns which requires proper monitoring and control procedures. The severity of the problem of air pollution can be understood from the point that 'emission control' was the major topic for the world leaders at Conference of Parties (COP 21) held in 2015 (Figure 6.1). In India, Air (Prevention and Control of Pollution) Act, was enacted in 1981 for prevention, control, and abatement of air pollution in the country.

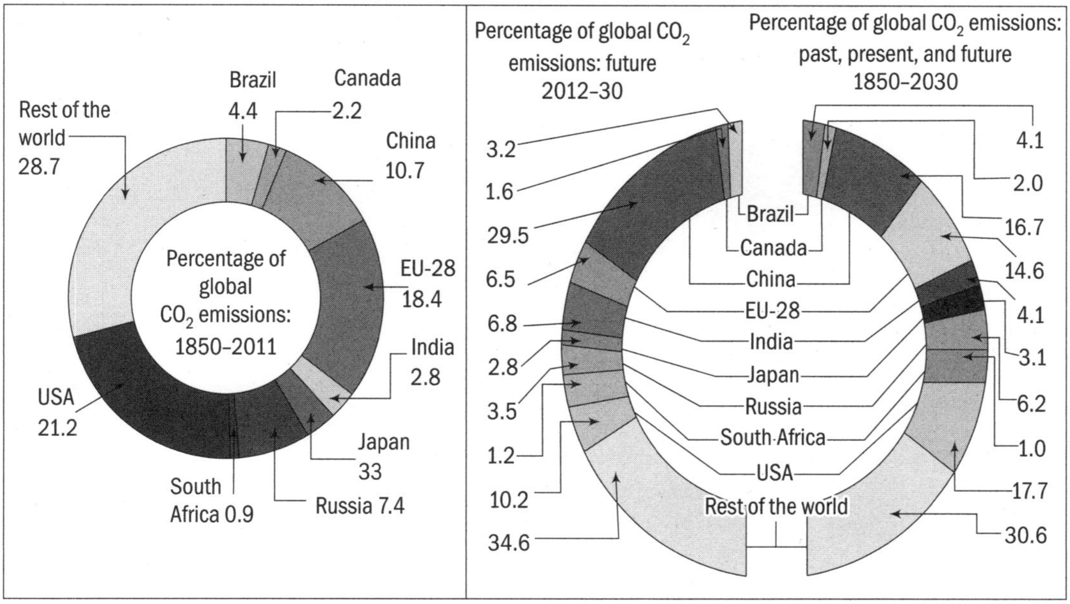

Figure 6.1 Global scenario of CO_2 emissions and predictions

6.2 AIR POLLUTION: CAUSES, SOURCES, AND HAZARDOUS EFFECTS

In India alone, indoor air pollution is responsible for causing 1.3 million deaths every year, which is among the highest in the world. Indoor air pollution is principally sourced from combustion, building material, and bioaerosols while sources of atmospheric air/pollution include industrial effluents and gases, waste burning in open, terrorist activities, burning of crackers, and aviation acts (Figures 6.2–6.4). In general, 'source' means any process, activity of LHGBRs which releases a greenhouse gas (GHG), an aerosol, or a precursor of GHG into the atmosphere.

Figure 6.2 Air pollution due to waste burning in open land/public places

Figure 6.3 Air pollution by industries located in residential areas

6.2.1 City Planning and High-rise Buildings

'Housing for all' is desirable to all LHGBRs. In fact, housing has an important part to play in economy, culture, and overall outlook of a country. Global space is limited, therefore, almost all developing cities and towns are paying due attention to construction of high-rise buildings in order to accommodate maximum number of LHGBRs. Development of high-rise apartments requires heavy civil engineering works that lead to considerable environmental degradation. This holds equally true for skyscraper offices of multinational companies (Figure 6.5).

Figure 6.4 Collapse of Twin Towers of The World Trade Center; as an act of terrorism; resulted in severe air pollution

(a) (b)

Figure 6.5 (a) High-rise residential apartments and (b) skyscraper offices

6.2.2 Industrial Effluents and Gases

On a global scale, all government agencies are paying attention towards prevention of atmospheric discharge of industrial effluents and gases as it has severe impacts on the environment.

In India, according to Air (Prevention and Control of Pollution) Act, 1981, air pollution can be defined as the presence of any solid, liquid, or gaseous substance in atmosphere in such concentrations that can be injurious to human beings or living creatures or plants or property or environment. Violation of this act, under Section 37 can lead to punishment with imprisonment for a term, which may extend to 3 months or a fine of ₹10,000 or both. Also under Section 46, no civil court will have jurisdiction to entertain any suit or proceeding in respect of any matter with an appellate authority constituted under this act is empowered by, or under this act to determine, and no injunction shall be granted by any other court or authority in respect of any action taken or to be taken in pursuance of any power conferred by or under this act.

The above act puts tremendous pressure on pollution-generating units/industrial activities. Some of the highlights of this act are briefed here:

(a) In India, air pollution control measures are primarily aimed at regulating pollution mainly caused by public and private vehicles, government-owned power stations, and small-scale industries.

(b) Under Section 5(2), the Chairman and Member-Secretary should be the persons having special knowledge or practical experience in respect of matters relating to environmental protection. Also three members of the state board are nominated to represent agriculture, fishery, trade, and industry.

(c) In case of discrepancy in analysis of sample collected by authorities under Sections 28 and 29, a third party should be appointed by mutual agreement (by the industry concerned and state authorities) to resolve the dispute.

6.2.3 Monitoring Objectives and Approach

Before considering any air quality monitoring programme, it is essential to establish objectives carefully, for example, it is generally not possible to use a network designed to monitor long-term trends of air pollution levels to investigate a specific complaint (Figure 6.6).

Principal objectives of the development of air quality monitoring programmes can be classified as under:

(a) Trend analysis

(b) Evaluate control strategies

(c) Activate episode controls

(d) Evaluate risk to human health

(e) Data for land-use planning

(f) Validate dispersion models

(g) Investigate complaints

(h) Initial assessment

Instruments for simultaneous measurements of meteorological data are given in Table 6.1.

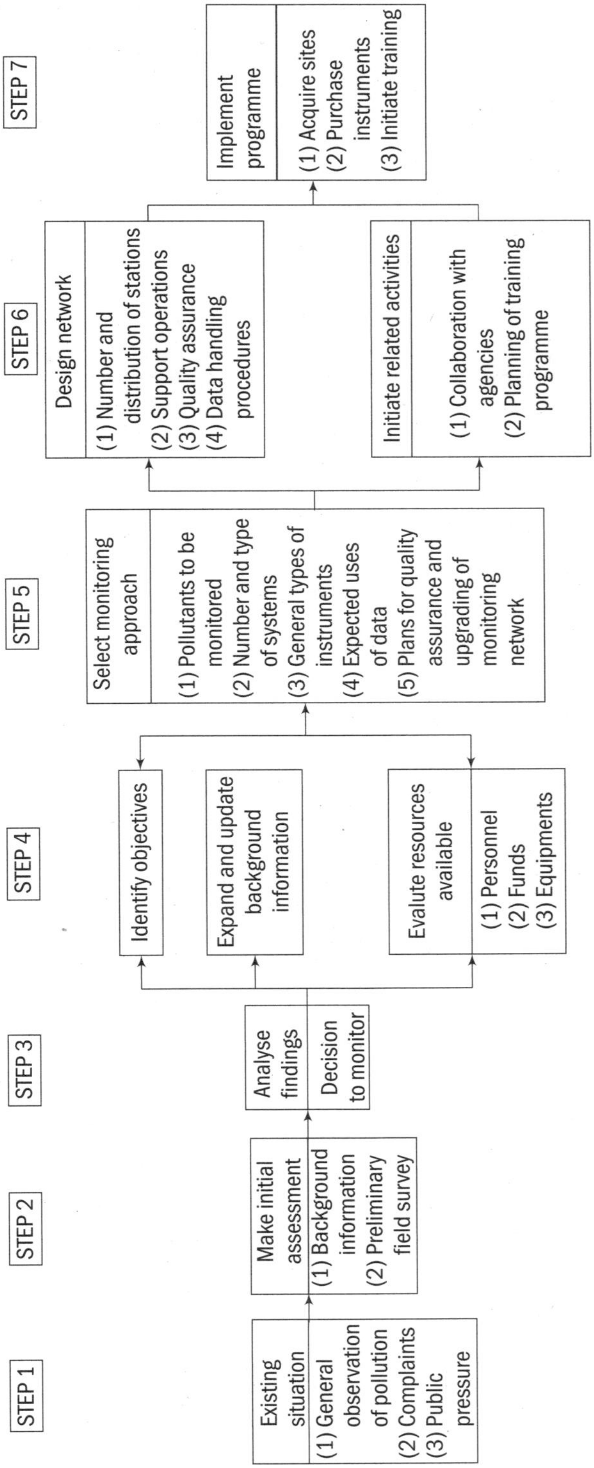

Figure 6.6 Steps involved in the development of an air quality monitoring programme

Table 6.1 Instruments for simultaneous measurements of meteorological data

Monitoring objective	Instruments
Initial assessment	Recording wind instruments
Trend analysis, evaluate risk to human health, evaluate risk of environmental damage	Recording wind instruments, thermo-hygrographs, and precipitation gauges
Evaluate control strategies	Recording wind instruments, thermo-hygrographs, precipitation gauges, vertical temperature profile observations (often based on radiowaves/ sound measurements from local meteorological services)
Validate dispersion models	Observations from a number of meteorological stations in the area. On one side should be a tower station for taking observations related to temperature, wind observations are made at different heights. Atmospheric stability conditions can be estimated by using ground-based parameters such as radiation intensity, cloudiness, and wind observations (consultation with a meteorological expert is recommended)
Activate episode control	Both wind and temperature observations can be made (collaboration with meteorological services essential)
Investigate complaints	Portable wind instruments, thermometers, and precipitation gauges
Database for land-use planning	Portable wind stations, useful if the topography of the area is complicated

6.2.4 Design of Air Quality Monitoring Programmes for Urban and Industrial Areas

The preliminary information, useful in designing air quality monitoring programmes, can be categorized into the following:

(a) Sources and emissions
(b) Health and population
(c) Meteorological data
(d) Topographical information
(e) Previous air quality data
(f) Land-use planning considerations

6.2.5 Selection of Air Pollutants

Major air pollutants include sulphur dioxide, oxides of nitrogen, carbon monoxide, fluorine, particulate matter, and hydrocarbons. Table 6.2 lists some of these major air pollutants with their sources. Air quality monitoring programmes are essentially developed after carefully studying the environmental effects of air pollutants.

6.2.6 Quality Assurance

Generally, air quality data gathered by various institutes within a country is accepted as a gospel truth. However, at times, data can be faulty and misleading. In order to safeguard the interest of industry and trade, co-rational programme on quality assurance

Table 6.2 Principal sources of pollutants in urban areas

Source	Fuel	Main pollutants
Domestic heating or cooking	Wood peat, dung, coal, light oil	Suspended particulate matter, carbon monoxide, oxides of nitrogen, sulphur dioxide
Industrial boilers, power plants	Coal, heavy oil	Sulphur dioxide, oxides of nitrogen, suspended particulate matter
Manufacturing industrial processes	Emissions	Specific pollutants related to nature of process, for example, sulphur dioxide, mercaptans from pulp mills, heavy metals from smelters, fluorides from aluminium smelters, iron oxide from steel works, dust from cement works
Transportation	Gasoline, diesel	Carbon monoxide, oxides of nitrogen, lead, hydrocarbons, oxidants*, suspended particulate matter, sulphur dioxide

*These are secondary pollutants, formed by photochemical reactions of primary pollutants.

Table 6.3 Common parameters for quality assurance

S. No.	Parameter
1	Sampling site selection
2	Sampling probe and manifold design
3	Data handling and reporting
4	Step-by-step description of sampling, analysis, and calibration
5	Methodology and operational matrix
6	Operational and auditing procedures

is essential. Objective of such a programme is to improve the quality of the data. It is also concerned with establishing and assessing data accuracy among organizations contributing to database. Regardless of the scope and magnitude of ambient air measurement systems, there are a number of common considerations; given in Table 6.3; pertinent to data collection methodology. A quality assurance programme should include a system for documenting and updating procedures. The essential elements and responsibilities of a quality assurance programme; in the form of quality assurance wheel; are shown in Figure 6.7.

6.2.7 Control Techniques for Gaseous Emissions

Gaseous pollutants can be grouped in two categories: (i) inorganic gases and (ii) organic vapours. Sulphur dioxide, hydrogen sulphide, nitrogen oxides, hydrogen chloride, silicon tetrafluoride, hydrogen fluoride, carbon monoxide, and ammonia are included in organic gases while hydrocarbons, mercaptans, alcohols, ketones, and esters belong to organic vapours.

Before considering control, global background concentrations and threshold value limits (as given in Table 6.4) are examined first. Standards for air quality in India are set by Central Pollution Control Board (CPCB). Indian Standards Institution (ISI), has

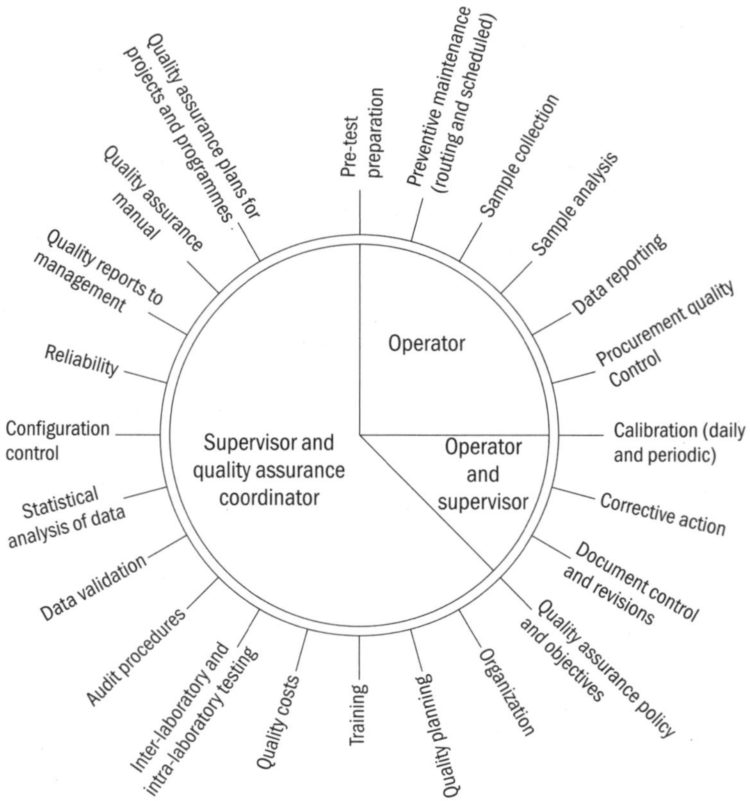

Figure 6.7 Quality assurance elements and responsibilities

Table 6.4 Global background concentration and threshold limit value of major pollutants (in ppm)

Pollutant	Background	Threshold limit value
Sulphur dioxide	0.2×10^{-3}	5
Hydrogen sulphide	0.2×10^{-3}	10
Nitrogen oxides	1×10^{-3}	5
Oxidants or ozone	3×10^{-2}	0.1
Methane	1.5	–
Carbon monoxide	0.1	–
Ammonia	–	50
Benzene	–	25
Formaldehyde	–	5
Hydrogen cyanide	–	10
Hydrogen flouride	–	3

ppm - parts per million

formulated a number of air quality standards (Table 6.5). The ambient air quality standards, recommended by United States Environmental Protection Agency (USEPA), adopted by ISI, are given in Table 6.6.

Table 6.5 A. Indian specification air quality standards

Description	IS No.	Year
Limits for gaseous emissions from sulphuric acid and phosphate fertilizer industries are set	8635	1977
Set limits for gaseous emission from nitric acid and nitrogenous fertilizer industries	9005	1978
Limits for gaseous emission from petroleum refineries are outlined	8636	1977

B. DRAET air quality standards

Description	IS No.	Year
Limits for gaseous pollutants from chlor-alkali industries are set	DOC : CDC 53(8353)	1980
Set limits for gaseous pollutants from thermal processes, for example, phosphoric acid industries	DOC : CDC 53(8347)	1980
Limits for emission of particulate matter from new cement industries are defined	DOC : CDC 53(8402)P	1979
Limits for gaseous emission from man-made fibre plants-part II synthetic fibre are described	DOC : CDC 53(8218)	1978
Limits for gaseous emissions from man-made fibre plants-part I cellulosic fibre are set	DOC : CDC 53(6737)F	1978
Code of practice for control of air pollution in iron and steel industries	DOC : CDC 53(8495)P	1979

Table 6.6 Ambient air quality standards, recommended by USEPA

Pollutant	Primary concentration (μg/m^3; ppm)	Secondary concentration (μg/m^3; ppm)	Type of measurement
SO_x	80; 0.03 365; 0.14	60; 0.02 260; 0.1 1300; 0.5	Annual arithmetic mean, maximum 24 h concentration, maximum 3 h concentration
Particulate	75; 260	60; 150	Annual geometric mean, maximum 24 h concentration
CO	10,000; 9 40,000; 35	Same as primary	Maximum 8 h concentration Maximum 1 h concentration
Photochemical oxidants	160; 0.08	Same as primary	Maximum 1 h concentration
NO_x	100; 0.05	Same as primary	Annual arithmetic mean
HC	160; 0.24	Same as primary	Maximum 3 h concentration (6–9 AM)

ppm - parts per million

Note Primary standards are concerned with public health while secondary standards are intended to protect public welfare.

6.2.8 Enforcement

The Air (Prevention and Control of Pollution) Act, 1981 clearly outlines under Section 18 that air pollution control, unlike water, is a subject of central government. To quote, it states that "every state board shall be bound by such directions in writing as the central board or the state government may give to it provided that where a direction given by the state government is inconsistent with the direction given by the central board, the matter shall be referred to the central government for its decision."

According to Section 23 of the Act, authorities should be informed whenever concentration of an air pollutant is found to exceed the acceptable levels. So far, no procedures have been laid down for enforcement. State water pollution control boards are responsible for implementing rules laid down in Air Pollution Act, 1981 in respective states.

Above-mentioned information can considerably help authorities in preparing recommendations for achieving air pollution control and abatement.

Some of the programmes, acts, and publications aimed at air pollution control, include the following:

(a) Air Monitoring Programme by WHO, Geneva (1977)
(b) Quality Assurance Handbook by USEPA (1976)
(c) Air Pollution Control and Control of Gaseous Emissions by USEPA
(d) Air (Prevention and Control of Pollution) Act, 1981 (India)

6.3 MAJOR CONTRIBUTORS TO AIR POLLUTION

Industrial exhaust gases, burning land wastes, and release of industrial toxic air/gases considerably contribute to air pollution. Figure 6.8 shows global-level overview of air pollution. In addition, burning of forest lands for agriculture and other purposes massively contributes to air pollution, with a substantial wildlife loss (Figure 6.9). Some of the other causes of air pollution are discussed in the ensuing section.

6.3.1 Festive Fireworks: A major source of air pollution

There are many factors responsible for increasing India's air pollution problem, burning of firecrackers is one of them. During festivals, firecrackers are burned on a large scale, this pollutes the air substantially. Every year during Diwali, firecrackers on a large scale are burned, which leads to the formation of the thick toxic smog over the affected area. This often results in forced closure of schools, power stations, and construction sites. In an attempt to fight air pollution, in 2017, Supreme Court of India banned sale of firecrackers in Delhi-NCR.

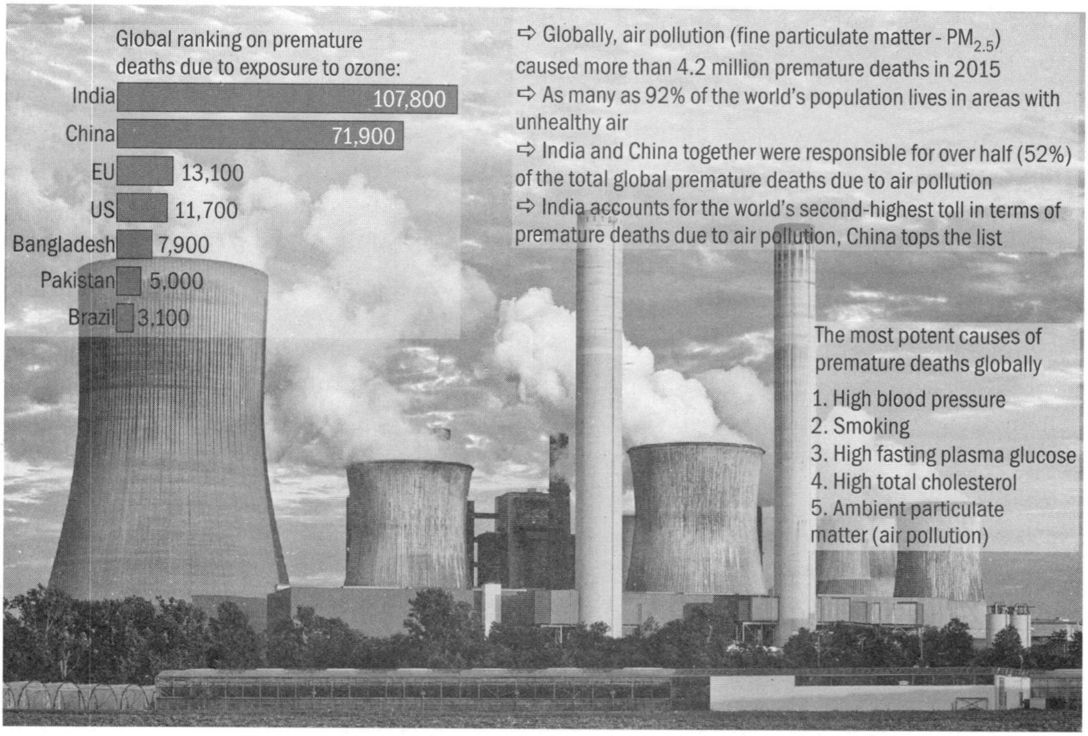

Figure 6.8 Global-level overview of air pollution

Intensity of the problem of environmental pollution can be understood from the fact that media extensively cover pollution in their areas of journalism and attempt towards spreading public awareness about the environmental conservation. Some of the notable headlines about air pollution levels in India from our national dailies are listed below:

(1) Even Inside Homes Delhites are Not Safe from Air Pollution
(2) Indoor Air Pollution Annually Kills 1.3 Million Indians
(3) Heat and Dust Raise Delhi's Air Toxins to Critical Levels

Figure 6.9 Burning of forest land for agricultural purposes results in severe air pollution and wildlife loss

(4) Feeling Breathless in 'Gas Chamber' In Delhi

(5) Bad Air Can Cripple Heart, Cause Cancer

(6) WHO: Air Pollution is the World's Top Environmental Risk

Air quality map (AQM) of Delhi suggests that increase in air pollution levels is also attributed to strong air/winds. According to CPCB, "air flowing to Delhi from the northwest is to blamed for high pollution level in Delhi during the month of April which carried particulate matter (PM) of 2.5 microgram (μg) size concentration to 1.23 $\mu g/m^3$ of air."

6.3.2 Air Pollution due to Aviation

Aviation industry, all over the world, contributes heavily to air pollution through emissions released from exhaust gases of aircrafts and helicopters (Figure 6.10). The exhaust gases contain CO_2, water vapours, hydrocarbons, carbon monoxide, nitrogen oxides, sulphur oxides, lead, and black carbon, thus, making atmosphere/ climate heavily polluted. Lowering CO_2 emissions is the first preventive step towards climate change.

6.3.3 Pollution due to Paints and Solvents

Worldwide, paint and solvent industry is growing at a fast pace. Paints are widely used for the purpose of providing protection or decoration or both. Modern paints are tailor-made or exterior coatings which are ready mixed. Paints consist of vegetable oil and a solvent dry. On drying, the solvent part vaporizes in the atmosphere, thus contributing to air pollution in form of volatile organic compounds (VOCs). Solvents such as mineral spirits or turpentine are added in various amounts to decrease the viscosity of the paints. Vapours of VOCs released into the atmosphere also contribute significantly to environmental pollution and thus need be tackled properly.

Figure 6.10 Aviation activities cause air pollution

6.3.4 Smog and Haze Hazards

In 2016, Delhi-NCR area was engulfed by a thick smog for several days. Readings of air quality index (AQI) during this period ranged in the severe 400–480. The status of air quality of Delhi and other metro cities is shown in Figure 6.11. In death toll due to air pollution, India is second after China. Together these two nations account for more than world's 50% total deaths occurring due to pollution. Smog and haze formation in many cities such as Hong Kong was so severe that it prompted world scientists to innovate methods to fight the problem of smog. In this regard, method of utilizing a new silver catalyst for reducing ill effects of smog came out promising. This catalyst purifies air by decomposing toxic gases into harmless components (The Times of India 2017).

According to the *State of Global Air 2017* report, released in Boston on 14 February 2017, as many as, 254,000 deaths occurred in 2015 on account of exposure to toxic gases such as ozone which resulted in chronic lung diseases. India accounts for the highest number of premature deaths in the world due to ozone pollution, its death toll is 13 times higher than Bangladesh's and 21 times higher than Pakistan's. Comparison of concentration of

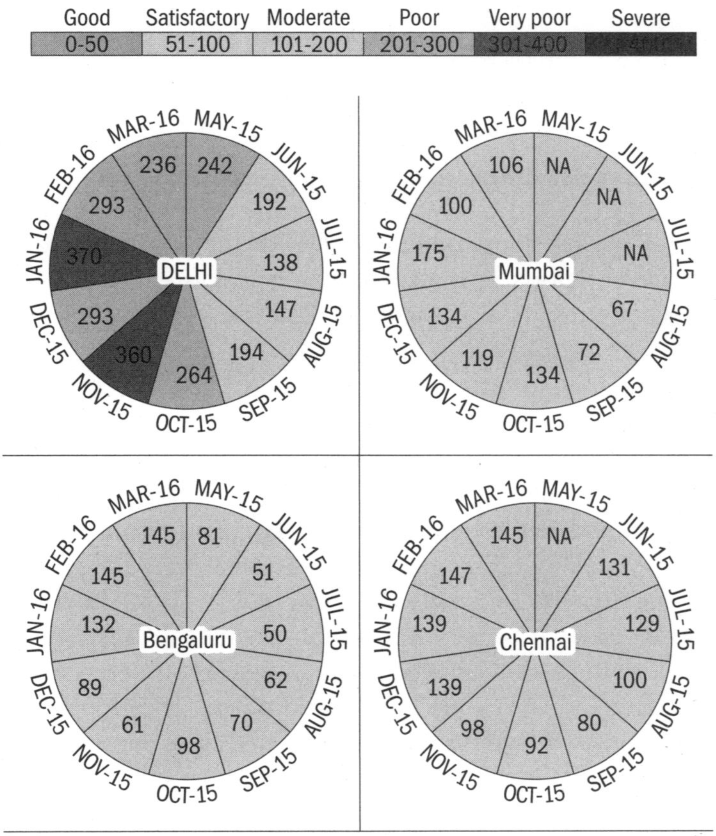

Figure 6.11 Delhi's air pollution is highest among metro cities

Figure 6.12 Delhi's air pollution is highest among metros cities (2011–15)

particulates in air of the ten most polluted cities of the world including Delhi is shown in Figure 6.12. However, efforts of the Government of India to reduce pollution levels for achieving better AQI figures had pushed Delhi down from the top of the world's most polluted cities' list (Figures 6.13 and 6.14, Box 6.1).

As per a CPCB's report, in 2016, Delhi, Varanasi, Lucknow, and Kanpur were among the most polluted cities of India whereas the coastal and southern areas were far better, in terms of air quality. But there is a huge disparity in the number of monitoring stations each city had. Delhi had nine monitoring stations, Mumbai had just one, and Bengaluru had five. Report says that Delhi had two months of 'very poor' and five months of 'poor' and five months of 'poor' air. Mumbai had no months in the 'poor' or 'very poor' categories, nor did Bengaluru or Chennai. Experts said meteorological factors play a crucial role in determining air quality. Coastal cities have better dispersal of pollutants. Cities such as Delhi, Kanpur, Varanasi, and others in the Indo-Gangetic Plain are worse off because of higher number of emission sources such as industries, brick kilns, and others. The report also indicates that the monitoring was very weak in certain parts with many cities showing 'no data' for several days. For instance, AQI of Gurgaon was under observation only for a day in the month of January. This indicates that the report failed to provide a precise picture of pollution levels of these cities.

City	Country	$PM_{2.5}$ (mg/m^3)	City	Country	$PM_{2.5}$ (mg/m^3)
Zabol	Iran	216.7	Baoding	China	126.0
Gwalior	India	176.1	Delhi	India	122.1
Allahabad	India	169.7	Ludhiana	India	121.9
Riyadh	Saudi Arabia	155.5	Dammam	Saudi Arabia	120.9
Al Jubail	Saudi Arabia	151.7	Shijiazhuang	China	120.6
Patna	India	148.9	Khanna	India	113.8
Raipur	India	143.7	Firozabad	India	113.3
Bamenda	Cameroon	132.0	Lucknow	India	112.9
Xingtai	Chaina	128.0	Handan	China	111.8
			Peshawar	Pakistan	111.0

Figure 6.13 Improved AQI figures of Delhi
Source WHO

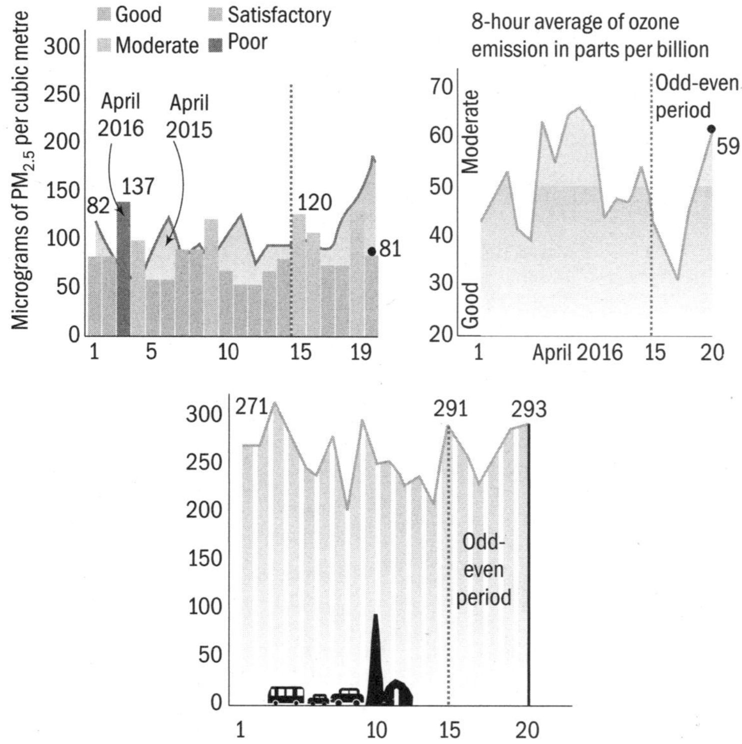

Figure 6.14 Government initiatives such as even–odd scheme have played a significant role in improving Delhi's air quality

Box 6.1 Delhi no longer world's most polluted city

According to WHO Urban Air Quality Database, released in 2016, Delhi which was once the world's most polluted city, now ranks 11th among 3000 cities in 103 countries on $PM_{2.5}$ levels and 25th in terms of PM_{10} levels.

As per WHO's 2014 database, Delhi was the most polluted city in the world, with a mean annual $PM_{2.5}$ level of 153 mg/m^3. That number came down to 122 in 2016. In the latest database, Zabol in Iran is the most polluted city in the world. Gwalior and Allahabad are at second and third positions, respectively. According to the analysis, $PM_{2.5}$ was associated with more serious health impacts than PM_{10}. Patna and Raipur ranked sixth and seventh, respectively. In India, because of temperature inversion, air pollution during winters is much higher in comparison to the air pollution in summers. India's air pollution trend can be visualized from Figure 6.15.

6.4 TECHNOLOGIES FOR ABATEMENT OF AIR POLLUTION

6.4.1 Chemical Technology

Scientists, engineers, and technologists (SETs) have confirmed the presence of many potentially toxic substances in ambient air that life forms including LHGBRs breathe. Modern technologies, for example, chemical technologies (CTs) and biotechnologies (BTs) are so precise that they can measure concentration of these toxic chemicals in the atmosphere, even

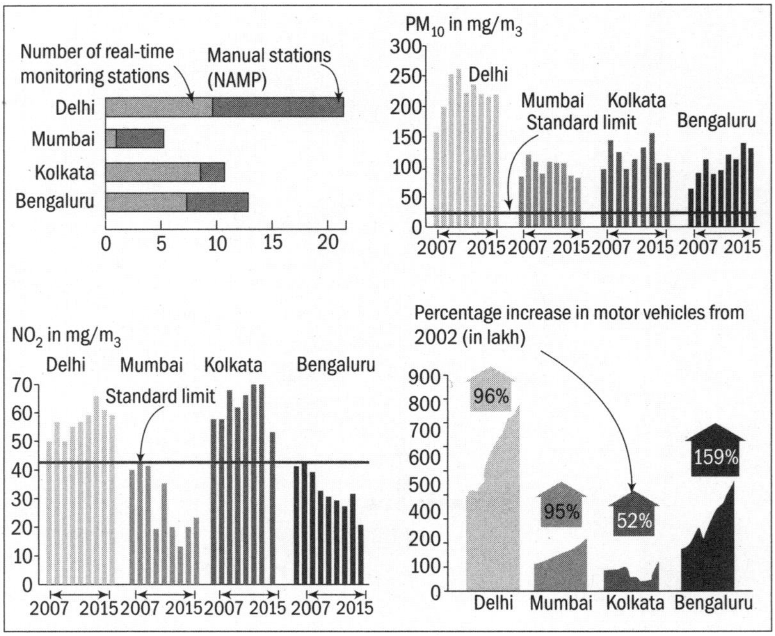

Figure 6.15 India's air pollution trend in mega cities

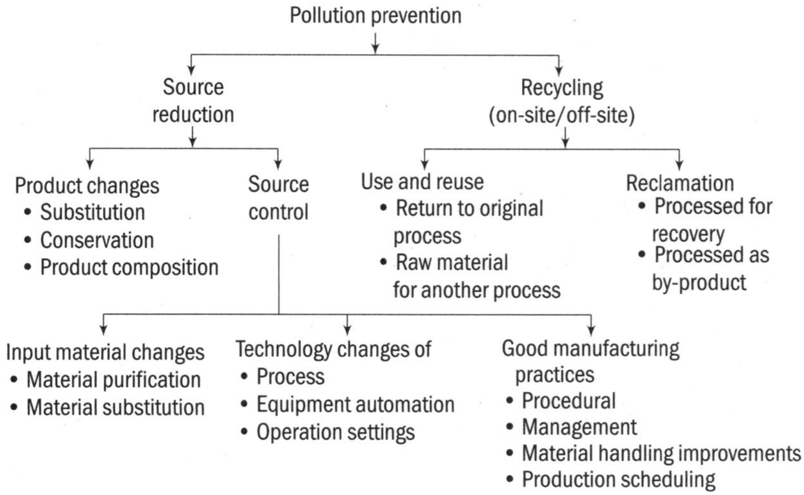

Figure 6.16 General pollution prevention techniques

when they are present in trace amounts. This was not possible in methods/technologies used earlier. These toxic pollutants of air are collectively called air toxics (ATs). ATs can enter the atmosphere from a variety of sources. They are present in motor vehicle exhaust, fuel vapours released from motor service stations, the smoke and exhaust from wood burning, heaters/ovens and the solvents used for dry cleaning, and spray painting industries. ATs are also emitted from point sources such as chemical plants, oil refineries, electroplaters, inclinators. Exposure to ATs normally occurs through inhalation but they may also contaminate water and soil and hence can be ingested with food. Therefore, pollution prevention becomes necessary. General pollution prevention techniques are primarily aimed at reducing pollutants' release from the source and recycling (Figure 6.16).

6.5 AIR POLLUTION CONTROL

Some of the strategies for achieving air pollution control include the following:

a. Cyclone collection
b. Wet scrubbing
c. Dry scrubbing
d. Electrostatic precipitation
e. Fibre-filtration
f. Adsorption
g. Plate column
h. Packed column
i. Combustion
j. Condensation

Each of these strategies has their own merits and demerits.

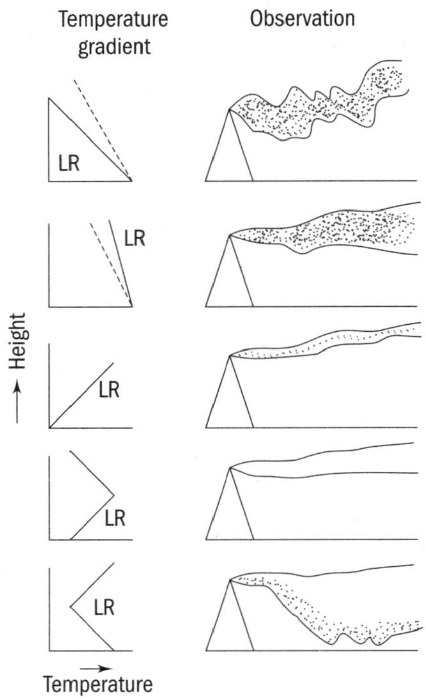

Figure 6.17 LR as a function of height of stack and temperature gradient of air dispersion.

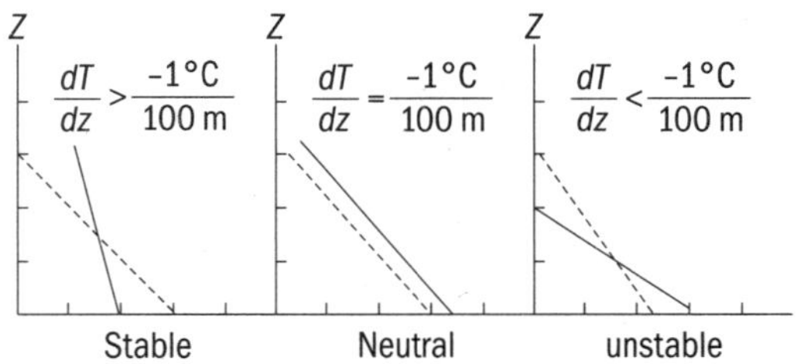

Figure 6.18 Stability as function of LR

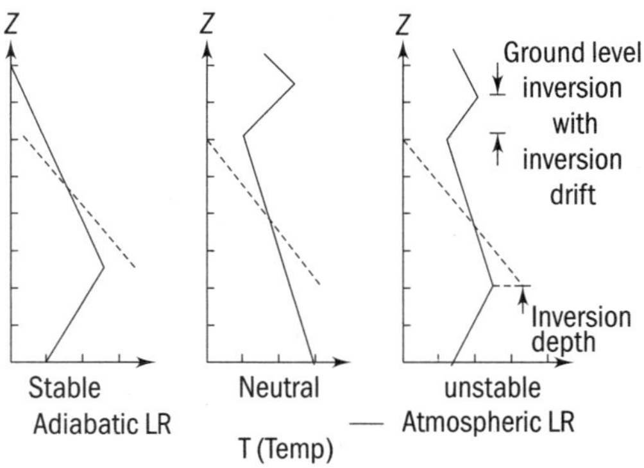

Figure 6.19 Behaviour of LR under inversion

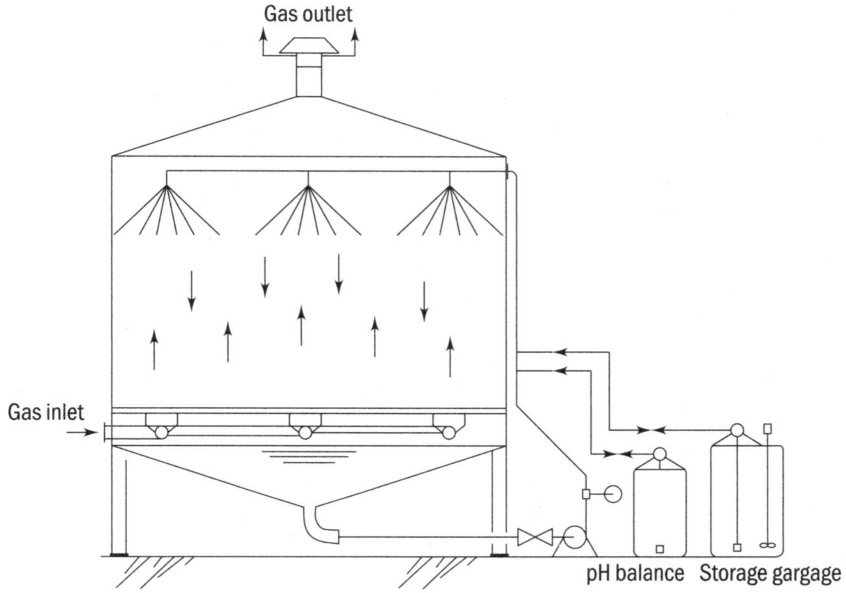

Figure 6.20 Bioscrubbers

Stack discharging is the most common method of discharging industrial waste gases into the atmosphere. Study of flow of air caused by wind is referred to as lapse rate (LR). There are some factors which influence result of the study such as obstruction by high-rise buildings, degree of turbulence, speed of diffusion/dispersion in the lower air, and variation in temperature. Dry adiabatic lapse rate (DALR) is the temperature change for a rising parcel of dry air. The DALR can be approximated by the following equations and is expressed as –1°C/100 m.

$$dt/dz = -10^{-z}\ °C/m \quad ...(6.1)$$

$$= -5.4°F/1000\ ft \quad ...(6.2)$$

6.5.1 Bio Routes of Achieving Air Pollution Control

6.5.1.1 Bioscrubbers

Bioscrubber mechanism consists of an absorption column and one or more bioreactors (Figure 6.20). Biological oxidation takes place in these bioreactors. The reaction tank is aerated and filled with a nutrient solution which is made to circulate continuously. The microbial mass mainly remains in the circulating liquor that passes through the absorption column. As the circulation rate is fast, this leaves very little scope for the formation of biofilm in the absorption column. The biofilm, if gets formed, is removed accordingly.

6.5.1.2 Biofilters

Some microorganisms can oxidize organic chemicals such as VOCs. This can be utilized to achieve biofiltration. Biofilters can be of open or closed type. Soil, compost, peat, heather, and bark are used in combination in biobeds. Uniformity and permeability of biobeds are

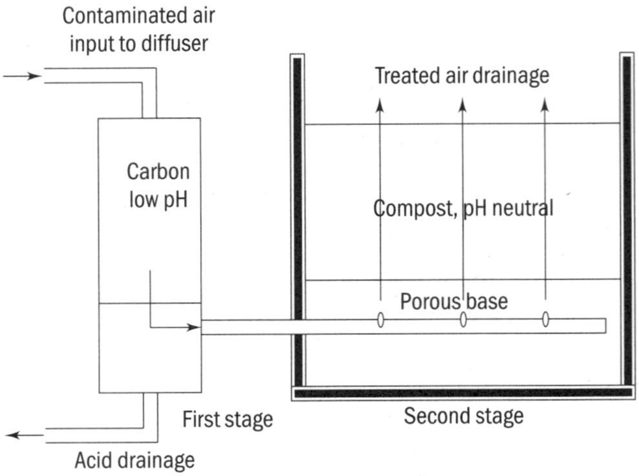

Figure 6.21 A two-stage biofilter

important factors in selection of the suitable gas treatment and also for avoiding bypassing or chocking. Biobeds require a lot of space; they are turned 2–3 times in a year. Proper drainage at the bottom is essential. Height of packing of bed is 1 m and flow rate of gas is 130 m^3/hm^2. Microorganisms used in biofilters are mesophilic. Temperature in 15–40°C range, 40%–60% moisture contents, and gas contact time of 10–30 s are maintained for optimum functioning of biofilters. A two-stage biofilter is shown in Figure 6.21.

6.5.1.3 Biotrickling filters

Biotrickling filters are composed of a packing system. Gases are passed through this packing system, water trickles down, dissolving the gas, and the pure air comes out (Figure 6.22). In trickling filters, the dissolved gas is unable to form acids.

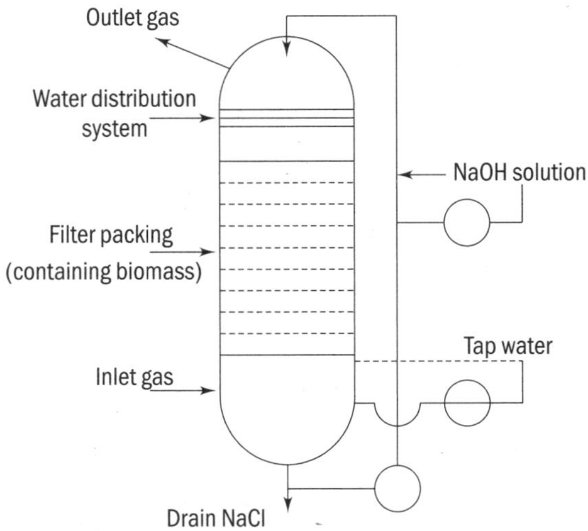

Figure 6.22 Biotrickling filter

Application of biosorption for treating polluted air has increased in the recent past. A large number of microorganisms have been observed to degrade a number of air pollutants. For example, pure culture of *Pseudomonas* sp. and genetically engineered *Escherichia coli* can efficiently degrade trichloroethylene (TCE). Biosorption can even decompose VOCs, releasing carbon dioxide, water, biomass, and inorganic salts as end products. This process of decomposing VOCs is cheaper than carrying out thermal and catalytic oxidation. *Thiobacillus ferrooxidans* can be used for treating hydrogen sulphide and sulphur dioxide.

6.6 POLLUTION-FREE AIR SUPPLY

6.6.1 Indoor/In-house

Indoor/ in-house air pollution is a leading cause of death and disability across the world. In India alone, indoor air pollution is responsible for 1.3 million premature deaths every year. Indoor air pollution is 10 times more dangerous than outdoor air pollution. In indoor air pollution, polluted air keeps on circulating within a residential unit and exposes whole family to pollutants which results in various allergies and chronic diseases/ailments such as asthma and respiratory disorders including lung cancer. Nowadays, air cleaning devices are available for improving indoor air quality. Eureka Forbes and Daikin are the leading indoor air purifier manufacturers. Likewise, indoor air quality of a car may be degraded if its air inlets are not properly clean. A car air purifier can be utilized for removing bad odour and airborne microparticles and gases. It also refreshes air by releasing negative ions.

Pollution-free air supply is a must for a safe habitation on earth. As per WHO, air pollution is world's top environmental health risk, responsible for occurrence of 1 out of 8

Table 6.7 Concentration of airborne microbes at various locations

Sampling location	Colonies/m^3 air
– Ground – (23 m above ground)	– ~10^2~ – 10 ~ 10^2
Classroom	~ 10^3 ~
Office	10^2 ~ 10^3
Factory	10^2 ~ 10^4
Bar/restaurant	10^3 ~ 10^4
Bedroom/dining room	10^2 ~ 10^3
Underground shelter	10^3 ~ 10^4
Sterile laboratory	10 ~ 10^2
Hospital (waiting room)	10^2 ~ 10^4
Hospital (lecture room/surgery room)	10^2 ~ 10^3

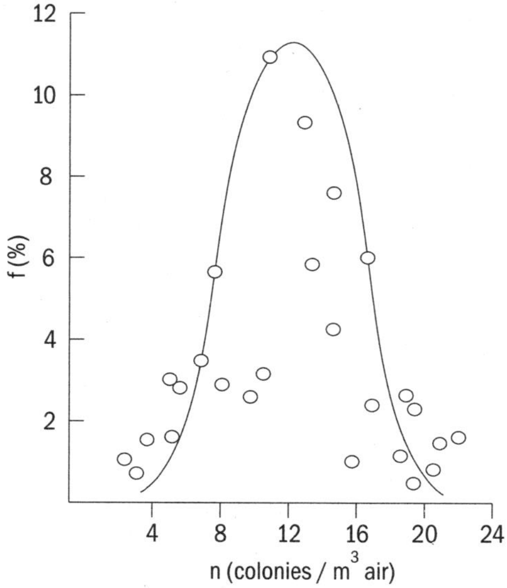

Figure 6.23 Frequency distribution of airborne microbes (operation period θ, aerosol concentration 10^6 cells/m^3/ V_s = 5 cm/s)

global deaths. Inhalation of polluted air can accelerate ageing of lungs which can lead to increased rate of heart diseases and cancers. Polluted air is composed of airborne microbes. Concentration of these microbes can vary with location (Table 6.7). The frequency distribution of airborne microbes is shown in Figure 6.23.

As already discussed, worldwide, air pollution is now recognized as a major environmental hazard, responsible for causing many diseases. Around 8.8% of premature deaths, due to air pollution, occurs in low- and middle-income countries, the percentage is much higher in the Western Pacific and South-East Asian regions.

Low-velocity effect

At low superficial air velocities

Particles are exerted by gravitational, diffusional,and electrostatic forces

Their effect is inversely proportional to air velocity.

High-velocity effect

At high superficial air velocities inertial force act on the particles. Inertial effect is such that below a certain critical air velocity collection due to inertial force is zero. This critical velocity

$$V_c = \frac{1.25\mu_a d_f}{C.\sigma_p.d_p^2}$$

In which

μ_a = Air viscosity

d_f = Fibre diameter

C = Slip flow factor

σ_a = Density of airborne organisms/particles

d_p =Particle diameter

Effectiveness

Penetration of microorganisms in aerosol into a fibrous filter is reported to be logarithmic in nature.

$$\ln\frac{N_1}{N_2} \propto \phi\ (L)$$

where N_1 is the number of organisms entering the filter, N_2 is those leaving a particular filter depth (L).

$$\text{So} = \ln\frac{N_1}{N_2} = \text{kL}$$

$$\text{Or} = \log\frac{N_1}{N_2} = \frac{k}{2.303}.L$$

where, K is filtration constant. It value depends on

- Air velocity
- Filter packing density
- Filter size
- Density of organisms to be removed

The filtration constant K is usually expressed in terms of depth of filter necessary to remove 90% of the entering organisms (that is, L_{90})

$$log\frac{N_1}{N_2} = \frac{2.303}{L_{90}}.L = kL$$

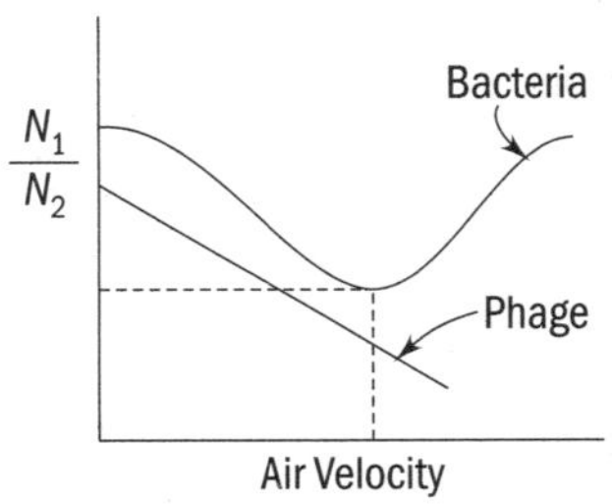

Figure 6.24 Effectiveness as a function of air velocity

Where

$$k = \frac{2.303}{L_{90}}$$

Thus $\quad log\frac{N_1}{N_2} = \frac{L}{L_{90}}$

Effectiveness as a function of air velocity can be understood with the help of the Figure 6.21.

Evaluation of air filter bed depth

$$\frac{N_1}{N_2} = \frac{CQt}{P}$$

Where N_1 = No. of organisms entering
N_2 = No. of organisms in exit
C = Concentration of m.o.s in incoming air
Q = Vol. flow rate of air
t = period of operation
p = allowable chance of penetrating the filter

So, $\frac{L}{L_{90}} = log\frac{N_1}{N_2} = log\frac{CQt}{P}$

Single-fibre efficiency

In collection of airborne particles/microbes by a single fibre the following mechanisms are presented and supported by experimentation to play major role.

(a) Inertial impaction
(b) Interception
(c) Diffusion

If η_0' is collection efficiency due to inertial impaction, then

$$\eta_0' = 0.16\left[N_R + \frac{1}{2}(1/2 + 0.8N_R)\Psi - 0.1052\ N_R\Psi^2\right]$$

where N_R^* = diametric ratio = $\frac{dp}{df}$

If η_0'' is collection efficiency by interception

$$\eta_0'' = N_R^2\ N_{Re}^{1/6} \text{ where, } N_{Re}^* = \frac{d_f\ \upsilon\ \sigma_a}{\mu_a\ *}$$

If η_0'' is collection efficiency by diffuse

$$\eta_0'' = N_R^{-2/3} N_{Re}^{-11/18} \text{ where, } N_{Re} = \frac{\rho_a}{D_{BM}}$$

These are applicable provided airflow around cylindrical fibre is laminar.

Thus, $\eta_0 = \eta'_0 + \eta_0'' + \eta_0''$

The airborne submicron particles are likely to display Brownian motion and it is conceivable that they will be collected by diffusion displacement. This collection efficiency of a single fibre due to diffusion has been computed as

$$\eta_0'' = \frac{1}{2(2 - ln\, N_{Re})}\left[2(1 + B)\,ln(1 + B) - (1 + B) + \frac{1}{(1 + B)}\right]$$

Particle upstream velocity has been assumed to be equal to its downstream velocity, that is, $V_o = V$ with particle diffusivity D_{BM}.

$$B = \frac{2(2 - ln\, N_{Re}) D_{BM}}{V_{df}}$$

And $D_{BM} = \dfrac{ckt}{3\pi\mu_a d_p}$

= Diffusivity of particles, cm^2/sec

In these relations

ψ = inertial parameter

$$= \frac{c\rho_p d_p^2 v_0}{18\mu_a d_f}$$

$$D_{BM} = \frac{ckt}{3\pi\mu_a d_p}$$

General equation for computation of η_{si} is provided below.

$$\eta_{si} = 0.01518\left(\frac{C_D N_{Re}}{2}\right) N_R^{1.5} + 0.16(0.25 + 0.4\ ) - 0.0268 N_R$$

Where $\quad N_I = \dfrac{\rho_p d_p^2\, v}{\pi\mu_a d_f}$

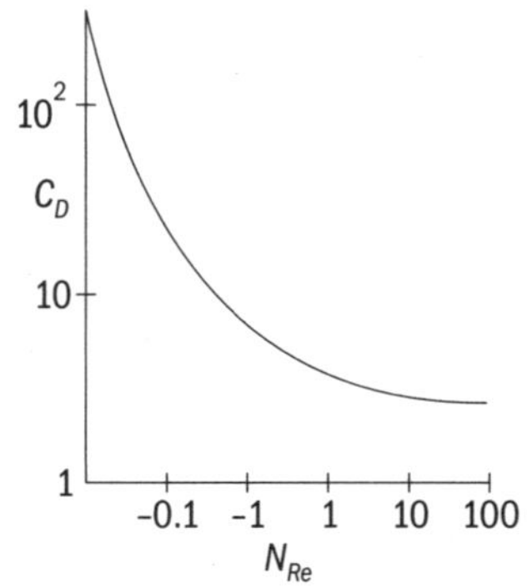

Figure 6.25 Standard C_D versus $^{N}R_e$

Using this the overall collection efficiency of the filter is

$$\eta_0 = 1 - \exp\left(\frac{4}{\pi}\eta_{si}\frac{\alpha \cdot L}{(1-\alpha)d_f}\right)$$

and pressure drop

$$\Delta p = \frac{pv^2}{2g}\cdot\frac{4L\alpha}{\pi d_f}C_D f^n(\alpha)$$

$$f^n(\alpha) = 1 + K\alpha$$

Value of K ranges between 15 and 80.

Assessment of log penetration law

$$\eta = \frac{N_1 - N_2}{N_1}$$

$$\eta_\alpha = \frac{\pi d_f(1-\alpha)}{4L\alpha}\, ln\frac{N_1}{N_2}$$

$$= \frac{\pi d_f(1-\alpha)}{4L\alpha}\, ln\left(\frac{1}{1-\eta}\right)$$

Also, $\eta_\alpha = \eta_0\,(1 + 4.5\,\alpha)$

Where η_0 is the collection efficiency of a single fibre and the value of being bound in the limit $o<\alpha<0.10$

6.6.2 Air Purification

6.6.2.1 *In hospitals*

In a hospital environment, airborne microbial load usually varies in 10^2–10^4 range. In such premises, installation of high-efficiency particle arrest (HEPA) filter air purifiers is essential for reducing spread of infections. For recombinant DNA research laboratories, the physical containment requirements are a must to prevent contamination.

6.6.2.2 *In Industries*

a. Non-biosystematic

Non-biosystematic industries, for example, steel manufacturing units release large amounts of effluent gases into the atmosphere. These effluent gases contain high concentration of CO_2 (a GHG). A rise in CO_2 concentration can lead to several environmental problems and thus survival of humans, animals, and other life forms on the earth will be threatened. Without putting a check on CO_2 emissions, climate change cannot be handled efficiently, and, therefore, keeping the global temperature increase to less than 2°C (above the pre-industrial era) will be hard to achieve. CO_2 levels in the atmosphere have already crossed an important threshold and are unlikely to come down, despite extensive environmental engineering actions. When fuels such as coal are burned in industrial set-ups or power plants, not all of the chemical energy contained in the fuel is utilized. The unused fraction of the fuel is discharged through the chimneys as unburnt gases, containing CO and polyaromatic (carcinogenic) compounds. In order to obtain clean gas/vapour from such mixed gas effluents, fibre-bed mist eliminators and spray tower systems for collecting air pollutants are utilized. Monsanto Enviro-Chem Co. is the leading company that provides engineering, consulting, and inspection services (to mitigate air pollution) to the industries engaged in fertilizer production, oil refining, and metal smelting.

b. Biosystematic

Industrial effluent gases can be treated to remove atmospheric pollutants by using process biotechnology (PB). Industrial effluent gases containing SO_2, nitrogen oxides, VOCs, and suspended particulates are responsible for causing several environmental hazards. Other two types of biological waste gas purification systems include biofilters and bio-trickling filters (discussed earlier).

6.7 CLIMATE CHANGE CONCERNS

World leaders met urgently in 2015 in Paris, not to discuss terrorism, but to discuss another major threat of the modern world, that is, climate change. It is clear that even temperature rise of less than 1°C can trigger devastating weather events. In India, due to climate change, farmers are hit again and again with unseasonal rain, hail, and extreme rain and temperatures, resultantly, crippling their livelihoods. Therefore, an agreement has to be made between governments to reduce GHG emissions to keep the world below the 2°C rise guardrail.

It is worth mentioning here that limiting temperature rise also limits the amount of carbon dioxide that can be emitted in the atmosphere. For this, amount of carbon that can be emitted to the atmosphere has to be fixed. In other words, nations have to share a global carbon budget in order to reduce carbon dioxide emissions. However, in spite of all the talk and rhetoric, the world still has not been able to delink growth from emissions. The stakes are high, so, terms of the agreement will change. Instead of requiring countries to cut their GHG emissions on the basis of their contribution, they should be asked how much GHG emission they can reduce for the sake of the environment. Climate change is a global problem, therefore, all countries need to come forward. Participation should not just be confined to the developed countries, mainly responsible for the bulk of GHG emissions. In this way, the firewall, which differentiates between developed countries and the underdeveloped and developing countries will be removed. But this is not all. The UN has accepted that the intended nationally determined contributions (INDCs)—actions submitted by all countries to reduce emissions by 2030—will take the world to at least 2.7°C rise. So, the world must agree to how it will ratchet up the commitments to cut emissions further. This is where the real problem of sharing the carbon emission lies. How world will share the carbon budget? According to the Intergovernmental Panel on Climate Change (IPCC), to keep the world below 2°C, with a 66% probability, the budget is nearly 2900 billion tonnes of carbon dioxide between 1870 and 2100. Industrialized nations have already emitted bulk of this quota (1900 billion tonnes) into the atmosphere. The world is left with only 1000 billion tonnes of carbon dioxide. The aim of the INDC is to surreptitiously appropriate the carbon budget. The USA, for instance, has already used up almost 21% of the used carbon budget. Between now and 2030, as per its lacklustre INDC, it will take up another 8%–10%. In this way, the INDC is not just a country's commitment to reduce emissions; it is its intention to occupy global carbon space. Once this space is taken, it is difficult to vacate.

The only way now to operationalize equity is to make sure that all countries take actions to reduce emissions, based on the fair share of the carbon budget. However, rich and industrialized countries are not in favour of such carbon budget as it bounds them to reduce their carbon emissions which can be a hindrance to their development process. This is why western media has started an orchestrated campaign to paint India, which is asking for fair share, as the climate villain. But this is why India must set the narrative straight. India too is concerned with the climate change and is taking the necessary actions to cut emissions. In fact, India's INDC is more ambitious than the USA's, in terms of moving towards non-fossil fuels. The USA in 2030, will have only 30% non-fossil in its energy mix, while India's commitment towards shifting to renewables is 40%. The USA even today has more per capita consumption of coal and its switch to natural gas means it is moving away from renewables. The fact is that natural gas is only marginally cleaner than dirty coal. It keeps the USA locked into fossil fuels. So, COP 21 is not just about an agreement to cut emissions, it is also about how the world will operationalize equity.

Currently the WHO has called for urgent action to avoid climate change by reducing emissions of black carbon, ozone, methane, and carbon dioxide. This is the first time that the agency has recommended steps that countries must take to reduce emissions that account for over seven million premature deaths annually.

The measures include encouraging people to eat plant-based foods and reduce methane emissions associated with some animal-sourced foods. WHO has also recommended reducing vehicular emissions, increasing investment and policy measures, and providing alternative cooking fuel to low-income households who primarily depend on wood, dung, and other solid fuels (Box 6.2).

Dr Flavia Bustreo, Assistant Director-General, WHO opined, "Actions that countries, health and environment ministries, and cities can take right now to reduce emissions, protect health and avoid illness and premature deaths, which often take the greatest toll on the most vulnerable."

The suggestions are part of a new report by the United Nations' health agency. The report calls for fast-track action to reduce global warming and highlights health risks associated with air pollution.

The report also talks about the relationship between improved diets, increase in physical activity, food security, and air pollution. For instance, the agency has encouraged governments to invest in dedicated rapid transit, such as buses and trains, and foster safe pedestrian and cycle networks. It could promote multiple benefits, including safer active travel and reduced health risks from air and noise pollution, physical inactivity, and road traffic injuries.

Box 6.2 Four ways suggested by WHO to reduce pollution and avoid the associated heath risks

1. Reducing vehicle emissions can improve air quality and reduce the disease burden attributable to outdoor air pollution.
2. Policies and investments that prioritize dedicated rapid transit, such as buses and trains, and foster safe pedestrian and cycle networks can reduce health risks posed by air and noise pollution, as well as physical inactivity.
3. Providing cleaner and more efficient stove and fuel alternatives to low-income households.
4. Encouraging consumption of plant-based foods could reduce the risk of heart diseases and some cancers, besides slowing methane emissions associated with some animal-sourced foods.

On climate change, India ratified the Paris Agreement on 2 October 2016. According to Prime Minister, Mr Narendra Modi, India will make sincere efforts towards controlling global warming. India, which accounts for 4.1%–4.5% of global GHG emissions, held up final steps of ratification to induce the USA to put pressure on China to let India into the nuclear suppliers group (NSG). NSG is a group of nuclear supplier countries that works towards nuclear proliferation by controlling the export of materials, equipment, and technology that can be used to manufacture nuclear weapons (The Hindustan Times 2016).

The International Civil Aviation Organization (ICAO) has also warned that a rise in sea level could threaten low-lying airports, while a higher number of clouds could reduce visibility, particularly, in West Asia and Latin America. Aviation is a significant cause of GHG emissions and the industry has faced criticism for its lack of action to address the problem. *ICAO Environmental Report 2016* has strongly suggested possibility of a warmer climate in the coming years. "Climate change poses a new set of risks that airports need to assess properly. The last decades have provided a glimpse of the future climate, but the main effects will be more evident three of four decades from now," it said.

"A rational response at all airports is therefore to carry out risk assessments of existing and new infrastructure in order to think ahead, reduce risks, minimize life cycle costs, and ensure the reliability and regularity of the aviation sector..." it added, "All actors in the aviation industry should carry put risk assessments, and decide if action is required." In order to control air pollution due to aviation, UN Development Policy has drawn attention of air transport regulation with respect to air transport industry as per ICAO norms.

6.8 NOISE-POLLUTED ATMOSPHERE

Owing to rapid growth of urbanization, industrialization, advancement of SETs, noise pollution along with air pollution has emerged as one of the major urban environmental problems. It has also threatened the quality of life of the people residing in urban settings where noise-like smog is giving rise to a number of health-related issues. These problems have led to the development of noise map to monitor annoying levels of noise.

The decibel levels are much higher than permissible limits in most parts of the modern cities. A noise map of Delhi could be a great aid for planners to counter the ever increasing din in the city. A noise map can be used for traffic planning and to manage and reduce noise levels in the vicinity of roads. More importantly, the maps can be used to test the effectiveness of different methods of reducing noise, whether by planning, as in re-routing traffic, or through noise-reduction techniques such as sound barriers.

In a noise map, the raucous areas resemble the hot areas of a weather map, the cooler portions depict the quieter sections of the city. In the words of Dr Nasim Akhtar, Principal Scientist, CSIR-CRRI, "It is usually done by calculations based on a computerized noise model of an area."

From 2009 till 2013, CRRI studied the noise levels at various areas of Delhi including East of Kailash, Nehru Nagar, Ashram Chowk, Lajpat Nagar, and Indraprastha Apollo Hospital. Results were surprising, for instance, silent zones such as hospitals actually experienced higher noise levels than prescribed limits. On some occasions, the magnitude of noise was equal to that in residential and commercial areas. While the noise range remained largely consistent during the day, each location showed two clear daily peaks, one in the morning rush hours and the other in the evening. The least noise was recorded during early morning hours. Even during night-time, some locations such as Moolchand, Lado Sarai, Asharam and Apollo Hospital had higher noise levels than other places. "This may be due to heavy traffic plying on the roads at night-time after the lifting of day time restrictions," observed Akhtar.

Traffic was found to be a major source of noise pollution in urban areas. As a general rule, the larger and heavier the vehicles, the noisier they are. Near bypass roads, the noise from heavy truck was found to be equivalent to that of 9.6 cars rushing by, with a single car logging 68–73 decibels. The noise generated by one bus was as much as that of 7.2 cars. As per WHO guidelines, noise levels above 70 decibels are considered severe.

In buildings close to a flyover, impact of noise pollution, even up to the seventh floor, was pronounced. During these four years, noise mapping at the DDA flats in East of Kailash was done twice. "An increase of 10 decibels in four years caused the noise to reach alarming levels on the higher floors," said Akhtar. Noise reflected by ceilings is primarily responsible for increase in sound intensity on high floors.

6.9 DUAL EFFECT OF AIR POLLUTANTS

According to a study, conducted by WHO, in 3000 cities around the world, the amount of microscopic particle pollution increased by 8%, between 2008 and 2013. The findings were particularly alarming in low- and middle-income countries, where around 98% of cities do not meet the WHO guidelines on air quality. The figure was 56% for high-income countries. More recent research shows that tiny particles of sulphur and nitrogen oxides, black carbon, and few other pollutants contribute to the death of over 3 million people each year. The smaller the particles, the more hazardous they are because they can lodge deep in the lungs or pass into the bloodstreams, contributing to asthma, cardiovascular diseases, cancers, and strokes. A recent study report published in *Nature*, a science journal, suggested that the global health burden due to air pollution is likely to double by 2050. However, interestingly, there is a silver lining accompanied with these dark pollutants, some of these toxic elements/particles help to cool the planet. This is particularly true of the sulphur compounds that are commonly emitted from cool-fired power plants. These particles reflect some light back into space, preventing it from reaching the earth's surface. On aggregate, the cooling effect of these particles is significant. According to a 2009 study, an estimated 47% of GHG-related warming is being masked by huge clouds of particles smaller than 1 μm in diameter (*MIT Technology Review* 2016).

6.10 HEALING BY NANOPARTICLES

Recently it has been established that nanoparticles can help in treating arthritis. According to Christine Pham from the Washington University School of Medicine, there may be a possibility of penetration of certain types of nanoparticles into an arthritic joint. If the introduced nanoparticles can remain up to a few hours in the inflamed joint, it may help in curing the pain. These nanoparticles are 10 times smaller than red blood cells and thus could penetrate tissues. The nanoparticles carry a peptide, derived from a protein called melittin. The peptide can bind itself to a molecule called small interfering RNA (siRNA). siRNA has anti-inflammatory effects on the diseased cells, thus, provides relief from the severe joint pain.

It is now well known that researchers could create single-cell protein (SCP) powder from renewable resources. They could also make food/feed from thin polluted air using electricity in a protein reactor. The SCP may not revolutionize food industry but it can provide a new platform for the future food. Academy of Finland is advancing in this direction. In order to combat world hunger, it is working on concept of processing CO_2 of the polluted air in a protein reactor to produce basic protein food.

Bibliography

Abeyratne, R. 2014. *Regulation of Air Transport: the slumbering sentinels.* Switzerland: Springer

Brown, L. R. 1972. *World Without Borders.* New Delhi: Affiliated East-West Press Pvt. Ltd

Cooney, C. 2018. Learn about fermentation and downstream processing, MIT Professional Short Course and MT Technology Review, July 23-27

Farand, C. 2017. Scientists make food from thin air and electricity, 2 August, p. 23. New Delhi: The Times of India

Feere Asuk Sobuj. 2015. *Feere Asuk Sobuj* 2 (3)

Green, D. W. and R. H. Perry. 2007. *Perry's Chemical Engineers Handbook,* 8th edn. New York: McGraw-Hill

Henry, T.B. 2015. *Crit. rev. tax.* 45: 643

Love, L. and E. Seokin. 1970. Air pollution and Human Health. *Sci.* Vol. 169: 3947

Ma, J., H. Huanges, C. Tian, and R. Kallenborn. 2011. *Nat. clim. chan.* 1: 255

Pinzonic, M., et al. 2015. *Environ. Res.* 142: 185

Rigel, F., *et al.* 2015. *Environ. pollut.* 11: 6

Santra, S. C., *et al.* 2007. In *Biotechnological Application,* Chapter 1. Kolkata: ENVIS Publication

Schulz, W. G. and Dodd (eds). 2007. C & EN, News of the Week. *Climate Change,* 26 March, p. 8

The Hindustan Times. 2015. 15 April. New Delhi

The Times of India. 2018. This device can purify air over 500 sq. m. p. 4, 26 Sept. New Delhi

The Times of India. 2016. 5 July. New Delhi, p. 5

UNDP. 2001. *The Stockholm Convention on Persistent Organic Pollutants.* Nairobi: UNEP

7 Wastes to Value-added Products

7.1 INTRODUCTION

Waste can be treated by biotechnology processes to form value-added products (VAPs). For this, principles of biological oxidation are utilized. Biological oxidation may occur in both mesophilic and thermophilic microorganisms. In their mode of operation, biooxidation processes can either be oxic or anoxic. In oxic thermophilic biodigestion (OTD) or biodegradation of wastes, microbial cells derive energy in the form of adenosine triphosphate (ATP) from the controlled oxidation reaction of hydrogen with oxygen to form valuable products such as biofuels (Figure 7.1).

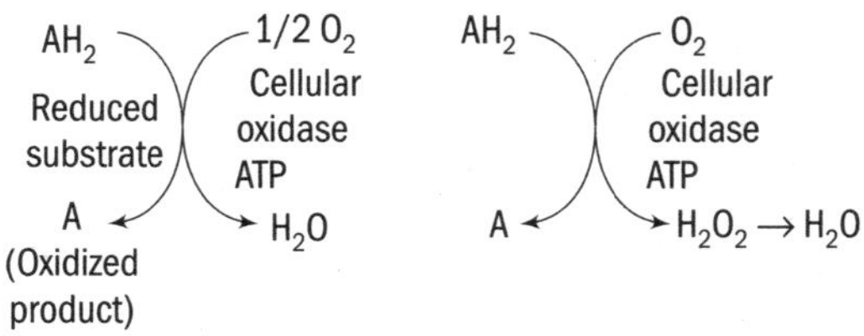

Figure 7.1 Oxic process biotechnology

7.2 BIOENERGY FROM BIOWASTES

7.2.1 Liquid Fuels

Agro-based solid feedstock and their wastes are generated in abundance throughout the world. These wastes are often discarded into solid, semi-solid, or liquid disposal systems which results into severe environmental pollution. Major advancements in biochemical engineering, process biotechnology, and chemical technology have taken place mainly in the last 5 years or so. It has also led to production of many VAPs such as liquid biofuels (for example, bioethanol), gaseous biofuels (for example, biomethane and biohydrogen) from carbohydrate wastes, leather, and petro-biochemical wastes. In addition, hydrolysed nanostructure composites can be produced successfully using these bioroutes.

7.3 BIOTECHNOLOGY, CHEMICAL TECHNOLOGICAL, AND BIOTECHNOLOGY–CHEMICAL TECHNOLOGY INTEGRATED PROCESSES

7.3.1 Liquid Biofuel Route

Anaerobic solventogenic cellulosic bioconversion, by fermentation technology (FT), under vacuum cycling, using *Clostridium thermocellum,* is an important biotechnology process. It produces bioethanol as main product and bio-hydrogen (bioH) gaseous fuel as co-product. Schematic diagram, as given in Figure 7.2, shows process of bioethanol production. Its production, however, is bound by many requirements. Two of the major requirements include:

(1) Benefits of bioethnol are realized in a sustainable way.

(2) Bioethanol production, in the long run, needs to overcome many challenges such as
   - Strategy of collection and size reduction
   - Proper storage of biomass
   - Cheaper pretreatment method(s)
   - Availability of high efficacy enzymes at low-cost process integration on industrial scale
   - Global support for development of integrated process biotechnologies

7.3.2 Liquid Biofuel other than Ethanol

7.3.2.1 Wastes to biodiesel from algae

Wastewaters are rich in nutrients and thus can promote algae growth, suitable for biodiesel production. Increase in diesel prices is the prime factor responsible for advancements made in biofuel technology. Various steps involved in biodiesel production from algae are shown in Figure 7.3.

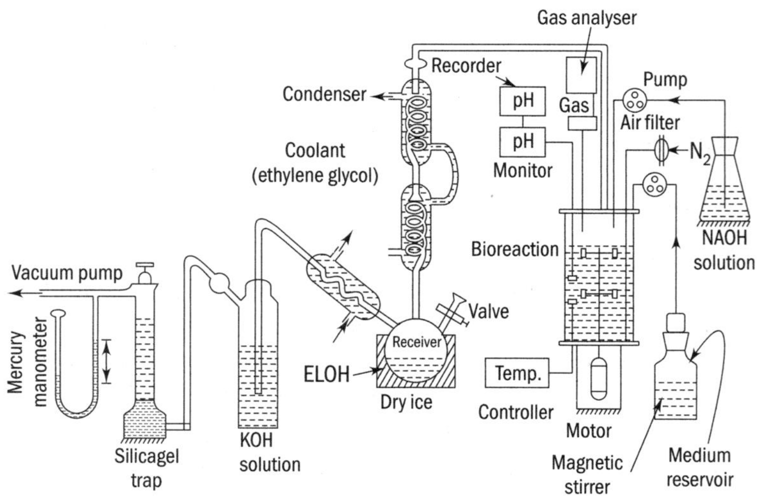

Figure 7.2 Experimental set-up of anaerobic solventogenic cellulose fermentation under vacuum cycling by *Clostridium thermocellum,* producing hydrogen gas as by-product

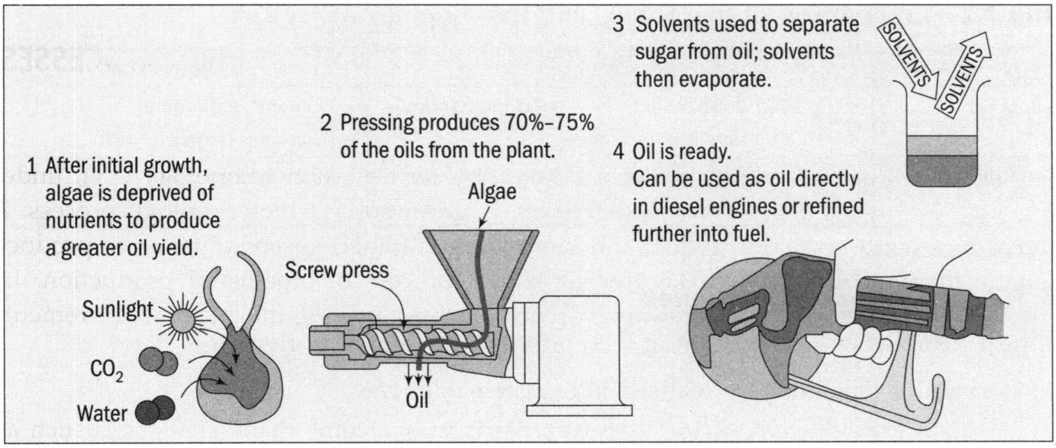

Figure 7.3 Biodiesel production from algae

7.3.3 Gaseous Biofuel Route

7.3.3.1 Biomethane

Many solid and semi-solid carbohydrate-based waste materials can be used in anaerobic digestion (AD) to generate biogas, containing biomethane as major gaseous biofuel. Biomethanation of biowastes is an age-old biotechnological process. In this process, from the sludge of the bioreactor, an isolated, pure culture of thermophilic, biomethane-forming microorganisms, is used in AD of various agro-residues/biowastes and their mixtures. It can produce a number of gaseous biofuels (Table 7.1). The AD can also be carried out using cow dung. In addition to biofuel gases, other value-added material, in the form of solid sludge, can be generated. The ratio of nitrogen, phosphorus, and potassium (NPK) in the sludge indicates its biomanural suitability. A comparison of mesophilic and thermophilic AD systems is given in Table 7.2.

Table 7.1 AD of solid/semi-solid biowastes (21 days)

1. Mesophilic

Waste		CH_4, H_2, CO_2 percentage			N:P:K of solid residue (approximately)
		CH_4	H_2	O_2	
a.	Cow dung	60	1.1	34.4	1.0.5:0.2
b.	a. + 1% ash	60.4	2.9	34.4	1.0.5:0.2
c.	a. + 0.4 sugar	57.6	2.1	38.4	1.0.5:0.2
d.	a. + 2.4% fresh legume leaves	61.6	4.0	32.0	1.0.5:0.2
e.	a. + 20% dry leaves (1.7% Kjeldahl nitrogen)	68.0	0.6	28	1.5:0.4:0.2
f.	a. + waste food	66.0	3.2	28.5	1.5:0.4:0.2

2. Thermophilic (at 55°C)

g.	a. + waste food	67.0	5.0	27.5	1.2:0.6:0.4

Table 7.2 Comparison of mesophilic and thermophilic AD systems

Type of AD	Advantages	Disadvantages
Mesophilic	– Digested solid residue is a good biomanure – Requires less energy input	– Contains pathogens – Promotes termite growth
Thermophilic	– Digested solid residue is a good biomanure – Pathogen free	– More energy input

AD - anaerobic digestion

7.3.3.2 Energy–pollution index

The heat obtained by combustion of CH_4 after bioconversion is given by

$$L_aQ = 113 \left[\frac{\lambda_{CH_4}}{M_{CH_4}}\right] \frac{\lambda_s \delta_s}{\lambda_{CH_4} \delta_{CH_4}} \quad ...(7.1)$$

$$\approx 9.14\ v_s \delta_{s\eta}\ (KJ/gs) \quad ...(7.1)$$

7.3.3.3 Relevance to pollution abatement

Considering the simple case

$$C_nH_aO_b{}^+ \left(\eta + \frac{a}{4} - \frac{b}{4}\right) O_2 \longrightarrow \eta CO_2 + \frac{a}{2} H_2O \quad ...(7.3)$$

Based on this balance

$$COD_+ = \frac{2}{3}\ v_s\ \delta_s\ g0_2/gs \quad ...(7.4)$$

$$BOD_t = 0.68\ COD_t \quad ...(7.5)$$

Municipal solid waste, waste from industrial set-ups, and effluents released from genetically modified organisms (GMO) are typically disposed of through sewer lines. These wastes and wastewaters contain pathogens and other harmful material and therefore should be treated prior to disposal. Pathogenic or inactivation procedures (approved) are best suited for treating such wastes. Treatment of such harmful/objectionable wastes with care is important from the perspective of biosafety norms. This is because recycling of biotech process wastes is purpose dependent. Wastes may be recycled for high-protein animal feed or for increasing agricultural yield by spreading sewage sludge or other lower-value organic matter on agricultural land or for carrying out AD (biomethanation). The use of sludge for increasing crop yield is a widely practised route of waste utilization. However, much attention has not been given to investigate molecular foundations of biomethanation reaction engineering in AD to integrate bioconversion engineering parameter (BEP), environmental engineering parameter (EEP), and genetic material parameter (GMP). These are important to quantitatively elucidate mixed cells systemic performance in AD. For this, objective development of concepts on the kinetic and dynamics of variation of genetic materials in AD process biotechnology is a prime necessity (Figure 7.4).

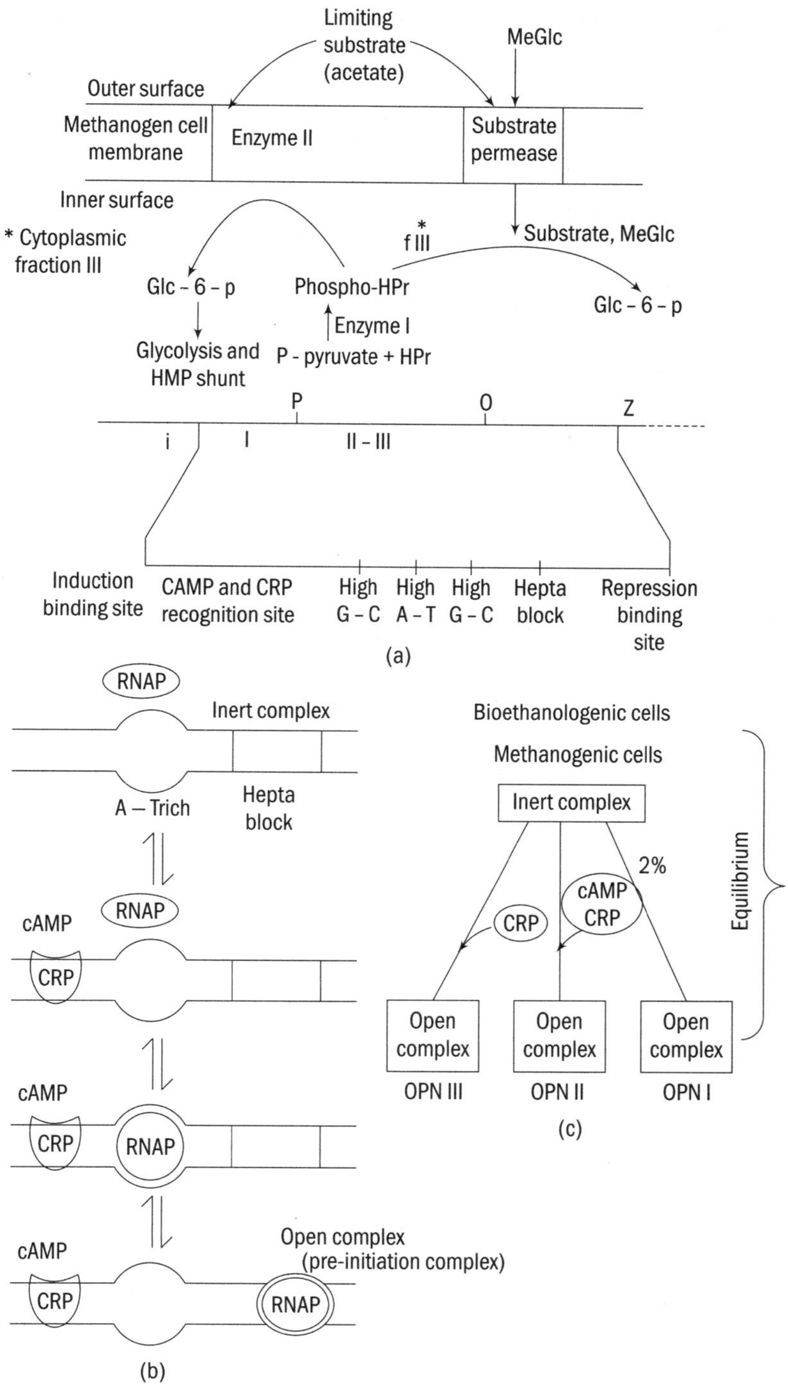

Figure 7.4 Methanogen: (a) simplified methanogenic cell; (b) DNAs involved with methanogens; and (c) conceptual representation for initiation of transcription in methanogens with operon systems

cAMP - cyclic AMP; CRP - cryptic DNA; RNAP - RNA polymerase

7.3.3.4 *Dynamic variation of genetic material in anaerobic digestion*

1. **Conceptual foundations** Dynamics of variation of ribosomal RNA (or RNA) in cells (eukaryotic and prokaryotic) is not unusual. Direct modification of the DNA (or RNA) molecules which carry the genetic material of cells may happen during processing by various mechanisms, for example, cell–cell conjugative interaction transformational modification. So, it can be stated that methane formation, by methanogens, in AD, highly depends on the kinetics and dynamics of genetic material of the flora, and the total contents of RNA/DNA in the waste to be treated. In order to treat wastes for biomethanation and pollution abatement, it is, therefore, necessary to consider total DNA level of the flora, the mechanisms of DNA transfer, and involvement and its subsequent expression.
2. **Genetic/cell biology foundations** (See Figure 7.4)
3. **Theoretical foundations**

A. Batch biomethonation

On the basis of conceptual schematics, the phenomenon of induction–repression, in genes of methanogens, in batch biomethanation system, can be explained with the help of reaction sequence schemes.

(1) Induction

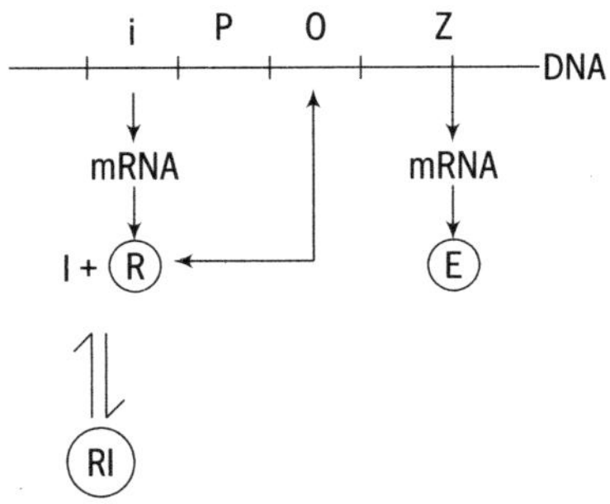

According to this scheme

$$R + nI \longrightarrow (RI)_n \qquad ...(7.6)$$

$$O + R \longrightarrow OR \qquad ...(7.7)$$

If $n = 1$,

$$Q = \frac{1 + nk_{Q1}I}{1 + k_{Q1}I + k_{2T}} \qquad ...(7.8)$$

Where

i = Initiation regulator gene of methanogen in methane-forming system

P = Promoter gene

O = Operator gene
Z = Methanogenic structural gene
R = Repressor in methanation
I = Inducer of methanation
RI = R–I complex
E = Rate-limiting enzyme in methanation
Q = Fraction of repressor-free operator
K_{Q1} and K_{Q2T} = constants

(2) Substrate-level repression reactions

RNAP + DNA INT: k_0 ...(7.9)

INT $\rightleftarrows$ OPN_I : k_I ...(7.10)

CAMP + CRP $\rightleftarrows$ CAMP.CRP : K_{AP} ...(7.11)

INT + CAMP.CRP $\rightleftarrows$ OPN_{II} : K_{II} ...(7.12)

INT + CRP $\rightleftarrows$ OPN_{III} : K_{III} ...(7.13)

$$[DNA]t = [DNA] + [INT] + \sum_{k=1}^{III} [OPEN_k] \quad ...(7.14)$$

Thus,

$$f = \frac{[OPN_I] + [OPN_{II}] + [OPN_{III}]}{[DNA]t} \quad ...(7.15)$$

Actually,

$$f = f_1 \cdot f_2 \quad ...(7.16)$$

in which

$$f_1 = [INT] + \sum_{k=1}^{III} [OPEN_k]$$

$$= 1 - \frac{[DNA]}{[DNA]t} \quad ...(7.17)$$

and

$$f_2 = \frac{\sum_{k=1}^{III} [OPEN_k]}{[INT] + \sum_{k=1}^{III} [OPEN_k]} \quad ...(7.18)$$

or

$$f_2 = \frac{k_I}{L} + \frac{k_{II}[CRP]}{L} + \frac{k_{AP}k_{III}[CAMP][CRP]}{L} \quad ...(7.19)$$

Here, $$L = 1 + K_I + K_{III}\,[CRP] + K_{AP}.K_{II}\,[CAMP]\,[CRP] \quad ...(7.20)$$

Thus, $$f_2 = \frac{k_I' + k_{II}'\,[CRP]}{1 + k_I' + k_{II}'\,[CAMP]} \quad ...(7.21)$$

where, $$k_{II}' = k_a + k_{II}.[CRP] \quad ...(7.22)$$

$$f_2 = \frac{k_I}{1 + k_I + k_{II}'\,[CAMP]} \quad ...(7.23)$$

or $$K_I = f_{2min}/1 - f_{2min} \quad ...(7.24)$$

Similarly,

$$f_{2_{max}} = \frac{k_I + k_{II}'\,[CAMP]max}{1 + k_I + k_{II}'\,[CAMP]max} \quad ...(7.25)$$

$\therefore$ $$F = \frac{f_2}{f_{2_{max}}} = \frac{k_I + k_{II}'\propto}{1 + k_I + k_{II}'\propto} \cdot \frac{1 + k_I + k_{II}'}{k_I + k_{II}'\propto} \quad ...(7.26)$$

in which

$$k_{II}'' + k_{II}'\,[CAMP]max \quad ...(7.27)$$

$$\propto = \frac{[CAMP]}{[CAMP]max} \quad ...(7.28)$$

(methanogens mRNA production rate)

$$= Const.F.Q.\,[RNA]_{total} \quad ...(7.29)$$

So, $$\propto = (1 - \propto_{GBC})\left(1 - \frac{G_B}{G_{BC}}\right) + \propto_{GBC} \text{ for } G_B \le G_{BC} \quad ...(7.30)$$

G_B = Biological growth promoter concentration

G_{BC} = Biological growth promoter critical concentration

$$= \propto^n{}_{GBC}\ \text{Exp}\{n - p\left(n\,\frac{G_B}{G_{BC}} - 1.0\right)\} \text{ for } G_B \le G_{BC} \quad ...(7.31)$$

$$[RNA]_{total} = \propto\mu + b \quad ...(7.32)$$

μ = specific methanogenic cell growth rate

b = a constant

B. Continuous biomethanation

Considering biomethanation in a continuous mixed bioreactor, the following rate equations (in transient phase) are possible:

$$\frac{dG_B}{dt} = D\,(G_{B\,in} - G_B) - \frac{C}{Y_G}\,\mu_G \quad ...(7.33)$$

$$\frac{dX_B}{dt} = D\,(X_{B\text{in}} - X_{B\text{in}}) - \frac{C}{Y_X}\,\mu_X\,W \quad ...(7.34)$$

$$\frac{dC}{dt} = -\,DC + \mu C \quad ...(7.35)$$

In the above correlations

$$\mu_G = \frac{\mu_{mB}.G_B}{K_{SG} + G_B + \dfrac{K_{SG}}{K_{IX}}.X_B} \quad ...(7.36)$$

$$\mu_X = \frac{\mu_{mX}.X_B}{K_{SX} + X_B + \dfrac{K_{SX}}{K_I G}.G_B} \quad ...(7.37)$$

$$W = X_B/X_B + G_B \quad ...(7.38)$$

$$\mu = \mu_G + \mu_X \quad ...(7.39)$$

So,

$$\frac{d(CM)}{dt} = -\,DCM + k_{tM}\,\text{FQ}\,(\propto \mu + b)\text{C} \quad ...(7.40)$$

and

$$\frac{d(CE)}{dt} = -\,\text{DCE} + k_{tM}\,\text{CM} \quad ...(7.41)$$

In these equations, the units of the factors are:

G_B, G_{Bin}, mg mL^{-1} (growth-limiting substrate)

X_B, X_{Bin}, mg mL^{-1} (or gL^{-1}) (cell concentration)

C, mg mL^{-1} (carbon source)

M, m mole (g cell)$^{-1}$ (methane)

E, mg mL^{-1} (rate-limiting enzyme)

D, h^{-1} t, h

7.4 VALUE-ADDED PRODUCTS FROM WASTES

7.4.1 Sewage Sludge to Make Concrete

Sewage treatment plants generate huge quantity of sewage sludge. In India, sewage sludge is often left untreated. However, this sludge could be dried by suitable methods. Dried sewage sludge can be recycled to form a number of VAPs. Dried sludge can also be added to cement to make concrete. Scientists and researchers at Universiti Teknologi MARA, Malayasia have developed methods to make concrete out of dried sewage sludge, however, more research is needed to yield a better quality product. In India, sewage sludge is being utilized for making bricks.

7.5 BIO-HYDRO METALLURGICAL EXTRACTION OF METALS FROM LOW-GRADE ORES

7.5.1 Methods for Removal of Heavy Metals Using Biological Treatments

Leaching is solid–liquid extraction. It is also called lixiviation. Microbial leaching or biohydrometallurgy is applied for the treatment of low-grade waste ores. From solid ores, metals are extracted by dissolution. Metabolic activities of bacteria are utilized for accelerating the process of dissolution. There are many factors that can influence the process of bioleaching, some of these factors including the following:

(1) Nature of the involved microorganisms
(2) Size and crystal structure of the ore
(3) Concentration of the metal present in the ore
(4) Thermal environment
(5) Acid environment
(6) Oxygen demand of the cell, the extent of air supply and presence of CO_2
(7) Inoculum size
(8) Exposure condition to sunlight
(9) Presence of surface active agents and shaking arrangement

Various microorganisms such as bacteria, fungi, algae/protozoa take part in biological treatment of sludge. Some of the important microorganisms include *T. thiooxidans*, *T. ferrooxidans*, *T. concretovorous*, *Ferrobacillus ferrooxidans*, and *F. thiooxidans* (Table 7.3). A laboratory set-up of microbial leaching is shown in Figure 7.5. For continuous microbial leaching with recycling of regenerated microbial solution, the process biotechnology is followed (Figure 7.6).

Table 7.3 Composition of nutrient solutions

T. thiooxidans (Waksman medium)		*F. ferrooxidans* (Lealhen's medium)	
	Per litre		Per litre
$(NH_4)_2\ SO_4$	0.2 g	$(NH_4)_2\ SO_4$	1.5 g
M_9SO_4, $7H_2O$	0.5 g	KCl	0.05 g
$CaCl_2$, $2H_2O$	0.01 g	K_2HPO_4	0.05 g
$FeSO_4$, $7H_2O$	0.01 g	$MgSO_4$, $7H_2O$	0.50 g
dH_2O	1000 mL	$Ca(NO_3)_2$	0.01 g
Colloidal sulphur 1 g/100 mL	dH_2O,$FeSO_4$,$7H_2O$	1000 mL 10% w/v solution	10 mL

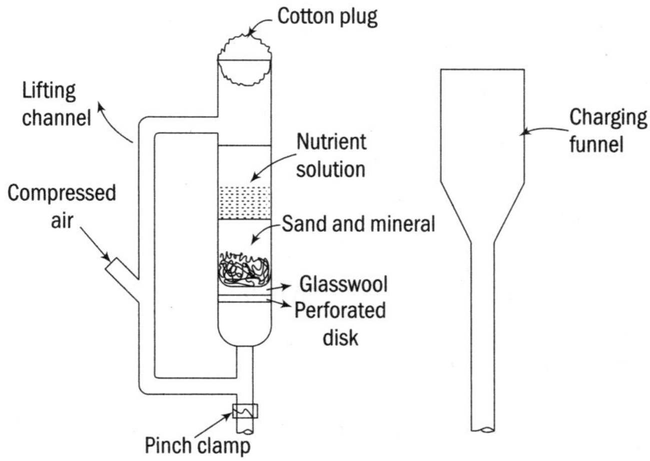

Figure 7.5 Percolator (airlift type)

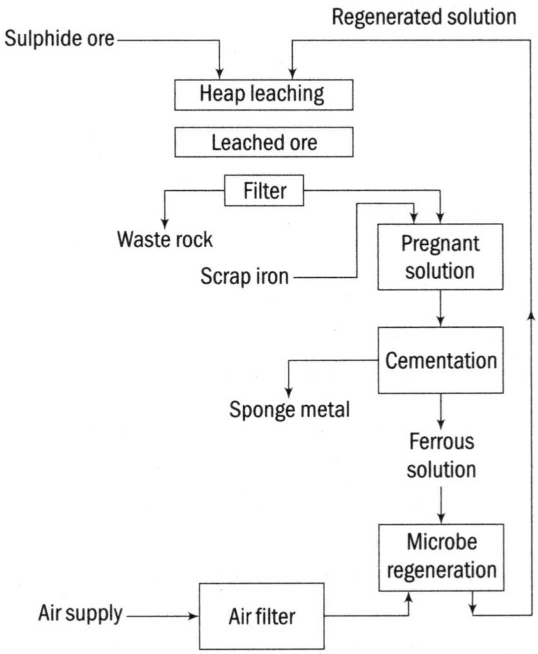

Figure 7.6 Recycling process biotechnology of microbial leaching

7.6 ENZYME AND MICROBIAL TECHNOLOGY

Bacteria mainly contribute to leaching through the following three ways:

(1) By direct oxidation of mineral sulphide
(2) By regeneration of ferric lixiviant which non-biologically oxidizes metal sulphides
(3) By depressing iron released from ore processing

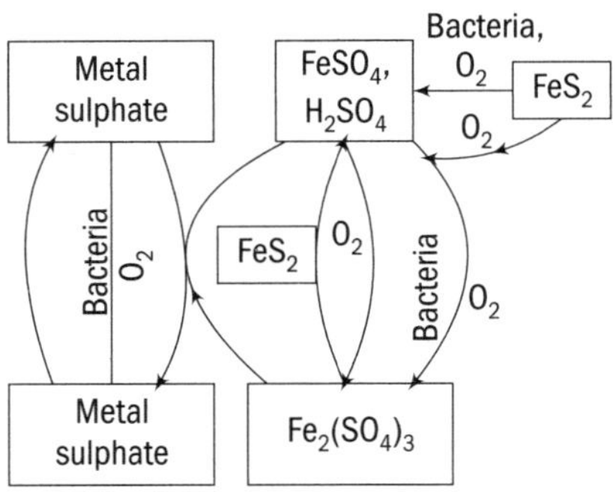

Figure 7.7 Mechanism of bacterial leaching

The mechanism of biooxidation for leaching is shown in Figure 7.7 and the extent of metals that can be extracted from bacterial leaching is given in Table 7.4.

Key agents responsible for oxidation of metals from mineral solutions are (i) iron-oxidizing bacteria, (ii) oxygen, and (iii) ferric sulphate.

In presence of moisture, the first reaction is

$$2\ FeS_2 + 7O_2 + 2H_2O \longrightarrow 2FeSO_4 + 2H_2SO_4$$

Second reaction

$$4\ FeSO_4 + O_2 + 2H_2SO_4 \xrightarrow{\textit{Thiobacillus}} 2Fe_2\ (SO_4)_3 + H_2O$$

Third reaction

$$MS + 2Fe^{3+} \longrightarrow M^{2+} + 2Fe^{2+} + S^{\circ}$$

For example

$$CuS + Fe_2\ (SO_4)_3 \longrightarrow CuSO_4 + 2FeSO_4 + 2S$$

$$UO_2 + Fe_2\ (SO_3)_3 \longrightarrow UO_2\ SO_4 + 2FeSO_4$$

$$FeS_2 + 7\ Fe_2(SO_4)_3 + 8H_2O \longrightarrow 15\ FeSO_4 + 8H_2SO_4$$

In metallurgical wastes, known as metal liquor suspended sludge (MLSS), substantial quantity of heavy metals are present. These metals can be recovered by applying suitable separation processes such as ion exchange. Table 7.5 lists some of these extractable metals present in MLSS.

Microbial sludge developed (in batch process) from synthetic wastewater containing peptone, glucose, KH_2PO_4 (50 mg L^{-1}), and NaCN is shown in Figure 7.8. From wastewater released from gold mining processes, the extraction/recovery of residual gold is carried out as per the flow sheet given in Figure 7.9. Table 7.6 lists the important impurities contained in wastewater. Table 7.7 lists impurity-wise percentage reduction.

7.6.1 Domestic Wastewater Treatment System

Owing to limited freshwater sources, awareness about reuse of domestic wastewater has increased in order to meet the ever increasing demand for drinking water, especially in urban areas. Domestic wastewater is one of the major components of wastewater (generated mainly

Table 7.5 Extractable heavy metals present in MLSS

MLSS (mgL^{-1})	Metals (mgL^{-1}) in MLSS		
	Zn	Cu	Fe
	With nutrients		
1060	75.47	10.38	4.53
740	77.57	2.43	2.97
840	107.14	9.76	2.55
	With sewage		
2840	35.21	9.37	1.07
2040	49.02	14.71	1.35
1560	51.92	22.18	1.13

MLSS - metal liquor suspended sludge

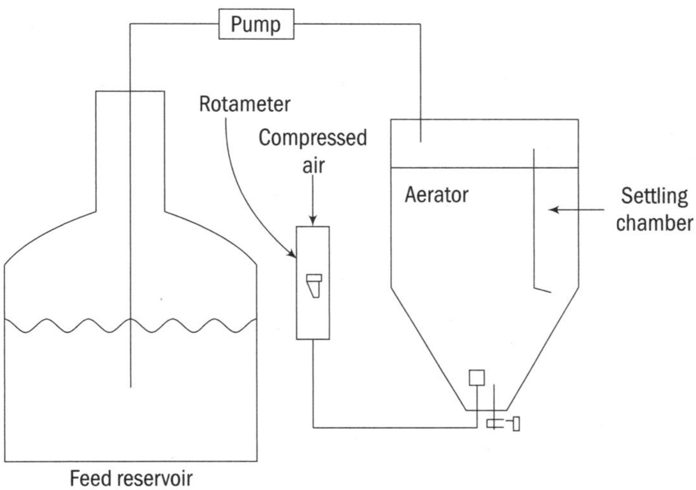

Figure 7.8 Microbial sludge development from synthetic wastewater

in cities) which is being released into the sewerage system. Domestic treatment and reuse of this wastewater has not attracted much attention in the past. Negligence on this front puts a great amount of pressure on freshwater resources. Wastage of freshwater could have been avoided if the concept of reuse of treated wastewater had been introduced earlier.

7.6.1.1 *Quantification of wastewater*

Much of the water used in a household ends up as wastewater. Wastewater produced from the bathrooms, kitchen sinks, and wash basins can be recycled and reused. Of the potable water supplied in Delhi, only less than a quarter is used for drinking and cooking purposes. The remaining 80% of water is discharged as wastewater into the sewerage system. At present, per capita water requirement for domestic use is 172 litres per capita

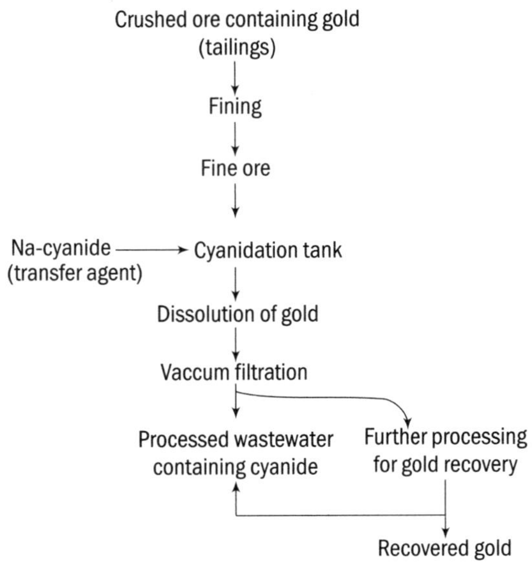

Figure 7.9 Recovery of gold from wastewater released from gold mining

Table 7.6 Impurities contained in wastewater

Contents	Concentration (mg L^{-1})
CN^-	20–35
COD	280–450
TOC	50–130
Alkalinity	210–290
Zn	6–8
Cu	7–9
Fe	0.3–0.5

COD - chemical oxygen demand; TOC - total organic carbon

Note pH of the wastewater is assumed to be 7.5.

daily (LPCD). Out of this, about 150 L is returned as wastewater. Leaving aside 50 LPCD water used for flushing toilets, the remaining 100 LPCD of wastewater generated from the kitchen, bathroom, and washbasins can be reused with a minor treatment for gardening and flushing purposes.

7.6.1.2 Benefits of wastewater re-cycling

It is possible to collect and reuse wastewater by filtering through a recycling process and use for irrigating kitchen garden, washing car, and flushing toilets. The main benefits of recycling wastewater include the following:

Table 7.7 Typical percentage reduction of impurities

Parameter	Reduction with nutrients (mg L^{-1})			Reduction with sewage (mg L^{-1})		
	Influent	Effluent	Reduction (%)	Influent	Effluent	Reduction (%)
CN'	21.5	2.5	80	21.2	4.4	79.24
COD	277	216	22.7	306	265	25.56
Zn	7.6	2.2	71.7	31	4.08	86.84
Cu	7.04	3.95	43.46	9.48	3.37	64.55
Fe	0.43	0.37	13.67	0.43	0.28	55.57

COD - chemical oxygen demand

Note pH for influents: 7.1 – 7.5, pH for effluents: 7.5 – 8.4

- Lesser quantity of potable water is used for purposes other than drinking.
- Stormwater and sewage discharges are reduced.
- Reuse of wastewater reduces pollution of waterbodies such as rivers.

7.6.1.3 Process of domestic wastewater treatment

Wastewater released from kitchen, bathroom, or washbasins, also known as domestic wastewater, contains suspended solids, oil, and soapy material. Treatment of such wastewater involves removal of impurities and stabilization of the pollutants. Overall process is rather simple and does not make use of intricate and complicated technologies. The principal steps involved in the treatment of domestic wastewater are outlined in Figure 7.10.

7.6.2 Construction of Simplified Domestic Wastewater Treatment System

This is basically a sedimentation tank in which sludge gets settled and stabilized. The chamber is divided into four compartments, with middle compartment having baffle walls

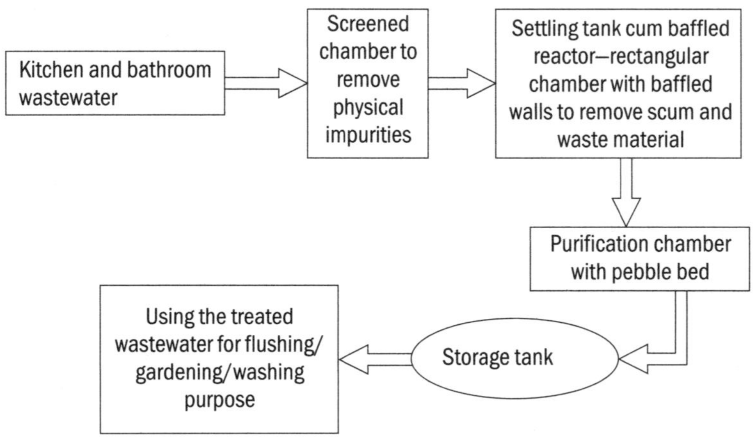

Figure 7.10 Domestic wastewater treatment

to divert the wastewater to flow top to bottom and again to top so that scum and suspended material is removed. This water is then passed through a pebble bed for purification. Finally, the treated water is stored in a tank for reuse. All the chambers are provided with one manhole for the purpose of maintenance. The dimensions of a treatment system are based on the volume of wastewater to be treated and pollution load of the wastewater. A chamber of 2 m length, 1 m width, and 1.2 m depth is sufficient to treat the wastewater produced by a small family (of four persons). Likewise, a chamber of 3 m length, 1 m width, and 1.2 m depth is enough to treat the wastewater generated by 10 to 15 persons (Figure 7.11).

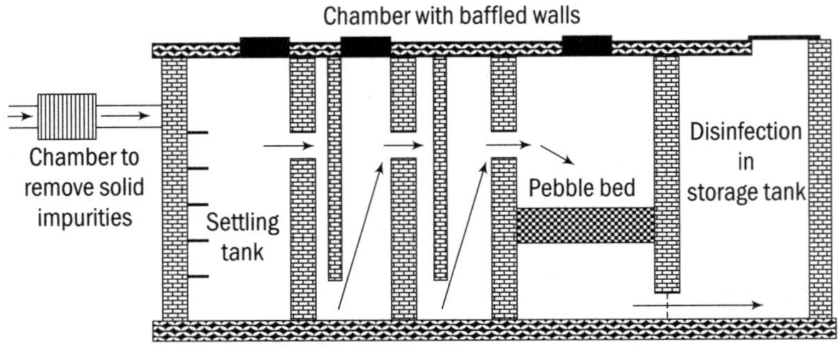

Figure 7.11 Domestic wastewater treatment system

Bibliography

Acharya, C. N. 1956. Your home needs a gas plant. *Indian Farming* 6(2):27–30

Brown, T. A. 2018. *Biochemistry*, 1st edn. Chapter 7. New Delhi: Viva Books Pvt. Ltd

Fisher, S. E. and M. Ridley. 2013. Culture, genes, and the human revolution. *Science* 340 (6135):929–30

Ghosh, S. and D. L. Klass. 1978. Two-phase anaerobic digestion. *Proc. Biochem.* 13:15–24

Kundu, S. 1998. *Microbial Conversion of Cellulose to Ethanol* (PhD thesis). Delhi: Indian Institute of Technology

Mishra, M. C. 1976. *Mass Culture of Algae* (PhD thesis). Kanpur: Indian Institute of Technology

Mukhopadhyay, S. N. 2017. *Food Engineering: process and technology.* New Delhi: Viva Books

Mukhopadhyay, S. N. 2017. Biochemical Engineering Analysis of Biomethanation: recent progress. Abstract JERET, January–March, p. 129. New Delhi

Mukhopadhyay, S. N. 2017. Lecture on Biofuels from Wastes (4 Aug.), Department of Biotechnology, IMS Engineering College, Ghaziabad

Mukhopadhyay, S. N. 2006. Technorama: a supplement to IEI News, 55 (T), March, pp. 10–20

National Academy of Sciences. 1977. *Methane Generation from Human, Animal, and Agricultural Wastes* (Report). Washington (DC): National Academy of Sciences

Pandey, A. (ed.). 1998. *Advances in Biotechnology*, pp. 503–508. New Delhi: Edu. Pub. and Distri.

Saha, N and S. N. Mukhopadhyay. 1997. *Studies on Thermophilic Conversion of Waste Material into Methane.* Proc. ICFB

Sathianathan, M. A. 1975. *Biogas Achievements and Challenges.* New Delhi: AVARD Publication

8 Waste Heat Treatments and Utilizations

8.1 INTRODUCTION

Essentialities of water in daily life, laboratories, process industries, and ecosystem are due to its attributes. Water is a unique and most abundant liquid chemical compound and resource material on earth. Its attributes are numerous. It is a universal drink for humans and animals. It is also necessary for aquatic life. Plants also need water for their growth and development. In all natural foods, water is present in a definite form. Water is essential for processes in which steam is used as a heating medium. If the steam in turn is not trapped in the process and goes away, it may serve to provide a natural application for heat pumps. Several such processes exist within chemical/biochemical process industries. A few examples include distillation, evaporation, thermal sterilization of food and fermentation medium, and to some extent drying unit operations. In the transportation industry, water provides aquatic avenues/routes. In the present industrialized age, water is a solvent in chemical and biochemical reactions and for product recovery and purification. Also, it is the heat transfer medium for both heating and cooling in which its properties play an immense role. Its property in relation to ecological engineering has also been highlighted in *Food Engineering: process and technology* (Mukhopadhyay 2017). In developing chemical technologies for better environment, the role of water is well known. Water management research considerably helps in generation of reports, based on water resources. Such reports primarily focus on water processing/treatment for purification with major emphasis on cheap physico-chemical and biological treatment processes. Thus, physical and chemical attributes of water make it a very important industrial solvent.

In recent years, there has been tremendous increase in awareness about water availability, quality, and attributes for consumption and S&T concerns. Various conventional and modern laboratories and industrial processes use several substances that are diluted with water. Processes requiring water are classified into two categories. In the first category of process water, demand is "quantity controlled". In this, the quantity of water is fixed. The second category of process water demand is "quality controlled". In processes involved with quality control, the amount of water required is a function of the quality of water provided. In food/chemistry, quantity controlled reaction is the addition of process water to a reactor charge.

However, the examples of the quality controlled chemical reaction process are the mass transfer controlled operation, medium preparation engineering in food and fermentation, and so on. For food and biotechnological process validation engineering attribute of quality of water is a key factor. If water is impure, process dynamics and quality of product become associated with side reaction and productivity goes down. In chemistry and chemical/ biochemical industry, hydrolysis reaction cannot occur without water attributes. Many redox-site interactions in chemistry are impossible without attributes of water. Likewise, water molecules near the active site of metalloproteins play a great attributory role in its function in biosystems. Chemical processing in water softening industry produces soft water in bulk for consumers. So, attributes of water with regard to chemistry, chemical/ biochemical laboratories, and related process industries play a vital role in designing better water networks and helps consumer living human genomic body/bioreactors (LHGBRs) or life forms (LFs) to remain free from ailments.

8.2 ATTRIBUTES OF WASTE HEAT WATER

8.2.1 Heat Pump Design Approaches

8.2.1.1 *General compression cycle*

A general steam-driven heat pump is shown in Figure 8.1. However, steam compression mode can be direct or indirect (Figure 8.2). For augmenting attributes of water in food and fermentation industries, the use of hot water is very common. Considerable quantities of waste heat of effluent liquid are found to be released in such research and development

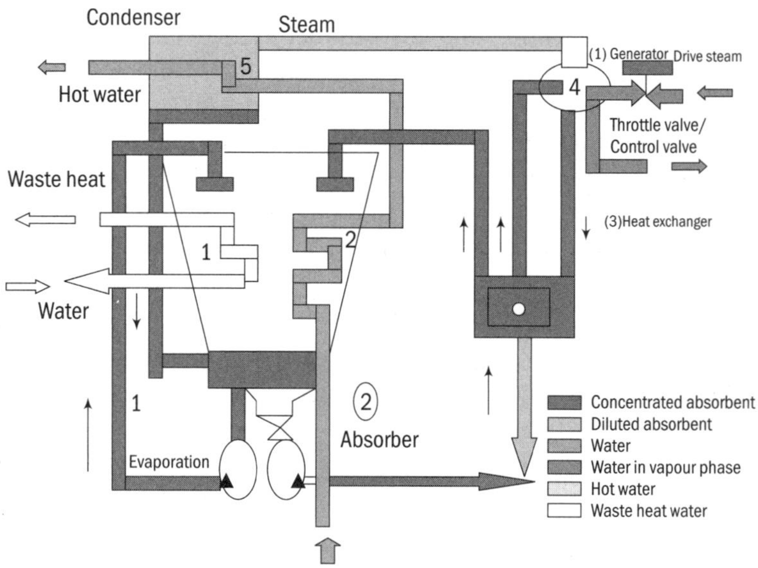

Figure 8.1 Absorption heat pump using steam as its driving medium

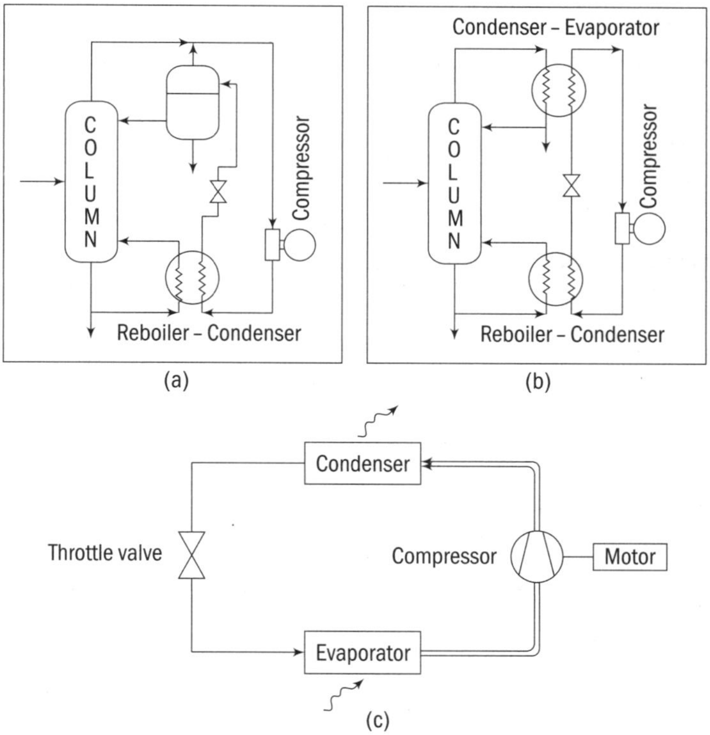

Figure 8.2 (a) Direct steam compression, (b) indirect heat pump, and (c) compress or heat pump cycle

laboratories or process industries. Generally, the medium in which this waste heat is carried is warm/hot water in quantities that greatly exceed internal site needs. This heat becomes available when processes are cooled. Here, the cooling device may be such that the temperature of the cooling water is quite low compared to the temperature that might be attained if large coolers were used. The heat, is therefore, carried by a large flow of medium/water at a relatively low temperature, which makes it difficult to use. Making use of heat pumps is one way of utilizing process water waste heat within the process industry. Some advanced countries such as Switzerland, Germany, and Sweden have utilized process water waste heat through the use of heat pumps for providing space heating and domestic hot water supplies in residential buildings and other recycling industrial premises.

8.2.1.2 Assumption in designing

Heat pump design for waste heat water recycling includes assumptions concerning mainly condenser and evaporator as well as a few other assumptions.

8.2.1.2.1 Condenser

(i) Fifteen degrees of superheat at the refrigerant inlet

(ii) Water leaving the condenser to reach the condensing temperature (as a result of assumption (i) and the sufficient heat transfer area)

(iii) Refrigerant subcooled to 10°C above the water inlet temperature

8.2.1.2.2 Evaporator

Evaporation temperature should be 5°C below the water outlet temperature.

8.2.1.2.3 Other assumptions

Sufficient suction (that is, evaporator) superheat must be supplied to meet the above assumption (i). This can be achieved using intercooling or waste engine (or motor) heat if the heat transfer from the waste stream is insufficient. Wastewater should be cooled to 20°C.

8.2.1.3 Theoretical aspect of heat pumps

Figure 8.3 shows the retrieval of heat from waste heat water in a process industry. The ratio between the final quantity of heat E_1 and the amount of energy supplied E_{01} is defined as the heat factor, φ. Thus,

$$\varphi = \frac{\frac{E_0}{T_0} \times T_1}{\frac{E_0}{T_0}(T_1 - T_0)} \qquad \text{...(8.1)}$$

So

$$E_1 = \varphi E_{01} \qquad \text{...(8.2)}$$

As thermodynamically, $E = ST$ (E is enthalpy),

$$\frac{E}{T} = S = \text{Entropy} \qquad \text{...(8.3)}$$

Therefore,

$$\phi = \frac{ST_1}{S(T_1 - T_0)} = \frac{T_1}{T_1 - T_0} \qquad \text{...(8.4)}$$

One may consider that at an absolute temperature T_0, energy E_a is available and should be used to produce heat at temperature T_1. E_a can be converted to heat or heat can be fetched from the surroundings and E_a can be used to raise its temperature to T_1. Thus, if E_1 is the heat produced, then

$$E_1 = \phi E_a = \frac{T_1}{T_1 - T_0} \times E_a \qquad \text{...(8.5)}$$

For example, if the ambient temperature in winter is 3°C and one has 2 kWh of electricity available to produce higher temperature (20°C) through the heat pump, then the heat produced will be

$$\frac{293}{293 - 276} \times 2 \approx 34 \text{ kWh}$$

8.2.2 Applications of Heat Pumps

8.2.2.1 Food and Pharmaceutical Industries

(a) **Flash steam system:** In many countries, a typical food-processing industry discharges about 10,000 tonnes/year of flash steam into the atmosphere. This steam could be utilized instead as the heating and drive energy source for an energy transformer to produce hot water. The energy transfer condenser could be cooled by a water network connected to a cooling tower.

(b) Cooling water systems

(i) **Dairies:** Heated cooling water in a closed cooling water network (for example, warm water in a milk evaporation/concentration industry) or waste heat water can be used as the heat source for producing hot water or preheating of cleaning water using a steam-driven absorption heat pump (Figure 8.3).

(ii) **Breweries:** In the brewing process, the temperature of returning hot water is very high. Heated cooling water from refrigeration condensers in a brewery, together with drainage water, can be used as the heat source for production of hot water through an absorption heat pump driven by hot water. Also, in breweries, compositional quality of water in beer production is of great concern.

(iii) **Canning:** Conditions have been found favourable for the use of heat pumps in the canning industry. Domestic hot water and space heating requirements are high in winter. Drainage water is again an interesting energy source for heat pump applications, coming not only from the process plant as a whole but also from the cooling circuits of autoclaves.

Figure 8.3 Use of heat pump in the dairy industry in Sweden

Table 8.1 Availability of discharge steam

Food industry	Quantity discharged (tonnes/year)
Milk processing/drying/concentrating	1,500–1,800
Fruit and vegetable processing	10,000
Breweries/distilleries	6,000–7,000
Antibiotic	20,000–20,500

Here, cooling device may be such that the temperature of the cooling water is quite low compared to the temperature that might be attained if large coolers were used. The heat is, therefore, carried by a large flow of medium/water at a relatively low temperature, which makes it difficult to use.

Making use of heat pumps is one approach of utilizing process water waste heat within the process industry. It can also be used for space heating and domestic hot water supplies in residential buildings and other recycling industrial premises.

Availability of discharge steam in a few food processing industries is given in Table 8.1.

Generally, the medium in which this waste heat is carried is warm/hot water. It is carried in quantities that largely exceed internal site needs. This heat becomes available when processes are cooled. The performance of a heat pump can be assessed by its pressure–enthalpy diagram (Figure 8.8).

8.2.2.2 Mining, Steel, Foundry, and Engineering Industries

In more recent years, there has been a wide use of heat pumps because of their potential in energy technology. Technical limitations constitute an important factor to be considered in mining industry as the requirement is that of robust heat pumps which could withstand abrasive water flow. Probably, the most profitable heat waste sources are water pumped out of the mining industry (temperature of about 10°C). It can be mixed with cooling water from the main compressor section and the heat can be used, say, for heating fresh air entering the mine's ventilation shaft. A typical example of the use of waste heat is that of the aluminium manufacturing industry in Sweden (Figure 8.4). However, heat pumps have yet to make a significant contribution to mining and other metallurgical branches of the industry.

There is potential for the use of heat pumps in waste heat recycling in iron and steel manufacturing industries, although, their economic viability is yet to be evaluated. More suitable arrangements could change this situation.

8.2.2.3 Chemical Industries

Processes in which steam is used as a heating medium and heat is not bound but eventually goes away provide a natural application for heat pumps. There are many examples of such processes within the chemical industry, including distillation, evaporation, and to some extent drying.

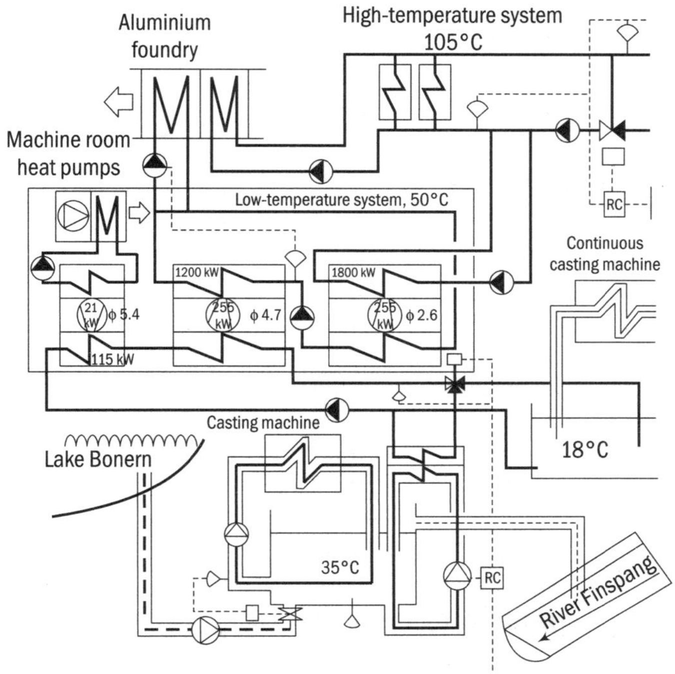

Figure 8.4 Energy conservation through heat pumps used in aluminium industry

8.2.2.4 Other Industries

(i) Pulp and paper industry: Most of the processes used in the pulp and paper industry require heat, generally in the form of steam. The use of fossil fuels in the pulp and paper industry can be eliminated through the application of heat pumps. In this industry, waste energy in process flows of steam, air, water, and fibres can be recovered with heat pumps. It is more difficult to determine whether this approach can be economically feasible. Heat pump applications in this industry are related to the following:

- Compression of flash system from continuous digesters
- Evaporation with mechanical vapour recompression
- Paper drying
- Heat recovery from waste

A typical heat pump application to the dryer section of a newsprint machine is shown in Figure 8.5.

(ii) Stone, wood, and textile industries: In Sweden, heat pumps are used with batch dryers in a large number of sawmills and woodworking plants. Heat pumps are used in the timber industry for timber drying. The most interesting application of heat pumps in textile finishing processes is to recover the waste heat from saturated air leaving the dryers after it is partially cooled by heat exchange with incoming fresh air.

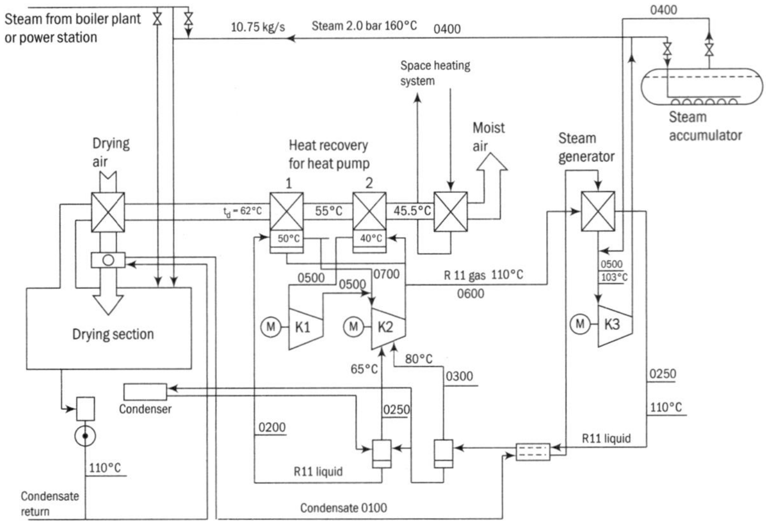

Figure 8.5 Heat pump in pulp, paper, and newsprint industry

8.3 POTENTIAL FOR IMPROVEMENT OF HEAT PUMPS

The efficiency of the compressor package (including the motor) in transforming electrical energy to refrigerant enthalpy is a critical factor in determining coefficient of performance (COP). Owing to the relatively low-saturation pressures of R114 (a refrigerant) compared to the more common refrigerants such as R12, R22, and R502 for which the compressor/motor combination was designed, the experimental power input figures were all well below the nominal full load figure. One expects, therefore, that the motor efficiency is below its optimum level. This supposition is confirmed by a plot of compressor efficiency versus loading (electrical) (Figure 8.6). Compressor efficiency is calculated by estimating the work required to compress the refrigerant between the known pressures and dividing the result by the actual electrical input. Motor loading is calculated by dividing actual electrical input by the full load rating. The average efficiency measured was 59%. The regression line

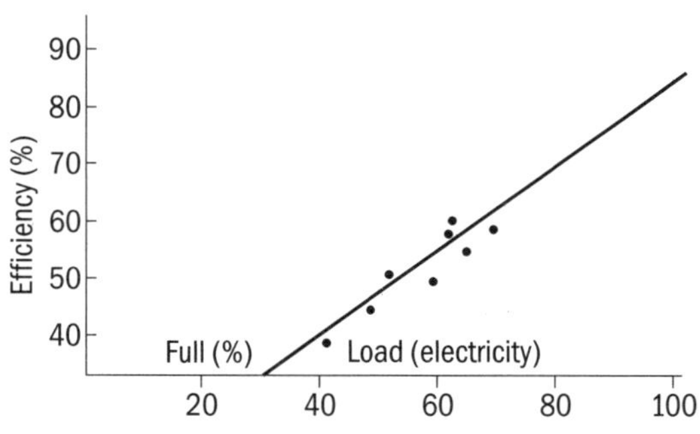

Figure 8.6 Compressor efficiency versus loading

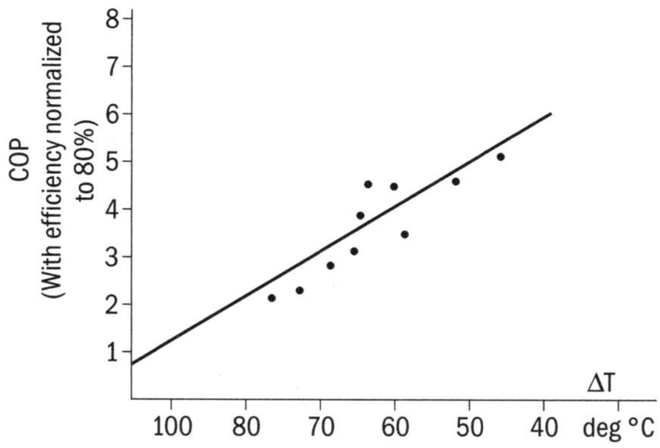

Figure 8.7 Differential temperature in relation to COP

indicates that an efficiency of 80% could be achieved with a loading in excess of 96.5%. In the design of an industrial heat pump, it is important to match the motor characteristics to the characteristics of the compressor under the required operating conditions. The effect on COP, operating at an efficiency of 80%, is shown in Figure 8.7.

8.4 PREDICTING HEAT PUMP PERFORMANCE

The optimum heat pump performance of a vapour compression cycle for a given fluid under particular evaporating and condensing conditions may be determined from the pressure–enthalpy diagram (Figure 8.8).

8.4.1 Performance Parameter

The COP (time) is determined from the following equation:

$$\text{COP} = \frac{Q_C}{Pe} \qquad ...(8.6)$$

It also gives the value of computed operation performance. Here,

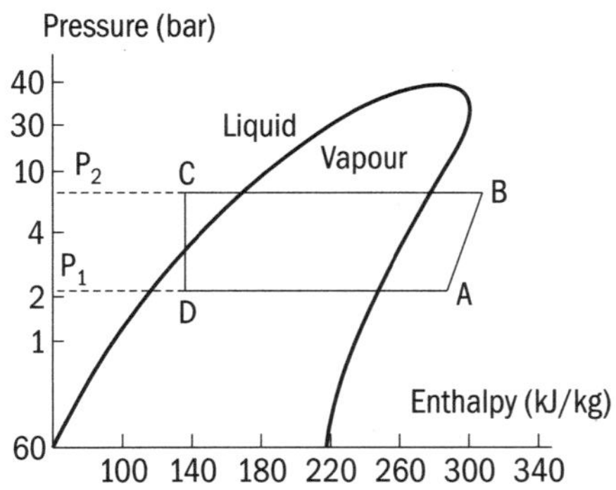

Figure 8.8 Pressure–enthalpy diagram (with halocarbon R114 refrigerant) of the heat pump

Table 8.2 Input/output conditions in a typical six-effect evaporator

Input	Temperature (°C)	Flow rate (tonnes/hour)
Feed (skimmed milk)	5	40
Cooling water	27	26.5
Steam	130	3.7
Outputs		
Concentrate	43	8.6
Cooling water	35.5	26.5
Condensate (with trace of the product)	60	31.0
Pure condensate (return of boiler)	–	3.7

$$Q_C = MCdt \qquad ...(8.7)$$

Here, M is the mass flow rate, C is the specific heat of water passing through the condenser, and t is the rise in temperature. P_e is electrical power input to condenser condensing temperature, T_c is the assumed saturation temperature at the compressor outlet pressure, P_2 is the evaporator pressure, and T_e is the assumed evaporator inlet temperature.

Six-effect evaporators are now in common use for concentrating milk. A typical input/output condition is given in Table 8.2.

8.4.2 Pilot-scale Heat Pump

An electrically driven vapour compression heat pump should be constructed with the following main components:

(a) **Compressor:** Prestcold semi-hermetic model LG 300–LG 0037 employing suction cooling of motor windings.
Displacement: 37 m^3/s
Nominal input power: 2.2 kW (three-phase)

(b) **Condenser:** Mueller model FC-300 water cooled spiral wound.
Heat transfer surface area: 5.05 m^2.

(c) **Expansion valve:** Danfoss TE-5

(d) **Evaporator:** Zelos model B coaxial tube water/refrigerant heat exchanger.
Heat transfer surface area: 0.26 m^2

(e) **Refrigerant:** Halocarbon R114 ($CCIF_2$–$CCLF_2$) was chosen from those suitable for use in the temperature range of 10–100°C. R114 exhibits good thermal stability and low toxicity and it previously gave satisfactory long-term performance.

The heat pump was operated under continuous flow conditions. Refrigerant temperatures were monitored at four points (condenser inlet/outlet and evaporator inlet/outlet). Suction and outlet compressor pressures were monitored. Water inlet/outlet temperatures and flow rates in the condenser and evaporator heat exchangers were monitored. The electrical power consumption of the compressor was measured. Water flow rates and source temperatures were varied, but only measurements pertaining to steady-state conditions were used for performance evaluation.

From the pressure–enthalpy diagram for the fluid (Figure 8.8),

$$COP_i = h_{BC}/h_{AB} \quad \text{...(8.8)}$$

Here, the path AB presents isentropic (ideal) compression. Points B and C are determined by the condensing pressure P_2 and the amount of superheating and subcooling expected in the condenser.

Point A is determined by following an isentropic line from B until it intersects pressure P_1. Alternatively, a simple approximation to the work of isentropic compression may be found which for R114 in the region of interest.

$$h_{AB} = 33 \log \pi \text{ kJ/kg} \quad \text{...(8.9)}$$

where π is the compression ratio (P_2:P_1). The practical COP_s (measured COP_s compensated to 80% compression efficiency) compared to ideal COP_s (from pressure–enthalpy diagram) and from the waste stream is insignificant.

Milk evaporator plants produce large volumes of warm condensate (in the temperature range of 40–60°C). This condensate goes to drain because it is considered to be too low a temperature for useful heat recovery (by simple heat exchange) to take place. This condensate, as well as exhausted cooling water that is also available, is an attractive source of heat for a heat pump that generates hot water in the range of 80–100°C. Figure 8.9 shows slope between COP_i and COP_P.

$$COP_P = 0.872 \times COP_i - 1.315 \quad \text{...(8.10)}$$

In predicting the performance of the heating, pumping schemes that follow this relation could be used. In each case, one could assume a source of waste heat at 45°C and a requirement of water at 85°C.

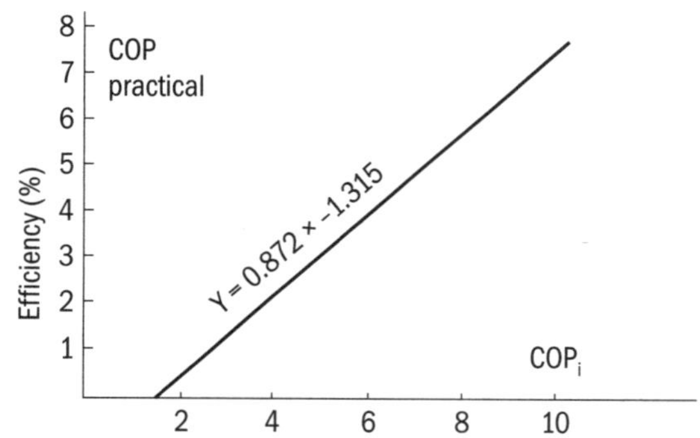

Figure 8.9 Ideal COP_i in relation to COP_P

8.4.2 Single-stage Scheme

For production of water at 85°C, 45°C stream can be split into two ways—one stream gives heat to the evaporator and the other is heated by the condenser to 85°C. The values of the various parameters are as follows:

P_1 = 1.5 bar, P_2 = 10.2 bar, ω = 6.8, H_{AB} = 27.5 kJ/kg, H_{BC} = 142 kJ/kg,
COP_i = 5.17, COP = 3.19.

Calculation: Let electrical input P_e = 1 MW and Q_C = 3.19 MW. Therefore,

$$3.19 \times 10^3 = m \times 4.2 \times (85 - 45)$$

Now m is 19 kg/s per MW of input power (m is the flow rate of water at 85°C).

8.4.3 Two-stage Scheme

A two-stage arrangement allows a reduction in the condensing/evaporating temperature difference, with the consequent improvement in COP.

Dots in Figure 8.10 show (experimentally) the condensing/evaporating temperature difference ($\Delta T = T_c - T_e$) on a $1/T$ scale. Value of COP_s lies in the range of 2.2–3.6 for atmospheric temperatures (39–66°C).

The COP_s can be compared with values obtained using the following formula

$$COP_C = \frac{T_C + 273}{T_C - T_e} \quad ...(8.11)$$

which gives the COP of an ideal heat pump operating on the Carnot cycle between the temperature limits T_c and T_e. The ratio between measured COP and COP_c has a mean value of 45.6% and a standard deviation of 7.6%.[1]

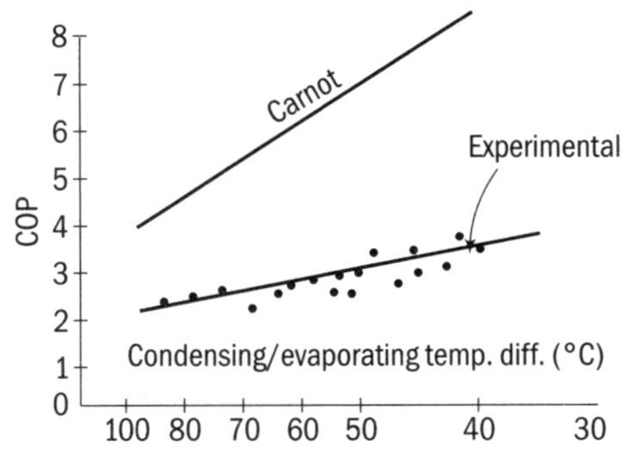

Figure 8.10 Plotting of COP against condensing/evaporating temperature difference

[1] The Carnot cycle is only a crude guide to the practical vapour compression cycle due to the effects of superheat and subcooling. It serves as a standard for comparison.

8.4.4 Influential Factors

Heat pumps find a major application in the food processing industry. There is a simultaneous need for both heating and cooling in the processing of foods in dairy industry. Costs are involved in the following cases:

- Drying
- Evaporation
- Beneficiary industry (for example, canning, dairy, meat processing, and breweries)
- Heat pump type (for example, normal temperature or high temperature)
- Variability concerns

Bibliography

Callaghan, D. J. O. and P. M. Kelly, Heat pump Applications in Dairy Plants. In *Engineering and Food*, B. M. McKenna (ed.), vol. 2, Elsevier, Applied Sci. Pubs, London, 1984, pp. 1059–1069

Mukhopadhyay, P. and S. N. Mukhopadhyay, Advances in Science and Technology Attributes of water for Process Industries, Proceed. Tech. Sessions – Transcriptions, vol. II, Ind. Chem. Engg. Congress, Calcutta, 2000, pp. WTR 29–32

Mukhopadhyay, S. N. 2017. *Food Engineering: process and technology*, 1st edn. New Delhi: Viva Books

Potter, N. N. and J. H. Hotchkiss. 1995. Food processing and environment. In *Food Science*. New York: Chapman & Hall

Reay, D. A. and D. B. Mac Michael. 1979. *Heat Pumps: Design and Application*, p. 204. Oxford: Pergamon Press

9 Theoretical, Numerical, and Multiple-choice Questions on Wastes and Treatments

9.1 INTRODUCTION

Throughout the world, waste treatments and environmental engineering are taught in colleges and universities. Practising scientists, engineers, and technologists (SET) dealing with problems associated with waste treatments and environmental engineering need to work out parameters to tackle pollution. Thus, the use of numerical solutions is as important as understanding the subject theoretically and experimentally. Knowledge of the methods used for determination of solid, liquid, and gaseous pollutants and their generations, characterizations, and computations is essential to understand all aspects of pollution related to hazards to human health and other life forms. This chapter deals with these methods. Also, nanosystemic particles of various sizes are on the rise in the environment and therefore require proper handling through suitable methods.

9.2 WASTES AND TREATMENTS

Living human genomic body/bioreactors (LHGBRs) and other life forms (LFs) function in a manner similar to the biotechnological process for removal of toxic/pollutants. These naturally acquired design engineering (NADE) and recombinatorially acquired design engineering (RADE) bodies are not designed to function without waste removal and treating organs such as kidneys, lungs, liver, intestines, and associated circulatory systems. Performance evaluation of LHGBRs and other LFs depends on their exposures to pollution-free environments. Pollutants generated by the activities of LHGBRs need judicious treatments so that clean environment is available to all life forms. In 2013, a study report, *Performance Evaluation of Sewage Treatment Plants,* funded under National River Conservation Directorate (NRCD) by the Central Pollution Control Board (CPCB) in 2013 showed that there is a huge gap between the sewage generated and the amount actually treated, indicating persistence of environmental pollution.

9.2.1 Theoretical Questions on Environment, Pollution, and Waste

1. What is environment?
2. What is meant by environmental pollution?
3. What are the general pollutants of the environment?
4. What are the differences between land, water, and air pollution?
5. How oxygen is important in assessing pollution?
6. How garbage/solid waste is defined? How garbage and pollution are related to each other?
7. Define the following indices used to measure/express the pollution strength of wastewaters.
   (i) Immediate oxygen demand (IOD)
   (ii) Total oxygen demand (TOD)
   (iii) Biochemical oxygen demand (BOD)
   (iv) Chemical oxygen demand (COD)
   (v) Ultimate BOD
   (vi) Theoretical oxygen demand (ThOD)

   Show graphically how they are related and explain how they are measured. How BOD and COD are quantitatively related? Show that COD is greater than BOD.
8. What is air pollution? Distinguish between haze and smog. What is meant by coefficient of haze (COH)? How it is expressed?
9. Write short notes on the following:
   (i) Role of saprophytic organisms in reduction of wastewater pollution
   (ii) Role of new silver catalyst to help fight smog
   (iii) Role of air pollution in chronic health hazards
10. Write down appropriate relations to describe anaerobic digestion (AD) dynamics when organisms recycle and flow reduction/increase is used as one of the control methods. Include a dynamic equation for the appropriate control variable stating your argument for its form.
11. In a trickling filter bed (TFB) treatment of wastewater, slime mold aggregation occurred. The aggregates eventually formed spores due to a gradient in the cAMP (cyclic adenosine monophosphate) attractant. Considering 'a' as the aggregate density and 'ρ' as cAMP attractant density, the following model was developed:

$$\frac{\partial a}{\partial t} = \frac{\partial}{\partial x}\left(\mu\frac{\partial a}{\partial x} - xa\frac{\partial \rho}{\partial x}\right)$$

$$\frac{\partial \rho}{\partial t} = fa - k\rho + D\frac{\partial^2 \rho}{\partial x^2}$$

Analysing this model, answer the following questions:

(a) Show from linearization of these equations that assumed solutions of the form $(\sin qx)e^{xt}$ for a and ρ lead to stability if

$$xa_0 f < \mu(k + D),\ q \neq 0$$

(b) Which wavelength perturbations lead to the earliest instability?

(c) Why aggregation is needed for sporulation?

9.3 NUMERICAL PROBLEMS ON WASTES AND TREATMENTS

9.3.1 Food Industry Waste Treatment

1. A portion of 5 mL sample of a food industry waste was diluted to 300 mL in a BOD bottle. The DO content of the mixture was 8 mg/L. After 5 days of incubation at 20°C, the DO content of the mixture was found to be 3 mg/L. Determine the 5-day BOD (20°C) of the wastewater. If the BOD rate constant (K) of this wastewater is 0.2 day^{-1}, determine the initial first stage BOD of the wastewater. If the initial first stage BOD of the wastewater is 450 mg/L, determine the BOD rate constant K.
2. In the above question, determine the 5-day BOD at 30°C of the wastewater.
3. The wastewater in the Question 1 is to be discharged to a stream. The volume of wastewater is 5 million gallons per day (MGD). The volume of flow through the stream is 25 MGD. On the basis of this information, determine the 5-day BOD of the stream when the stream is mixed and temperature is maintained at (a) 20°C and (b) 25°C.

9.3.2 Food Microbiology Industry Waste

1. Foodborne bacteria have cilium. They use energy at each point. This energy is derived from the hydrolysis of adenosine triphosphate (ATP) and used at the rate of 5 mol/(s m^3). The cilium is 50 μm long and has the cross-sectional area of 0.03 μm^2. Using these details, calculate the minimum concentration that must be maintained at the proximal end of the bacteria to keep it moving and the amount of ATP in the cilium. Assume the diffusion coefficient of ATP to be 2.5×10^{-10} m^2/s.

9.3.3 Wastewater Disposal

1. A stream having the flow of 36 MGD was carrying a 5-day BOD (20°C) of 8 mg/L. A wastewater having a flow rate of 4 MGD and a 5-day BOD (20°C) of 400 mg/L is to be discharged to the stream. The temperature of the stream is maintained at 20°C. If the initial first stage BOD in the stream water is not to increase more than 18 mg/L, estimate the percentage of BOD removal required by the treatment plant (K = 0.1, 20°C).

2. The treated effluent of a food waste was discharged into a stream having a flow of 36 MGD. The wastewater before treatment and the stream water were examined and the following results were obtained.

Temperature	Stream water	Wastewater
5-day BOD (20°C)	25°C	25°C
K_1 at 20°C	0.25 mg/L	400 mg/L
K_2 at 20°C	36 MGD	4 MGD

Determine the percentage treatment if the initial first stage BOD of the water is not to exceed 15 mg/L.

3. A town of 20,000 people is to discharge treated domestic sewage to a stream with minimum flow of 0.127 m^3/s and BOD 2 mg/L. The sewage daily water flow (DWF) is 135 litres per person per day and the per capita BOD contribution is 0.068 kg/day. If the BOD in the stream below the discharge is not to exceed 4 mg/L, determine the maximum permissible effluent BOD and the percentage purification required in the treatment plant.
4. A stream with a flow of 0.75 m^3/s and BOD 3.3 mg/L is saturated with DO (9.17 mg/L at 20°C). It receives an effluent discharge of 0.25 m^3/s, BOD 20 mg/L, and DO 5.0 mg/L. Determine the DO deficit at a point 35 km downstream if the average velocity of the flow is 0.2 m/s. Assume that the temperature is 20°C throughout, K_1 for effluent/water mixture is 0.10 per day, and K_2 for stream is 0.40 per day.
5. Evaluate L_a and K from the BOD data given in the following table.

Y_t	Y_1	Y_2	Y_3	Y_4	Y_5	Y_6	Y_7	Y_8	Y_9	Y_{10}
BOD (mg/L)	45	65.4	80	91.5	100	108.5	114	119	123	125

6. A cider plant waste is produced at the rate of 0.94 MGD with the BOD of 980 ppm, available N_2 content of 5 ppm, and the phosphorous content of 200 m. Flow rate and BOD are assumed to be constant. The required removal efficiency is 90%. Specify the design data for the aeration tank in terms of (a) volume of tank, (b) air requirement, and (c) nutritional demands.
7. The characteristics of the composite food processing waste and the plant capacity of several industries are tabulated below. Evaluate the population equivalent of various wastes.

Industry	Meat processing		Citrus industries	Potato processing
	Slaughterhouse	Stockyard		
Capacity, per day	1,000 hogs	3,000 acre	100 tonnes	10,000 lb
Wastewater volume/gallon	200 per hog	23,000 per acre	1,000 per tonne	2 per lb
5-day BOD at 20°C (mg/L)	900	64	2,100	1,250
Total suspended solids (mg/L)	645	173	7,200	1,650
Volatile suspended solids (mg/L)	580	132	–	–
Organic nitrogen (mg/L)	113	11	–	–
Ammonia–nitrogen (mg/L)	–	8	–	–
Population equivalent	____________ Per hog	________ Per acre	________ Per tonne	__________ Per 1,000 lb

The basic domestic load of 4 g of BOD and 90 g of suspended solids of non-domestic sewage may be expressed as population equivalents.

8. Suppose the above industries (mentioned in the table of Question 7) are located in a city, having a population of 100,000, and there is plan to install a waste treatment plant in the city. The domestic sewage along with industrial wastes is to be treated in the same plant (joint treatment). The amount of wastewater produced by an individual is 40 gallons/capita/day (GPCD). On the basis of the data available, estimate the following:
   (i) Population equivalent of the plant
   (ii) Total daily flow of the wastewater in million gallons
   (iii) 5-day BOD (20°C) of the wastewater in mg/L
   (iv) Total suspended solid concentration of the wastewater in mg/L
9. Estimate the values of deoxygenation constant and the first stage BOD from the 20°C BOD data given in the following table.

Time, day	2	4	6	8
BOD (mg/L)	178	280	334	364

9.3.4 Denitrification of Wastewater

1. Using the following data and relations, determine the concentration of organic carbon and nitrate-nitrogen left in the microbial denitrifier operated as the continuous stirred tank bioreactor:
   (i) Dilution rate = 1 day^{-1}

(ii) Growth kinetic relation

$$\mu = \mu_m \left(\frac{S_1}{K_1 + S_1}\right)\left(\frac{S_2}{K_2 + S_2}\right)$$

Here, S_1 is the nitrate–nitrogen, S_2 is the organic carbon, μ_m = 5 day^{-1}, K_1 = 25 mg/L, K_2 = 75 mg/L, influent nitrogen = 40 mg/L, influent organic carbon = 120 mg/L, Y_1 (yield coefficient for NO_3^- – N) = 0.68, Y_2 (yield coefficient for organic carbon) = 0.17.

9.3.5 Solid-state Cultivation of Food on Wastes

1. After solid-state fermentation (SSF) cultivation of mushrooms, the residual mushroom compost was examined to find its suitability as a fertilizer in pot cultivation of vegetables. Using the given table, array the weights, represent them graphically, and calculate the following:
   (a) Sample mean deviation
   (b) Sample standard deviation
   (c) Sample variance of mean
   (d) Deviation of the estimated mean from that of the sample population
   (e) Value of chi-square

Also, comment on the results based on the calculated statistical information.

Plot number	Brinjal			Ladies' finger		
	FI (g)	FMC (g)	FC (g)	FI (g)	FMC (g)	FC (g)
1	853	790	660	202	165	120
2	840	795	650	200	160	125
3	830	785	645	210	170	122
4	825	780	630	205	172	188
5	810	780	632	204	168	115
6	840	782	670	198	162	130
7	820	770	688	195	160	125
8	832	762	625	208	158	110
9	845	750	630	202	155	115
10	835	805	650	192	160	108
11	830	810	645	205	165	115
12	852	812	642	190	170	120
13	815	775	678	212	162	125
14	805	768	675	192	160	130
15	840	778	668	195	165	122
16	830	786	662	205	168	125
17	858	792	642	208	170	120
18	832	798	645	210	172	115

FI – inorganic fertilizer; FMC – mushroom compost; FC – without fertilizer

9.3.6 Wastewater Treatments

1. The concentration of total suspended solids (TSS) of a wastewater is estimated to be 250 mg/L. If the *K* value of the system is 0.6, calculate the particulate BOD concentration of the wastewater.
2. In a rotating biological contactor (RBC), inflow of wastewater is 2.2 MGD, BOD concentration is 170 mg/L, and suspended solid (SS) concentration is 140 mg/L. Assuming *K* value to be 0.7, calculate how much lb/day of soluble BOD enters in RBC.
3. An RBC treats a primary effluent flow rate of 0.233 MGD. Calculate its hydraulic loading rate (HLR) in gallons per day/square foot (GPD/ft^2) considering the medium surface area (MSA) to be 96,600 ft^2.
4. An RBC has a total media surface area of 102,500 ft^2 and receives a primary effluent flow rate of 0.269 MGD. Considering the soluble BOD concentration in influent to RBC to be 159 mg/L, calculate the organic loading rate (OLR) in lb/1000 ft^2.
5. The inflow to an activated sludge process (ASP) bioreactor has BOD_5 (20°C) of 220 mg/L and the outlet must be below 15 mg/L. Its characteristics are given in the following table.

Characteristics	Waste treatment
Temperature (°C)	10–30
Variation	±5
Rheology	Newtonian
Viscosity variation, cP	1–10
Substrate concentration, g/L	0.1–5
Loading (BOD, ppm)	100–5,000
Bioreactor size, gallons	50,000–400,000
P/V, hp/10^3 gallons	0.05–0.50
μ (growth rate), h^{-1}	0.05–0.1
Loading (lb BOD/1000 ft^3)	10–1,000

Complete the design of ASP bioreactor. Active solid concentration is 3000 mg/L, recycle ratio (*R*) is 0.46, and flow (*F*) is 4.5 × 10^6 gallons/day. The values of *Y*, μ_{max}, K_s, K_d are given in the table below.

Y	μm (day^{-1})	Ks (mg/L)	Kd (day^{-1})	Coefficient basis	Temperature (°C)
0.5	–	–	0.055	BOD5	20

Considering all nutrients except BOD to be in excess, calculate sludge age, bioreactor volume needed, concentration of active solids in recycle line, and daily aeration rate, assuming 7.5% oxygen consumption. Also, evaluate $R_{optimum}$ (make suitable assumption) and the ratio of costs between digester and settler.

9.3.7 Hospital Waste Treatment for Value-added Product

From a hospital waste containing placental fluid a value-added product (VAP), that is, a food grade enzymic protein concentrate was recovered. It contained 0.001 kg/m^3 protease enzyme and was separated at 10°C using foam fractionation in a column with diameter 0.04 m and foam height 0.42 m. Considering the volume of feed to be 100 mL, surface tension of the fluid to be given by the ratio of mg protein/2L_o, and effluent protein reading to be 334.1 mg over a strip length (L_o) of 0.25 m, calculate the following:

(i) Surface tension of the placental fluid
(ii) Fold of enzyme purification
(iii) Heat of desorption
(iv) Mass transfer coefficient
(v) Average diameter of bubbles in the foam
(vi) Number of transfer units (NTU) required
(vii) Per cent deviation in the height of transfer unit (HTU) calculation

9.4 NUMERICAL PROBLEM ON AIR POLLUTION

A cigarette smoker exhales carbon monoxide at the level of 400 ppm. What is its value in per cent by volume and its concentration in μg/m^3 at 25°C and 1 atm pressure?

9.5 MULTIPLE-CHOICE QUESTIONS

1. In water pollution control, BOD test is used to measure
   (a) the total dissolved solids (b) sludge volume index
   (c) strength of water pollution (d) suspended solids
2. A portion of 5 mL sample of a wastewater was diluted to 300 mL in a BOD bottle. The dissolved oxygen (DO) content of the mixture was 8 mg/L. If DO of the mixture is 3 mg/L after 5 days of incubation at 20°C, what will be the BOD_5(20°C) of the wastewater?
   (a) 50 mg/L (b) 100 mg/L
   (c) 200 mg/L (d) 300 mg/L
3. A trickling filter involves a method of
   (a) primary wastewater treatment
   (b) secondary wastewater treatment

(c) combined wastewater treatment
(d) All of the above

4. In a BOD fuel cell, the microbial reaction is used to produce electrical energy. This reaction produces 673 kcal/mol of glucose. Considering 1 kcal = 1.16 Wh (watt hour), the amount of electrical energy produced by this fuel cell is
   (a) 1.5 Wh (b) 2.0 Wh
   (c) 3.0 Wh (d) 6.0 Wh
5. The ecological unit composed of organisms and their physical environment is called *a*(*n*)
   (a) niche (b) population
   (c) ecosystem (d) community
6. In the ecological environment, the organisms that break down compounds of dead organisms are called
   (a) phagotrophs (b) parasites
   (c) saprotrophs (d) producers
7. Which of the following must be present in an ecosystem if the ecosystem is to be maintained?
   (a) Producers and carnivores (b) Producers and decomposers
   (c) Carnivores and decomposers (d) Herbivores and carnivores
8. A sequence of species, in an environment, in which predator and prey are related to one another is called a(n)
   (a) trophic level (b) ecosystem
   (c) food chain (d) climax
9. Which of the following is a non-renewable natural resource?
   (a) Forests (b) Field crops
   (c) Iron (d) Animal fur
10. Which of the given material is the most resistant to anaerobic digestion/decomposition?
    (a) Proteins (b) Wood
    (c) Coal (d) Paper
11. Which of the following countries caused maximum ozone pollution in 2017?
    (a) Iran (b) India
    (c) Indonesia (d) Ireland
12. One of the ways of managing garbage/trash for pollution control is
    (a) Burning (b) Incineration
    (c) Bio-mining (d) Landfilling

13. For fighting smog to purify air pollution, the catalytic treatment involves
    (a) new silver catalyst
    (b) gold–silver catalyst
    (c) palladium–silver catalyst
    (d) None of the above
14. Gold refineries in Switzerland generate approximately 43 kg of gold in wastewater every year that has value of approximately
    (a) \$1.8 m
    (b) \$2.7 m
    (c) 0.036
    (d) None of the above
15. Biodrying of semi-solid biowastes under sun may produce
    (a) Biomanure
    (b) Biocake
    (c) Biofuel
    (d) All of the above

CONCLUSION

The contents and problem-solving practice questions given in the chapter can help prepare schemes for promotion of academic and research collaboration (SPARC) for national and global KI to create innovative minds of HOE.

Bibliography

Aiba, S., A. E. Humphrey, and N. F. Millis. 1973. *Biochemical Engineering,* 2nd edn. Tokyo: University of Tokyo Press

Bailey, J. E. and D. F. Ollis. 1986. *Biochemical Engineering Fundamentals,* 2nd edn. New York: McGraw-Hill

The Times of India. 2019. Will only IITs lead the way in SPARC scheme (S. Banchariya), p. 1. 18 March. New Delhi

Eckenfelder, Jr., W. W. 1956. *CEP* 52: 286

Metcaff and Eddy, Inc. 1989. *Wastewater Engineering: Treatment, Disposal, Reuse,* 2nd edn, revised by G. Tchobanoglous, New Delhi: McGraw-Hill

Mukhopadhyay, S. N. 2018. *Fundamental of Process Biotechnology,* 4th edn. New Delhi: Viva Books

Mukhopadhyay, S. N. 2012. *Process Biotechnology: Theory and Practice.* New Delhi: TERI

Rich, L. G. 1973. *Environmental Systems Engineering.* New York: McGraw-Hill

Spellman, F. R. 2016. *Handbook of Environmental Engineering.* Boca Raton, FL: CRC Press, Taylor & Francis Group

Vedajananda, S. 2000. PhD thesis on *Studies on the separation of proteolytic enzymes: trypsin and chymotrypsin using affinity ultrafiltration technique.* Jadavpur University, Kolkata

Appendix

APPENDIX A: SCIENTIFIC SYMBOLS, CONSTANTS, AND DATA

Avogadro number (N_A) = 6.022×10^{23} mol^{-1}

Velocity of light (c) = 3×10^{10} cm/s

1 electron volt (eV) = 1.602×10^{-12} erg = 3.827×10^{-20} cal = 1.002×10^{-19} J

1 erg = 2.389×10^{-8} cal = 6.242×10^{11} eV

Table A.1 Henry's law constants for gases in pure water at low pressure

Temperature (°C)	Oxygen[a]	Nitrogen[a,b]	Carbon dioxide	Air[a]
0	2.55	5.29	728	4.32
5	2.91	5.97	876	4.88
10	3.27	6.68	1040	5.49
15	3.64	7.38	1220	6.07
20	4.01	8.04	1420	6.64
25	4.38	8.65	1640	7.20
30	4.75	9.24	1860	7.71
35	5.07	9.85	2090	8.23
40	5.35	10.4	2330	8.70
45	5.63	10.9	2570	9.11
50	5.88	11.3	2830	9.46
60	6.29	12.0	3410	10.1
70	6.63	12.5	–	10.5
80	6.87	12.6	–	10.7
90	6.99	12.6	–	10.8
100	7.01	12.6	–	10.7

[a] Values × 10^4

[b] Includes argon

Note The table gives the value of H (Henry's law solubility constant) for the expression $x_a = p/H$, where x_a is the mole fraction of gas in solution and p_a is the partial pressure, in atmosphere, of component "a" in the vapour at equilibrium. Solute gas must have less than 1 atm partial pressure.

Source International Critical Tables, vol. 3, pp. 256–260

Radioactivity: 1 Ci = 103 m Ci (Ci, Curie)

1 Da = 1.6724×10^{-24} g (Da, Dalton)

Charge of an electron (e) = 4.8×10^{-10} esu = -1.6×10^{-19} C (coulomb)

Mass of an electron = 9.1×10^{-28} g

Mass of a proton = 1.672×10^{-24} g

Charge of a proton = 4.8×10^{-10} esu = 1.6×10^{-19} C (coulomb)

Mass of a neutron = 1.675×10^{-24} g

Charge of a neutron = 0 (electrically neutral)

Survival time of god particles < 3×10^{-24} s

Table A.2 Henry's law constants for high partial pressures

Partial pressure, p (mmHg)	Oxygen[a]		Nitrogen[a]	
	23°C	25.9°C	19.4°C	24.9°C
800	–	4.79	–	–
900	4.58	–	8.24	9.08
2000	4.59	4.80	8.32	9.15
3000	4.60	4.83	8.41	9.25
4000	4.68	4.88	8.49	9.38
5000	4.73	4.92	8.59	9.49
6000	4.80	4.98	8.74	9.62
7000	4.88	5.05	8.86	9.73
8100	–	–	9.04	–
8150	4.98	–	–	–
8200	–	5.16	–	9.91

[a] Value × 10^4

Source International Critical Tables, vol. 3, p. 257

Table A.3 Viscosity and specific weight of water

Temperature (°F)	Specific weight (lb/ft^3)	Absolute viscosity (lb s/ft^2)[a,b]	Kinematic viscosity (ft^2/s)[b,c]
32	62.42	3.746	1.931
40	62.43	3.229	1.664
50	62.41	2.735	1.410
60	62.37	2.359	1.217
70	62.30	2.050	1.059
80	62.22	1.799	0.930
90	62.11	1.595	0.826
100	62.00	1.424	0.739
130	61.55	1.069	0.558

[a] To convert to centipoise, divide by 2.088×10^{-5}.
1 centipoise = 10^{-2} g/(cm s)

[b] Value $\times 10^{-5}$

[c] To convert to centistoke, divide by 1.075×10^{-5}.
1 centistoke = 10^{-2} cm^2/s

Table A.4 Composition of dry air

Component	Volume per cent
Nitrogen	78.09
Oxygen	20.95
Argon	0.93
Carbon dioxide	0.03

Source Steiner, L. E. and J. A. Campbell. 1955. *General Chemistry.* New York: Macmillan

Table A.5 Vapour pressure of pure water

Temperature (°C)	*p*(mmHg)	Temperature (°C)	*p*(mmHg)
0	4.579	55	118.0
5	6.543	60	149.4
10	9.209	65	187.5
15	12.79	70	233.7
20	17.54	75	298.1
25	23.76	80	355.1
30	31.82	85	433.6
35	42.18	90	525.8
40	55.32	95	633.9
45	71.88	100	760.0
50	92.51		

Source Hodgman, C. D. 1954. *Handbook of Chemistry and Physics*, 36th ed. Chemical Rubber Publishing Co.

Table A.6 Solubility of air and oxygen in pure water in equilibrium with wet air at 1 atm of pressure

Temperature (°C)		Concentration (mg/L)
	Air	Oxygen
0	37.0	14.54
5	32.7	12.71
10	29.0	11.27
15	26.9	10.08
20	23.7	9.07
25	21.7	8.25
30	20.0	7.52
35	18.5	6.95
40	17.2	6.46
45	16.0	6.00
50	15.0	5.57
60	12.8	4.76
70	10.6	3.90

Source Perry, J. H., *et al.* 1963. Energy's law coefficients and vapor pressure. In *Chemical Engineers' Handbook*. New York: McGraw-Hill

1 IA	2 IIA	3 IIIB	4 IVB	5 VB	6 VIB	7 VIIB	8 VIIIB	9 VIIIB	10 VIIIB	11 IB	12 IIB	13 IIIA	14 IVA	15 VA	16 VIA	17 VIIA	18 VIIIA
1 **H** Hydrogen 1.008 1																	2 **He** Helium 4.0026 2
3 **Li** Lithium 6.94 2-1	4 **Be** Beryllium 9.0122 2-2											5 **B** Boron 10.81 2-3	6 **C** Carbon 12.011 2-4	7 **N** Nitrogen 14.007 2-5	8 **O** Oxygen 15.999 2-6	9 **F** Fluorine 18.998 2-7	10 **Ne** Neon 20.180 2-8
11 **Na** Sodium 22.98976928 2-8-1	12 **Mg** Magnesium 24.305 2-8-2											13 **Al** Aluminium 26.982 2-8-3	14 **Si** Silicon 28.085 2-8-4	15 **P** Phosphorus 30.974 2-8-5	16 **S** Sulfur 32.06 2-8-6	17 **Cl** Chlorine 35.45 2-8-7	18 **Ar** Argon 39.948 2-8-8
19 **K** Potassium 39.0983 2-8-8-1	20 **Ca** Calcium 40.078 2-8-8-2	21 **Sc** Scandium 44.955908 2-8-9-2	22 **Ti** Titanium 47.867 2-8-10-2	23 **V** Vanadium 50.9415 2-8-11-2	24 **Cr** Chromium 51.9961 2-8-13-1	25 **Mn** Manganese 54.938044 2-8-13-2	26 **Fe** Iron 55.845 2-8-14-2	27 **Co** Cobalt 58.933 2-8-15-2	28 **Ni** Nickel 58.693 2-8-16-2	29 **Cu** Copper 63.546 2-8-18-1	30 **Zn** Zinc 65.38 2-8-18-2	31 **Ga** Gallium 69.723 2-8-18-3	32 **Ge** Germanium 72.630 2-8-18-4	33 **As** Arsenic 74.922 2-8-18-5	34 **Se** Selenium 78.971 2-8-18-6	35 **Br** Bromine 79.904 2-8-18-7	36 **Kr** Krypton 83.798 2-8-18-8
37 **Rb** Rubidium 85.4678 2-8-18-8-1	38 **Sr** Strontium 87.62 2-8-18-8-2	39 **Y** Yttrium 88.90584 2-8-18-9-2	40 **Zr** Zirconium 91.224 2-8-18-10-2	41 **Nb** Niobium 92.90637 2-8-18-12-1	42 **Mo** Molybdenum 95.95 2-8-18-13-1	43 **Tc** Technetium (98) 2-8-18-13-2	44 **Ru** Ruthenium 101.07 2-8-18-15-1	45 **Rh** Rhodium 102.91 2-8-18-16-1	46 **Pd** Palladium 106.42 2-8-18-18	47 **Ag** Silver 107.87 2-8-18-18-1	48 **Cd** Cadmium 112.41 2-8-18-18-2	49 **In** Indium 114.82 2-8-18-18-3	50 **Sn** Tin 118.71 2-8-18-18-4	51 **Sb** Antimony 121.76 2-8-18-18-5	52 **Te** Tellurium 127.60 2-8-18-18-6	53 **I** Iodine 126.90 2-8-18-18-7	54 **Xe** Xenon 131.29 2-8-18-18-8
55 **Cs** Caesium 132.90545196 2-8-18-18-8-1	56 **Ba** Barium 137.327 2-8-18-18-8-2	57-71 Lanthanides	72 **Hf** Hafnium 178.49 2-8-18-32-10-2	73 **Ta** Tantalum 180.94788 2-8-18-32-11-2	74 **W** Tungsten 183.84 2-8-18-32-12-2	75 **Re** Rhenium 186.21 2-8-18-32-13-2	76 **Os** Osmium 190.23 2-8-18-32-14-2	77 **Ir** Iridium 192.22 2-8-18-32-15-2	78 **Pt** Platinum 195.08 2-8-18-32-17-1	79 **Au** Gold 196.97 2-8-18-32-18-1	80 **Hg** Mercury 200.59 2-8-18-32-18-2	81 **Tl** Thallium 204.38 2-8-18-32-18-3	82 **Pb** Lead 207.2 2-8-18-32-18-4	83 **Bi** Bismuth 208.98 2-8-18-32-18-5	84 **Po** Polonium (209) 2-8-18-32-18-6	85 **At** Astatine (210) 2-8-18-32-18-7	86 **Rn** Radon (222) 2-8-18-32-18-8
87 **Fr** Francium (223) 2-8-18-32-18-8-1	88 **Ra** Radium (226) 2-8-18-32-18-8-2	89-103 Actinides	104 **Rf** Rutherfordium (267) 2-8-18-32-32-10-2	105 **Db** Dubnium (268) 2-8-18-32-32-11-2	106 **Sg** Seaborgium (269) 2-8-18-32-32-12-2	107 **Bh** Bohrium (270) 2-8-18-32-32-13-2	108 **Hs** Hassium (277) 2-8-18-32-32-14-2	109 **Mt** Meitnerium (278) 2-8-18-32-32-15-2	110 **Ds** Darmstadtium (281) 2-8-18-32-32-17-1	111 **Rg** Roentgenium (282) 2-8-18-32-32-17-2	112 **Cn** Copernicium (285) 2-8-18-32-32-18-2	113 **Nh** Nihonium (286) 2-8-18-32-32-18-3	114 **Fl** Flerovium (289) 2-8-18-32-32-18-4	115 **Mc** Moscovium (290) 2-8-18-32-32-18-5	116 **Lv** Livermorium (293) 2-8-18-32-32-18-6	117 **Ts** Tennessine (294) 2-8-18-32-32-18-7	118 **Og** Oganesson (294) 2-8-18-32-32-18-8

57 **La** Lanthanum 138.91 2-8-18-18-9-2	58 **Ce** Cerium 140.12 2-8-18-19-9-2	59 **Pr** Praseodymium 140.91 2-8-18-21-8-2	60 **Nd** Neodymium 144.24 2-8-18-22-8-2	61 **Pm** Promethium (145) 2-8-18-23-8-2	62 **Sm** Samarium 150.36 2-8-18-24-8-2	63 **Eu** Europium 151.96 2-8-18-25-8-2	64 **Gd** Gadolinium 157.25 2-8-18-25-9-2	65 **Tb** Terbium 158.93 2-8-18-27-8-2	66 **Dy** Dysprosium 162.50 2-8-18-28-8-2	67 **Ho** Holmium 164.93 2-8-18-29-8-2	68 **Er** Erbium 167.26 2-8-18-30-8-2	69 **Tm** Thulium 168.93 2-8-18-31-8-2	70 **Yb** Ytterbium 173.05 2-8-18-32-8-2	71 **Lu** Lutetium 174.97 2-8-18-32-9-2
89 **Ac** Actinium (227) 2-8-18-32-18-9-2	90 **Th** Thorium 232.04 2-8-18-32-18-10-2	91 **Pa** Protactinium 231.04 2-8-18-32-20-9-2	92 **U** Uranium 238.03 2-8-18-32-21-9-2	93 **Np** Neptunium (237) 2-8-18-32-22-9-2	94 **Pu** Plutonium (244) 2-8-18-32-24-8-2	95 **Am** Americium (243) 2-8-18-32-25-8-2	96 **Cm** Curium (247) 2-8-18-32-25-9-2	97 **Bk** Berkelium (247) 2-8-18-32-27-8-2	98 **Cf** Californium (251) 2-8-18-32-28-8-2	99 **Es** Einsteinium (252) 2-8-18-32-29-8-2	100 **Fm** Fermium (257) 2-8-18-32-30-8-2	101 **Md** Mendelevium (258) 2-8-18-32-31-8-2	102 **No** Nobelium (259) 2-8-18-32-32-8-2	103 **Lr** Lawrencium (266) 2-8-18-32-32-8-3

Periodic Table of Elements

Table A.7 Atomic weight of elements (^{12}C = 12 amu)

Element	Symbol	Atomic number	Atomic weight	Element	Symbol	Atomic number	Atomic weight
Actinium	Ac	89	[227]	Mendelevium	Md	101	[256]
Aluminium	Al	13	26.9815	Mercury	Hg	80	200.59
Americium	Am	95	[243]	Molybdenum	Mo	42	95.94
Antimony	Sb	51	121.75	Neodymium	Nd	60	114.24
Argon	Ar	18	39.948	Neon	Ne	10	20.179
Arsenic	As	33	74.9216	Neptunium	Np	93	237.0482
Astatine	Ai	85	[210]	Nickel	Ni	28	58.7
Barium	Ba	56	137.34	Niobium	Nb	41	92.906
Berkelium	Bk	97	[249]	Nitrogen	N	7	14.0067
Beryllium	Be	4	9.0122	Nobelium	No	102	[254]
Bismuth	Bi	83	208.980	Osmium	Os	76	190.2
Boron	B	5	10.81	Oxygen	O	8	15.9994
Bromine	Br	35	79.904	Palladium	Pd	46	106.4
Cadmium	Cd	48	112.40	Phosphorus	P	15	30.9738
Calcium	Ca	20	40.08	Platinum	Pt	78	195.09
Californium	Cf	98	[251]	Plutonium	Pu	94	242
Carbon	C	6	12.01	Polonium	Po	84	[210]
Cerium	Ce	58	140.12	Potassium	K	19	39.096
Cesium	Cs	55	132.905	Praseodymium	Pr	59	140.9077
Chlorine	Cl	17	35.453	Promethium	Pm	61	[145]
Chromium	Cr	24	51.996	Protactinium	Pa	91	231
Cobalt	Co	27	58.9332	Radium	Ra	88	226
Copper	Cu	29	63.54	Radon	Rn	86	[222]
Curium	Cm	96	[247]	Rhenium	Re	75	186.2
Dysprosium	Dy	66	162.50	Rhodium	Rh	45	102.905
Einsteinium	Es	99	[254]	Rubidium	Rb	37	85.4678
Erbium	Er	68	167.26	Ruthenium	Ru	44	101.07
Europium	Eu	63	151.96	Samarium	Sm	62	150.35
Fermium	Fm	100	[253]	Scandium	Sc	21	44.956
Fluorine	F	9	18.9984	Selenium	Se	34	78.96
Francium	Fr	87	[223]	Silicon	Si	14	28.086
Gadolinium	Gd	64	157.25	Silver	Ag	47	107.868
Gallium	Ga	31	69.72	Sodium	Na	11	22.9898
Germanium	Ge	32	72.59	Strontium	Sr	38	87.62
Gold	Au	79	196.967	Sulphur	S	16	32.06

Contd...

Table A.7 *Contd...*

Element	Symbol	Atomic number	Atomic weight	Element	Symbol	Atomic number	Atomic weight
Hafnium	Hf	72	178.49	Tantalum	Ta	73	180.948
Hahnium	Ha	105	[260]	Technetium	Tc	43	98.9062
Helium	He	2	4.0026	Tellurium	Te	52	127.60
Holmium	Ho	67	164.930	Terbium	Tb	65	158.9254
Hydrogen	H	1	1.0079	Thallium	Tl	81	204.37
Indium	In	49	114.82	Thorium	Th	90	232.038
Iodine	I	53	126.9045	Thulium	Tm	69	168.934
Iridium	Ir	77	192.22	Tin	Sn	50	118.69
Iron	Fe	26	55.847	Titanium	Ti	22	47.90
Krypton	Kr	36	83.80	Tungsten	W	74	183.85
Kurchatovium	Ku	104	[257]	Uranium	U	92	238.03
Lanthanum	La	57	138.91	Vanadium	V	23	50.94
Lawrencium	Lr	103	[257]	Xenon	Xe	54	131.30
Lead	Pb	82	207.2	Ytterbium	Yb	70	173.04
Lithium	Li	3	6.941	Yttrium	Y	39	88.9059
Lutetium	Lu	71	174.97	Zinc	Zn	30	65.38
Magnesium	Mg	12	24.305	Zirconium	Zr	40	91.22
Manganese	Mn	25	54.9380	–	–	–	–

Note The value given in square brackets denotes the mass number of the longest lived or best known isotope.

USEFUL SCIENTIFIC CONSTANTS AND DATA

Physical Constants

Avogadro constant = 6.0225×10^{23} mol^{-1}

Gas constant (R) = 8.3143×10^{7} erg/(K mol) = 1.9872 cal/(K mol) = 82.054 cm^3 atm/(K mol)

Faraday constant (F) = 96,487 C/equivalent

Speed of light in vacuum (c) = 2.99792×10^{10} cm/s

Boltzmann's constant (k) = 1.38×10^{-16} g cm^2/(K s) = 1.37×10^{-16} erg/K = 2.234×10^{-27} kcal/K

Conversion Factors

1 in. = 2.54 cm

1 Ib = 453.59 g

1 gal (US) = 3.78 L

1 mg/L = 8.34 lb/million gal (US) water

1 cal = 4.187 J = 0.003968 Btu

1 hp (metric) = 7.35×10^{9} erg/s = 735 W

Fractions and Multiples

atto	a	10^{-18}	milli	m	10^{-3}	kilo	k	10^{3}
femto	f	10^{-15}	centi	c	10^{-2}	mega	M	10^{6}
pico	p	10^{-12}	deci	d	10^{-1}	giga	G	10^{9}
nano	n	10^{-9}	deka	da	10	terra	T	10^{12}
micro	μ	10^{-6}	hecto	h	10^{2}			

h = Planck's constant = 6.62×10^{-27} erg/s

1 erg = 2.389×10^{-8} cal = 6.242×10^{11} eV

Density of air (20°C) = 1.2×10^{-3} g/cc

Viscosity of air (20°C) = 1.7×10^{-4} g/(cm^2 s) = 0.017 cP

1 Å = 1×10^{-10} m = 10^{-8} cm

1 μm (micron) = 10^{4} Å = 10^{-4} cm

1 mil = 0.00254 cm = 0.001 in.

1 N (newton) = 10^{5} dyn = 1 kg m/s^2

1 Torr = 1 mmHg = 133.3 N/m^2

1 tonne = 10^{3} kg

Table A.8 Perfect gas law constant ($PV = nRT$)

P	V	n	R	T	V_{STP}[a]
atm	L	g-mol	0.082057	[b]°K	22.4
kg/m^2	m^3	g-mol	0.8478	°K	0.0224
mmHg	L	g-mol	62.361	°K	22.4
dyn/cm^2	m^3	g-mol	8.316×10^{7}	°K	0.0224
atm	ft^3	lb-mol	0.7302	[c]°R	359
atm	gal	lb-mol	5.463	°R	2686
psi	ft^3	lb-mol	10.73	°R	359
ftH_2O[d]	ft^3	lb-mol	24.73	°R	359
in.H_2O[d]	ft^3	lb-mol	296.8	°R	359
in.Hg[e]	ft^3	lb-mol	21.83	°R	359
atm	L	g-mol	0.0456	°R	22.4
mmHg	L	g-mol	34.64	°R	22.4
mmHg	ft^3	lb-mol	555.0	°R	359
atm	ft^3	g-mol	16.08×10^{-4}	°K	0.791
atm	ft^3	g-mol	28.97×10^{-4}	°K	0.791
in.Hg	ft^3	g-mol	86.68×10^{-3}	°K	0.791

[a] Volume (in indicated units) of 1 mol at 1 atm and 0°C.

[b] °K = 273.1 + °C

[c] °R = 459.7 + °F

[d] Water at 39.1°F

[e] Mercury at 0°C

APPENDIX B: CONVERSION TABLES OF UNITS

Multiply	By	To obtain
Acre	43,560	Square foot
Acre	4,047	Square metre
Acre	1.562×10^{-3}	Square mile
Acre	4,840	Square yard
Acre-foot	43,560	Cubic foot
Acre-foot	325,851	Gallon
Acre-foot	1,233.49	Cubic metre
Atmosphere	76.0	Centimetres of mercury
Atmosphere	29.92	Inches of mercury
Atmosphere	33.90	Feet of water
Atmosphere	10,333	kg/m^2
Atmosphere	14.70	$lb/in.^2$
Barrels—oil	42	Gallons—oil
British thermal units	0.2520	Kilogram-calorie
British thermal units	777.9	Foot-pound
British thermal units	3.927×10^{-4}	Horsepower hour
British thermal units	107.5	Kilogram-metre
British thermal units	2.928×10^{-4}	Kilowatt-hour
Btu/min	12.97	ft-lb/s
Btu/min	0.02356	Horsepower
Btu/min	0.01757	Kilowatt
Btu/min	17.57	Watt
Centimetre	0.3937	Inch
Centimetres of mercury	0.01316	Atmosphere
Centimetres of mercury	0.4461	Foot of water
Centimetres of mercury	136.0	kg/m^2
Centimetres of mercury	27.85	lb/ft^2
Centimetres of mercury	0.1934	$lb/in.^2$
cm/s	1.969	ft/min
cm/s	0.03281	ft/s
cm/s	0.036	km/h
cm/s	0.6	m/min
cm/s	0.02237	miles/h
cm/s	3.728×10^{-4}	miles/min
cm/s	21,206	$gal/day/ft^2$
cm/s^2	0.03281	ft/s^2

Contd...

Appendix B *Contd...*

Multiply	By	To obtain
Cubic centimetre	3.531×10^{-5}	Cubic foot
Cubic centimetre	6.102×10^{-2}	Cubic inch
Cubic centimetre	10^{-6}	Cubic metre
Cubic centimetre	1.308×10^{-6}	Cubic yard
Cubic centimetres	2.642×10^{-4}	Gallon
Cubic foot	2.832×10^{4}	Cubic centimetre
Cubic foot	1728	Cubic inch
Cubic foot	0.02832	Cubic metre
Cubic foot	0.03704	Cubic yard
Cubic foot	7.48052	Gallon
Cubic foot	28.32	Litre
ft^3/min	472.0	cm^3/s
ft^3/min	0.1247	gal/s
ft^3/min	0.4720	L/s
ft^3/s	0.646317	Million gallons per day
ft^3/s	448.831	gal/min
Cubic inch	16.39	Cubic centimetre
Cubic inch	5.787×10^{-4}	Cubic foot
Cubic inch	1.639×10^{-5}	Cubic metre
Cubic inch	2.143×10^{-5}	Cubic yard
Cubic inch	4.329×10^{-3}	gal
Cubic inch	1.639×10^{-2}	Litre
Cubic metre	10^{6}	Cubic centimetre
Cubic metre	35.31	Cubic foot
Cubic metre	61,023	Cubic inch
Cubic metre	1.308	Cubic yard
Cubic metre	264.2	Gallon
Cubic metre	10^{3}	Litre
Cubic yard	7.646×10^{5}	Cubic centimetre
Cubic yard	27	Cubic feet
Cubic yard	46,656	Cubic inch
Cubic yard	0.7646	Cubic metre
Cubic yard	202.0	Gallon
Cubic yard	764.6	Litre
Cubic yard per minute	0.45	ft^3/s
Cubic yard per minute	3.367	gal/s

Contd...

Appendix B *Contd...*

Multiply	By	To obtain
Cubic yard per minute	12.74	L/s
Degrees (angle)	60	Minute
Degrees (angle)	0.01745	Radian
Degrees (angle)	3600	Second
Degrees per second	0.01745	rad/s
Degrees per second	0.1667	Revolutions per minute
Degrees per second	0.002778	Revolutions per second
Dram	27.34375	Grain
Dram	0.0625	Ounce
Dram	1.771845	Gram
Fathom	6	Foot
Foot	30.48	Centimetre
Foot	12	Inch
Foot	0.3048	Metre
Foot	1/3	Yard
Foot of water	0.02950	Atmosphere
Foot of water	0.8826	Inches of mercury
Foot of water	304.8	kg/m^2
Foot of water	62.43	lb/ft^2
Foot of water	0.4335	$lb/in.^2$
ft/min	0.5080	cm/s
ft/min	0.01667	ft/s
ft/min	0.01829	km/h
ft/min	0.3048	m/min
ft/min	0.01136	miles/h
ft/min	10,772	$gal/day/ft^2$
ft/s	30.48	cm/s
ft/s	1.097	km/h
ft/s	0.5921	Knots
ft/s	18.29	m/h
ft/s	0.6818	miles/h
ft/s	30.48	cm/s^2
ft/s	0.3048	m/s^2
Foot-pounds	1.286×10^{-3}	British thermal units
Foot-pounds	5.050×10^{-7}	Horsepower-hour
Foot-pounds	3.241×10^{-4}	Kilogram-calorie

Contd...

Appendix B *Contd...*

Multiply	By	To obtain
Foot-pounds	0.1383	Kilogram-metre
Foot-pounds	3.766×10^{-7}	Kilowatt-hour
Foot-pounds per minute	1.286×10^{-3}	Btu/min
Foot-pounds per minute	0.01667	Foot-pounds per second
Foot-pounds per minute	3.030×10^{-5}	Horsepower
Foot-pounds per minute	3.241×10^{-4}	Kilogram-calorie per minute
Foot-pounds per minute	2.260×10^{-5}	Kilowatt
Foot-pounds per second	7.717×10^{-2}	Btu/min
Foot-pounds per second	1.818×10^{-3}	Horsepower
Foot-pounds per second	1.945×10^{-2}	Kilogram-calorie per minute
Foot-pounds per second	1.356×10^{-3}	Kilowatt
Gallon	3785	Cubic centimetre
Gallon	0.1337	Cubic foot
Gallon	231	Cubic inch
Gallon	3.785×10^{-3}	Cubic metre
Gallon	4.951×10^{-3}	Cubic yard
Gallon	3.765	Litre
Gallon	8	Pints (liquid)
Gallon	4	Quarts (liquid)
US Gallon	1.20095	US Gallon
US Gallon	0.83267	Imperial Gallon
Gallons of water	8.3453	Pounds of water
gal/min	2.228×10^{-3}	ft^3/s
gal/min	0.06308	L/s
gal/min	8.0208	ft^3/h
gal/min	1.440×10^{-3}	Million gallons per day
gal/min-ft^2	4.716×10^{-5}	cm/s
gal/min-ft^2	9.283×10^{-5}	ft/min
Grain (troy)	1	Grain (avoir.)
Grain (troy)	0.06480	Gram
Grain/US Gallon	17.118	Parts/million
Grain/ US Gallon	142.86	lb/million gallons
Grain/Imperial Gallon	14.286	Parts/million
Gram	980.7	Dynes
Gram	15.43	Grains
Gram	0.03527	Ounces

Contd...

Appendix B *Contd...*

Multiply	By	To obtain
Gram	0.03215	Ounces (troy)
Gram	2.205×10^{-3}	Pounds
g/cm	5.600×10^{-3}	Pounds/inch
g/cm^3	62.43	Pounds per cubic foot
g/cm^3	0.03613	Pounds per cubic inch
g/L	58.417	Grains/gallon
g/L	8.345	Pounds per 1000 gal
g/L	0.062427	Pounds per cubic foot
Hectare	2.471	Acres
Hectare	1.076×10^5	Square foot
Horsepower	42.44	Btu/min
Horsepower	33.000	Foot pounds per minute
Horsepower	550	Foot pounds per second
Horsepower	1.014	Horsepower (metric)
Horsepower	10.70	kg-calories/minute
Horsepower	0.7457	Kilowatt
Horsepower	745.7	Watt
Horsepower (boiler)	33.479	Btu/h
Horsepower (boiler)	9.803	Kilowatt
Horsepower hour	2547	British thermal units
Horsepower hour	1.98×10^6	Foot-pounds
Horsepower hour	641.7	Kilogram-calorie
Horsepower hour	2.737×10^5	Kilogram-metre
Horsepower hour	0.7457	Kilowatt-hour
Inch	2.540	Centimetre
Inches of mercury	0.03342	Atmosphere
Inches of mercury	1.133	Feet of water
Inches of mercury	345.3	kg/m^2
Inches of mercury	70.73	lb/ft^2
Inches of mercury	0.4912	$lb/in.^2$
Inches of water	0.002458	Atmosphere
Inches of water	0.07355	Inches of mercury
Inches of water	25.40	kg/m^2
Inches of water	0.5781	$Ounces/in.^2$
Inches of water	5.202	lb/ft^2
Inches of water	0.03613	$lb/in.^2$

Contd...

Appendix B *Contd...*

Multiply	By	To obtain
Kilogram	980,665	Dyne
Kilogram	2.205	Pound
Kilogram	1.102×10^{-3}	Ton (short)
Kilogram	10^3	Gram
Kilogram-calorie	3.968	British thermal units
Kilogram-calorie	3086	Foot-pounds
Kilogram-calorie	1.558×10^{-3}	Horsepower-hour
Kilogram-calorie	1.162×10^{-3}	Kilowatt-hour
Kilogram-calorie per minute	51.43	Foot-pounds per second
Kilogram-calorie per minute	0.09351	Horsepower
Kilogram-calorie per minute	0.06972	Kilowatt
kg/m	0.6720	lb/foot
kg/m^2	9.678×10^{-5}	Atmosphere
kg/m^2	3.281×10^{-3}	Foot of water
kg/m^2	2.896×10^{-3}	Inch of mercury
kg/m^2	0.2048	lb/ft^2
kg/m^2	1.422×10^{-3}	$lb/in.^2$
Kilometre	10^5	Centimetre
Kilometre	3281	Foot
Kilometre	10^3	Metre
Kilometre	0.6214	Mile
Kilometre	1094	Yard
km/h	27.78	cm/s
km/h	54.68	ft/min
km/h	09113	ft/s
km/h	0.5396	Knot
km/h	16.67	m/min
km/h	0.6214	mile/h
Kilowatt	56.92	Btu/min
Kilowatt	4.425×10^4	Foot-pound per minute
Kilowatt	737.6	Foot-pound per second
Kilowatt	1.341	Horsepower
Kilowatt	14.34	Kilogram-calorie per minute
Kilowatt	10^3	Watt
Kilowatt-hour	3415	British thermal units
Kilowatt-hour	2.655×10^6	Foot-pounds

Contd...

Appendix B *Contd...*

Multiply	By	To obtain
Kilowatt-hour	1.341	Horsepower-hour
Kilowatt-hour	860.5	Kilogram-calorie
Kilowatt-hour	3.671×10^5	Kilogram-metre
Litre	0.03531	Cubic foot
Litre	61.02	Cubic inch
Litre	10^{-3}	Cubic metre
Litre	1.308×10^{-3}	Cubic yard
Litre	0.2642	Gallons
L/min	5.886×10^{-4}	ft^3/s
L/min	4.403×10^{-3}	gal/s
Metre	3.281	Foot
Metre	39.37	Inch
Metre	1.094	Yard
m/min	1.667	cm/s
m/min	3.281	ft/min
m/min	0.05468	ft/s
m/min	0.06	km/h
m/min	0.03728	mile/h
m/s	196.8	ft/min
m/s	3.281	ft/s
m/s	3.6	km/h
m/s	0.06	km/min
m/s	2.237	mile/h
m/s	0.03728	mile/min
Microns	10^{-6}	Metre
Mile	1.609×10^5	Centimetre
Mile	5280	Feet
Mile	1.609	Kilometre
Mile	1760	Yard
M/h	44.70	cm/s
M/h	88	ft/min
M/h	1.467	ft/s
M/h	1.609	km/h
M/h	0.8684	Knots
M/h	26.82	m/min
M/min	2682	cm/s

Contd...

Appendix B *Contd...*

Multiply	By	To obtain
M/min	88	ft/s
M/min	1.609	km/min
Millimetre	0.03937	Inch
Million gallons per day	1.54723	ft^3/s
Million gallons per day	694.4	gal/min
Minutes (angle)	2.909×10^{-4}	Radian
Ounce	16	Dram
Ounce	437.5	Grain
Ounce	0.0625	Pound
Ounce	28.349527	Gram
Ounce	2.790×10^{-5}	Tonne (long)
Ounce	2.835×10^{-5}	Tonne (metric)
Ounces (fluid)	1.805	Cubic inch
Ounces (fluid)	0.02957	Litre
Ounces /in.^{2}	0.0625	lb/in.^{2}
Parts/million	0.0584	Grains/US Gallon
Parts/million	0.07016	Grains/ Imperial Gallon
Parts/million	8.345	lb/million gallon
Pound	16	Ounce
Pound	7000	Grain
Pound	0.0005	ton (short)
Pound	453.5924	Gram
Pounds of water	0.01602	Cubic foot
Pounds of water	27.68	Cubic inch
Pounds of water	0.1198	Gallon
Pounds of water per minute	2.670×10^{-4}	ft^3/s
Pounds per cubic foot	0.01602	g/cm^3
Pounds per cubic foot	16.02	kg/m^3
Pounds per cubic foot	5.787×10^{-4}	lb/in.^{3}
Pounds per cubic inch	27.68	g/cm^3
Pounds per cubic inch	2.768×10^4	kg/m^3
Pounds per cubic inch	1728	lb/ft^3
Pounds per foot	1.488	kg/m
Pounds per inch	178.6	g/cm
Pounds/ft^2	0.01602	Foot of water
Pounds/ft^2	4.883	kg/m^2

Contd...

Appendix B *Contd...*

Multiply	By	To obtain
Pounds/ft^2	6.945×10^{-3}	lb/in.^{2}
Pounds/in.^{2}	0.06804	Atmosphere
Pounds/in.^{2}	2.307	Foot of water
Pounds/in.^{2}	2.036	Inch of mercury
Pounds/in.^{2}	703.1	kg/m^2
Quadrant (angle)	90	Degree
Quadrant (angle)	5400	Minute
Inch	1.571	Radian
Quarts (dry)	67.20	Cubic inch
Quarts (liquid)	57.75	Cubic inch
Radian	57.30	Degree
Radian	3438	Minute
Radian	0.637	Quadrant
rad/s	57.30	Degrees/s
rad/s	0.1592	Revolutions/s
rad/s	9.549	Revolutions/min
rad/s^2	573.0	Revolutions/min^2
rad/s^2	0.1592	Revolutions/s^2
Revolutions	360	Degree
Revolutions	4	Quadrant
Revolutions	6.283	Radian
Revolutions/min	6	Degree/s
Revolutions/min	0.1047	rad/s
Revolutions/min	0.01667	Revolutions/s
Revolutions/min^2	1.745×10^{-3}	rad/s^2
Revolutions/min^2	2.778×10^{-4}	Revolutions/s^2
Revolutions/s	360	Degree/s
Revolutions/s	6.283	rad/s
Revolutions/s	60	Revolutions/min
Revolutions/s^2	6.283	rad/s^2
Revolutions/s^2	3600	Revolutions/min^2
Seconds (angle)	4.84×10^{-6}	Radian
Square centimetre	1.076×10^{-3}	Square foot
Square centimetre	0.1550	Square inch
Square centimetre	10^{-4}	Square metre
Square centimetre	100	Square millimetre

Contd...

Appendix B *Contd...*

Multiply	By	To obtain
Square foot	2.296×10^{-5}	Acre
Square foot	929.0	Square centimetre
Square foot	144	Square inch
Square foot	0.09290	Square metre
Square foot	3.587×10^{-8}	Square mile
Square foot	1/9	Square yard
Square inch	6.452	Square centimetre
Square inch	6.944×10^{-3}	Square foot
Square inch	645.2	Square millimetre
Square kilometre	247.1	Acre
Square kilometre	10.76×10^{6}	Square foot
Square kilometre	10^{6}	Square metre
Square kilometre	0.3861	Square mile
Square kilometre	1.196×10^{6}	Square yard
Square metre	2.471×10^{-4}	Acre
Square metre	10.76	Square foot
Square metre	3.861×10^{-7}	Square mile
Square metre	1.196	Square yard
Square mile	640	Acre
Square mile	27.88×10^{6}	Square foot
Square mile	2.590	Square kilometre
Square mile	3.098×10^{6}	Square yard
Square millimetre	0.01	Square centimetre
Square millimetre	1.550×10^{-3}	Square inch
Square yard	2.066×10^{-4}	Acre
Square yard	9	Square foot
Square yard	0.8361	Square metre
Square yard	3.228×10^{-7}	Square mile
Temperature (°C) + 273	1	Absolute temperature (K)
Temperature (°C) + 17.78	1.8	Temperature (°F)
Temperature (°F) + 460	1	Absolute temperature (°R)
Temperature (°F) – 32	5/9	Temperature (°C)
Tonne (long)	1016	Kilogram
Tonne (long)	2240	Pound
Tonne (long)	1.12000	Ton (short)
Tonne (metric)/tonne	10^{3}	Kilogram

Contd...

Appendix B *Contd...*

Multiply	By	To obtain
Ton (metric)	2205	Pounds
Ton (short)	2000	Pound
Ton (short)	32,000	Ounce
Ton (short)	907.18486	Kilogram
Ton (short)	0.89287	Ton (long)
Ton (short)	0.90718	Ton (metric)
Torr	1	mmHg
Watt	0.05692	Btu/min
Watt	44.26	Foot-pounds/min
Watt	0.7376	Foot-pounds/s
Watt	1.341×10^{-3}	Horsepower
Watt	0.01434	kilogram-calorie/min
Watt	10^{-3}	Kilowatt
Watt-hour	3.415	British thermal units
Watt-hours	2655	Foot-pound
Watt-hour	1.341×10^{-3}	Horsepower-hour
Watt-hour	0.8605	Kilogram-calorie
Watt-hour	367.1	Kilogram-metre
Yard	91.44	Centimetre
Yard	3	Foot
Yard	36	Inch
Yard	0.9144	Metre

Note 1000 tons = 1 tonne

CONVERSION FACTORS FOR SOME COMMON UNITS

Length

*1 in.	=	25.4 mm
*1 ft	=	0.3048 m
*1 yd	=	0.9144 m
1 mile	=	1.6093 km
*1 Å (angstrom)	=	10^{-10} m

Time

*1 min	=	60 s
*1 h	=	3.6 ks

*1 day	=	86.4 ks
1 year	=	31.5 Ms

Area

1 ft^2	=	0.092903 m^2
1 yd^2	=	0.836127 m^2
1 acre	=	4046.9 m^2
1 arc	=	100 m^2
1 $mile^2$	=	2.590 km^2

Volume

1 $in.^3$	=	16.387 cm^3
1 ft^3	=	0.02832 m^3
1 yd^3	=	0.76453 m^3
1 UK Gallon	=	4546.1 cm^3
1 US Gallon	=	3785.4 cm^3
1 litre	=	10^{-3} m^3

Mass

1 oz	=	28.352 g
*1 lb	=	0.45359237 kg
1 cwt	=	50.8023 kg
1 ton	=	1016. 06 kg
1 tonne	=	10^3 kg

Force

1 pdl	=	0.13826 N
1 lbf	=	4.4482 N
1 kgf	=	9.8067 N
1 tonf	=	9.9640 kN
*1 dyn	=	10^{-5} N

Temperature difference energy (work, heat)

*1 °F (°R)	=	5/9 °C (K)
1 ft-lbf	=	1.3558 J
1 ft-pdl	=	0.04214 J
*1 cal	=	4.1868 J
1 erg	=	10^{-7} J

1 Btu	=	1.05506 kJ
1 Chu	=	1.9004 kJ
1 hp h	=	2.6845 MJ
*1 kWh	=	3.6 MJ
1 therm	=	105.51 MJ
1 thermie	=	4.1855 MJ

Calorific value

1 Btu/ft^3	=	37.259 kJ/m^3
1 ft/s	=	0.3048 m/s
1 mile/h	=	0.44704 m/s
1 knot	=	4.394 3 m/s

Metric flow

1 ft^3/s	=	0.028316 m^3/s
1 ft^3/h	=	7.8658 cm^3/s
1 UK gal/h	=	1.2628 cm^3/s
1 US gal/h	=	1.0515 cm^3/s

Flow

1 lb/h	=	0.12600 g/s
1 ton/h	=	0.28224 kg/s

Per unit area

1 lb/in.^{2}	=	703.07 kg/m^2
1 lb/ft^2	=	1.882 4 kg/m^2
1 ton/ mile^2	=	392.30 kg/m^2
1 lb/in.^{3}	=	27.680 g/cm^3
1 lb/ft^3	=	16.019 kg/m^3
1 lb/UK gal	=	99.776 kg/m^3
1 lb/US gal	=	119.83 kg/m^3
1 g/cm^3	=	1000 kg/m^3

Pressure

1 lbf/in.^{2}	=	6.8948 kN/m^2
1 tonf/in.^{2}	=	15.444 MN/m^2
1 lbf/ft^2	=	47.880 N/m^2
*1 standard atm	=	101.325 kN/m^2

*1 atm (1 kgf/cm^2)	=	98.0665 kN/m^2
*1 bar	=	10^5 N/m^2
1 ft water	=	2.9891 kN/m^2
1 in. water	=	249.09 N/m^2
1 in.Hg	=	3.3864 kN/m^2
1 mm Hg (1 Torr)	=	133.32 N/m^2

Power (heat flow)

1 hp (British)	=	745.70 W
1 hp (metric)	=	735. 50 W
1 erg/s	=	10^{-7} W
1 ft lbf/s	=	1.3558 W
1 Btu/h	=	0.29307 W
1 Chu/h	=	0.527 54 W
1 ton of refrigeration	=	3516.9 W

Moment of inertia

1 lb ft^2 = 0.042140 kg m^2

Momentum

1 lb ft/s = 0.13826 kg m/s

Angular momentum

1 lb ft^2/s = 0.042 140 kg m^2/s

Viscosity, dynamic

*1 P (poise)	=	0.1 N s/m^2
1 lb/(ft h)	=	0.41338 mN s/m^2
1 lb/(ft s)	=	1.4882 N s/m^2

Viscosity, kinematic

*1 S (Stokes)	=	10^{-4} m^2/s
1 ft^2/h	=	0.25806 cm^2/s

Surface energy (surface tension)

1 erg/cm^2 (1 dyn/cm) = 10^{-3} J/m^2 (10^{-3} N/m)

Mass flux density

1 lb/(h ft^2) = 1.3562 g/(s m^2)

Heat flux density

1 Btu/(h ft^2)	=	3.1546 W/m^2
1 Chu/(h ft^2)	=	5.678 4 W/m^2
*1 kcal/(h m^2)	=	1.163 W/m^2

Heat transfer coefficient

1 Btu/(h ft^2 °F)	=	5.6783 W/(m^2 K)
1 Chu/(h ft^2 °C)	=	5.678 W/(m^2 K)

Specific enthalpy

*1 Btu/lb	=	2.326 kJ/kg

Heat capacity (specific heat)

*1 Btu/(lb °F)	=	4.1868 kJ/(kg K)

Thermal conductivity

1 Btu/(h ft °F)	=	1.7307 W/(m K)
1 kcal/(h m °C)	=	1.163 W/(m K)

Note (*) denotes an exact relationship.

SOME COMMON PHYSICAL CONSTANTS

Avogadro's number (N_A)	=	6.023×10^{23} mol^{-1}
Boltzmann constant (k)	=	1.3805×10^{-23} J/k
Planck's constant (h)	=	6.626×10^{-34} J s
Gravitational acceleration (g)	=	9.807 m/s^2
Universal gas constant (R)	=	8.314 J/(mol K)
Stefan–Boltzmann constant (σ)	=	5.6697×10^{-8} W/(m^2 K^4)
Universal constant of gravitation (G)	=	6.67×10^{-11} N m^2/kg^2
Electronic change (e)	=	1.662×10^{-19} C
Electron volt (eV)	=	1.602×10^{-19} J
Velocity of light in vacuum (c)	=	2.99795×10^8 m/s
Volume of 1 kmol of ideal gas at STP	=	22.41 m^3
Standard temperature and pressure at STP	=	1.013×10^5 N/m^2 and 273.15 K

APPROXIMATE VALUES FOR SOME COMMON PROPERTIES

Density of water	= ~10^3 kg/m^3
Viscosity of water (18°C)	= ~10^{-3} N s/m^2
Specific heat capacity of water	= ~4.2 kJ/(kg K)
Thermal conductivity of water	= ~0.6 W/(m K)
Latent heat of boiling water	= ~2.3 MJ/kg
Density of air at STP	= ~1.3 kg/m^3
Viscosity of air at STP	= ~1.7×10^{-5} N s/m^2
Specific heat capacity of air at STP	= ~1 kJ/(kg K)
Thermal conductivity of air at STP	= ~0.024 W/(m K)

APPENDIX C: ABBREVIATIONS AND CONVERSION SCALE

Amber mutation = am
Covalently closed circles = ccc
Dihydrofolate reductase = DHFR
Gene for DNA ligase = *lig*
Kilobases = kb
Megadaltons = MDa
Molecular weight = mol. wt
Plaque-forming unit = pfu
Temperature-sensitive (mutation) = ts

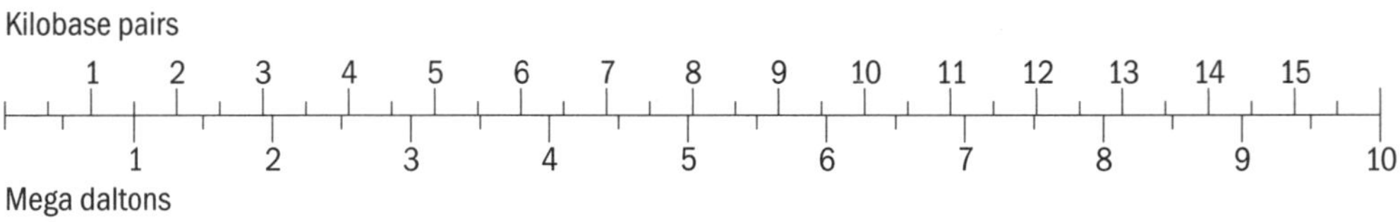

Figure C.1 Scale for conversion between kilobase pairs of duplex DNA and molecular weight

APPENDIX D: IMPORTANT BIO-FORMULAE OF GENETIC ENGINEERING

UNITS

1 μg = 10^{-6} g
1 ng = 10^{-9} g
1 pg = 10^{-12} g

1 kb of DNA = 6.5×10^5 Da of duplex DNA (sodium salt) = 3.3×10^5 Da of single-strand DNA (sodium salt) = 3.4×10^5 Da of single-strand RNA (sodium salt)

Average molecular weight of a deoxynucleotide base	= 324.5 Da
Average molecular weight of a deoxynucleotide base pair	= 649 Da
1 μg/mL of DNA	= 3.08 μM phosphate
1 μg/mL of 1 kb of DNA	= 3.08 nM 5′ ends
1 mol of pBR322	= 2.83×10^6 g
1 pmol of linear pBR322, 5′ ends	= 1.4 μg
1 A_{260} unit of duplex DNA	= 50 μg
1 A_{260} unit of single-strand DNA	= 37 μg
1 kb of DNA = 333 amino acids of coding capacity	= 37, 000 Da
1 Morgan	= 30×10^{-5} A°

FORMULAE

(i) Buoyant density (Θ) versus fraction GC (f_{GC})[1]

$f_{GC} = 10.12\Theta - 16.74$

$\Theta = 0.0988 f_{GC} + 1.6541$

(ii) DNA melting temperature (T_m) versus fraction GC (f_{GC}) and total salt concentration (C_S)[2]

$T_m = 16.6 \log C_S + 41 f_{GC} + 81.5$

(iii) Sedimentation velocity ($S_{20,w}$) of sodium DNA versus molecular weight (M) (solutions in 1 M sodium salt)[3]

Native, double-strand DNA

$S_{20,w} = 0.0882\ M^{0.346}$

$\log M = 2.89(\log S_{20,w} + 1.055)$

---

[1] Woodward, R. S. and J. Lebowitz. 1980. A revised equation relating DNA buoyant density to guanine plus cytosine content. Biochem. Biophys. Methods 2: 307

[2] Cantor, C. R. and P. R. Schimmel. 1980. Biophysical Chemistry. San Francisco: W. H. Freeman

[3] Studier, F.W. 1967. Sedimentation studies of the size and shape of DNA. J. Mol. Biol. 11: 373

- Neutral, denatured (single-strand) DNA

  $S_{20,w} = 0.0105\ M^{0.549}$

  $\log M = 1.82\ (\log S_{20,w} + 1.979)$

- Alkaline, denatured DNA

  $S_{20,w} = 0.0528\ M^{0.400}$

  $\log M = 2.50\ (\log S_{20,w} + 1.277)$

Table D.1 Codon dictionary

5′-OH terminal base	Middle base				3′-OH terminal base
	U	C	A	G	
U	Phe	Ser	Tyr	Cys	U
	Phe	Ser	Tyr	Cys	C
	Leu	Ser	STOP	STOP	A
	Leu	Ser	STOP	Trp	G
C	Leu	Pro	His	Arg	U
	Leu	Pro	His	Arg	C
	Leu	Pro	Gln	Arg	A
	Leu	Pro	Gln	Arg	G
A	Ile	Thr	Asn	Ser	U
	Ile	Thr	Asn	Ser	C
	Ile	Thr	Lys	Arg	A
	Met	Thr	Lys	Arg	G
G	Val	Ala	Asp	Gly	U
	Val	Ala	Asp	Gly	C
	Val	Ala	Glu	Gly	A
	Val[a]	Ala	Glu	Gly	G

[a] Codes for Met if in the initiator position.

Table D.2 Physical constants of the nucleotide triphosphates

Compound	MW	λ_{max}(pH 7.0)	$\lambda_{max}(\times 10^{-3})$ (pH 7.0)	Absorbance ratio (280/260)	pK_a(base)
ATP	507.2	259	15.4	0.15	4.1
CTP	483.2	271	9.1	0.97	4.8
GTP	523.2	252	13.7	0.66	3.3 9.3
UTP	484.2	262	10.0	0.38	9.5
dATP	491.2	259	15.2	0.15	4.1
dCTP	467.2	271	9.1	0.97	4.8
dGTP	507.2	253	13.7	0.67	3.3 9.3
dTTP	482.2	267	9.6	0.73	9.3

Table D.3 DNA fragment that migrates with dyes in polyacrylamide gels (base pairs)

Percentage of gel	Bromophenol blue	Xylene cyanol
3.5	100 bp	460 bp
5.0	65	260
8.0	45	160
12.0	20	70
20.0	12	45

Source Maniatis, T., E. F. Fritsch, and J. Sambrook. 1982. *Molecular Cloning: A Laboratory Manual.* New York: Cold Spring Harbor Laboratory

Table D.4 DNA fragment that migrates with dyes in denaturing polyacrylamide gels (bases)

Polyacrylamide (%)	Bromophenol blue	Xylene cyanol
5	35 b	130 b
6	26	106
8	19	70–80
10	12	55
20	8	28

Source Maniatis, T., E. F. Fritsch, and J. Sambrook. 1982. *Molecular Cloning: A Laboratory Manual.* New York: Cold Spring Harbor Laboratory

Table D.5 Approximate agarose gel concentration for the separation of DNA fragments of various sizes

Amount of agarose in gel (%)	Efficient range of separation of linear DNA molecules (kb)
0.3	60–5
0.6	20–1
0.7	10–0.8
0.9	7–0.5
1.2	6–0.4
1.5	4–0.2
2.0	3–0.1

Source Maniatis, T., E. F. Fritsch, and J. Sambrook. 1982. *Molecular Cloning: A Laboratory Manual.* New York: Cold Spring Harbor Laboratory

Table D.6 Estimate of polyacrylamide concentration required to strand separate DNA fragments of different sizes

Polyacrylamide (percentage)	DNA fragment size (in nucleotides)
4	>200
5	80–200
8	40–100
12	10–50

Source Maniatis, T., E. F. Fritsch, and J. Sambrook. 1982. *Molecular Cloning: A Laboratory Manual.* New York: Cold Spring Harbor Laboratory

Table D.7 Minimum detection limits of single-stranded and double-stranded DNA in acrylamide gels by various staining procedures

Stain	Detection limit[a]	
	ssDNA in 20% acrylamide, 7 M area	dsDNA in 5% acrylamide
UV shadowing[b]	2 μg	2 μg
Ethidium bromide	0.25 μg	15 ng
Stains-all	0.10 μg	50 ng
Methylene blue	1 μg	300 ng
Silver stain		2 ng

[a] In a 1 cm wide well of 0.5 mm thick gel.

[b] Place plastic wrapped gel on fluorescent indicator TLC sheet. Visualize bands with handheld UV source (250–280 nm). Other staining procedures are available upon request.

Source Maniatis, T., E. F. Fritsch, and J. Sambrook. 1982. *Molecular Cloning: A Laboratory Manual.* New York: Cold Spring Harbor Laboratory

Index

W